Crown Compounds: Toward Future Applications

EDITED BY

Stephen R. Cooper

Inorganic Chemistry Laboratory
University of Oxford

Stephen R. Cooper, Ph.D.
Inorganic Chemistry Laboratory
University of Oxford
Oxford OX1 3QR
England

Library of Congress Cataloging-in-Publication Data

Crown compounds : toward future applications / [edited by] Stephen R.
  Cooper.
    p. cm.
    Includes index.
    ISBN 1-56081-024-6
    1. Crown ethers. 2. Macrocyclic compounds. I. Cooper, Stephen R.
  QD305.E7C74   1992
  661'.84—dc20   92-26923
                                                       CIP

Printed in the United States of America
ISBN 1-56081-024-6
ISBN 3-527-28073-1

Printing History:
10 9 8 7 6 5 4 3 2 1

Published jointly by

VCH Publishers, Inc.   VCH Verlagsgesellschaft mbH   VCH Publishers (UK) Ltd.
220 East 23rd Street   P.O. Box 10 11 61   8 Wellington Court
New York, New York 10010   D-6490 Weinheim   Cambridge CB1 1HZ
                                 Federal Republic of Germany   United Kingdom

To my parents, and most of all, to L.T.

Preface

In the 25 years since its inception, the chemistry of crown ethers and crown-like compounds has undergone remarkable development. Intensive study of crown synthetic, structural, and solution chemistry has provided a wealth of information about crown compounds based upon oxygen, nitrogen, and sulfur donor atoms. These fundamental advances have been summarized in a number of excellent books and reviews. On the other hand, extensive investigation on the basic level contrasts with the less-developed extent of practical exploitation, for which crown compounds would now seem ripe. It remains fair to say that the practical application of crown compounds is still in its infancy.

This book represents an experiment to redress this situation. Unlike the retrospective viewpoint of its predecessors, this book attempts a *prospective* view. It attempts to answer the question "Where are these research fields *going?*" instead of explaining where they have been. Internationally prominent researchers were asked to survey their own areas briefly, speculate on where likely future applications might arise, and to identify the research (both basic and applied) necessary to bring their predictions to fruition. This latter section might resemble the introductory section of a grant proposal, for example; it addresses the question "What applications of this work are likely to be in use 20 years from now?"

I hope that *Crown Compounds: Toward Future Applications* will not only stimulate future research but help to guide it. As a compendium of fruitful areas of research in crown compounds as identified by leaders in the field, it was intended to stand on the present frontier of research and illuminate directions for future exploration, or, in another metaphor, to serve as a vector with its base in the present and its arrow in the future. I should like particularly to thank each of the contributors for making time in busy schedules to contribute their perspectives to this book.

Oxford
S.R.C.
February 1992

Contents

Contributors

ADAM, KENNETH R. Department of Chemistry and Biochemistry, James Cook University of North Queensland, Townsville, Queensland 4811, Australia

BELL, THOMAS W. Department of Chemistry, SUNY Stony Brook, Stony Brook, NY 11794-3400, U.S.A.

BRADSHAW, JERALD S. Department of Chemistry, Brigham Young University, Provo, Utah 84602, U.S.A.

CHEN, ZHIHONG Department of Chemistry, University of Miami, Coral Gables, Florida 33124, U.S.A.

COOPER, STEPHEN R. Inorganic Chemistry Laboratory, University of Oxford, Oxford OX1 3QR, England

DE SANTIS, GIANCARLO Dipartimento di Chimica Generale, Universita di Pavia, Via Taramelli 12, Pavia 27100, Italy

ECHEGOYEN, LUIS Department of Chemistry, University of Miami, Coral Gables, Florida 33124, U.S.A.

FABBRIZZI, LUIGI Dipartimento di Chimica Generale, Universita di Pavia, Via Taramelli 12, Pavia 27100, Italy

GOKEL, GEORGE W. Department of Chemistry, University of Miami, Coral Gables, Florida 33124, U.S.A.

HANCOCK, ROBERT D. Department of Chemistry, University of Witwatersrand, Johannesburg, South Africa

IZATT, REED M. Department of Chemistry, Brigham Young University, Provo, Utah 84602, U.S.A.

KADEN, THOMAS A. Department of Inorganic Chemistry, University of Basel, Spitalstrasse 51, Basel 4051, Switzerland

KELLOGG, RICHARD M. Department of Organic Chemistry, University of Groningen, Nijenborgh 16, Groningen 9747 AG, The Netherlands

KIMURA, EIICHI Hiroshima University School of Medicine, 1-2-3 Kasumi, Minami-ku Hiroshima 734, Japan

LICCHELLI, MAURIZIO Dipartimento di Chimica Generale, Universita di Pavia, Via Taramelli 12, Pavia 27100, Italy

LINDOY, LEONARD F. Department of Chemistry and Biochemistry, James Cook University of North Queensland, Townsville, Queensland 4811, Australia

MARTELL, ARTHUR E. Department of Chemistry, Texas A&M University, College Station, Texas 77843-3255, U.S.A.

NAKANO, AKIO Department of Food and Technology, Toa University, Ichinomiya-Gakuencho 2-1, Shimonoseki, Yamaguchi 751, Japan

NEWKOME, GEORGE R. The Graduate School, FAO 126, University of South Florida, Tampa, Florida 33620-7901, U.S.A.

PALLAVCINI, PIERSANDRO Dipartimento di Chimica Generale, Universita di Pavia, Via Taramelli 12, Pavia 27100, Italy

PARKER, DAVID Department of Chemistry, University of Durham, South Road, Durham DH1 3LE, England

SEEL, CHRISTIAN Institut für Organische Chemie und Biochemie, Universität Bonn, D-5300 Bonn 1, Federal Republic of Germany

SUTHERLAND, IAN O. Department of Organic Chemistry, University of Liverpool, Liverpool L69 3BX, England

VÖGTLE, FRITZ Institut für Organische Chemie und Biochemie, Universität Bonn, D-5300 Bonn 1, Federal Republic of Germany

ZHU, CHENG Y. Department of Chemistry, Brigham Young University, Provo, Utah 84602, U.S.A.

CHAPTER 1

Feeble Forces and Flexible Frameworks

George W. Gokel and Akio Nakano
Department of Chemistry
University of Miami
Coral Gables, Florida 33124 U.S.A.

Department of Food and Technology
Toa University
Ichinomiya-Gakuencho 2-1
Shimonoseki, Yamaguchi 751
Japan

1.1 Introduction

During the past 25 or so years, the crown ether[1] field has evolved into the crown ether and cryptand[2] field and more generally into the host–guest[3] field, for which the name "supramolecular chemistry"[4] has been offered. Of late, the efforts in this area of chemistry have been placed under the broad rubric of molecular recognition. During this evolutionary process, a number of research philosophies have emerged. These are as different as the individuals who espouse them but can be categorized into a few broad approaches.

One approach is to imitate or duplicate the structures of biological systems. An alternative to this method is to understand biological function in terms of structural relationships, which can then be duplicated in "unnatural products" usually having simpler and more compact superstructures than the biomolecules they emulate. An important goal of this work is to define the key functions and structural relationships in a given chemical reaction (for example, ester hydrolysis) and then to design a structural framework that supports these functional groups as rigidly as possible. Making the host molecules very rigid has an advantage in that the structural relationships are fixed. Further, the cost of organizing the structure is paid during the synthesis. This concept or approach has proved successful in a number of examples. The difficulty with this strategy is, in a sense, precisely its advantage: the cost is paid during synthesis.

Benner has differentiated the academic timescale from nature's timescale by contrasting the two as "Ph.D. time" versus "geological time."[5] The message inherent in this distinction is very important for us because nature can do simple syntheses quickly and then wait for evolution to make decisions about the correctness of functional groups, hydrophobicity, stereochemical disposition, and so on. Organic or bioorganic chemists rarely enjoy the luxury of waiting many years for structures to sort themselves out. Still, evolution is an important principle in the field of molecular recognition. Unlike natural product synthesis, the final melting point does not signify the end of the program. In contrast, how the structure performs its intended function gives information about the next direction for the work or the next alteration in the structure to be undertaken.

The biological mimics that have been prepared based on a detailed analysis of structural relationships have in many cases been magnificent indeed. The synthetic efforts required for their production have sometimes been heroic. Of course, an additional challenge in such syntheses is to devise straightforward routes from simple and readily available starting materials. Thus the synthetic complexity may be reduced.

Rigidity is extremely appealing to the organic chemist who has a tendency to think enthalpically (in terms of ΔH). The analysis of host–guest interactions usually focuses on the forces at work directly at the interaction site rather than on the overall conformational changes, solvent reorganization, van der Waals forces, or any of the other subtle forces that have an important, but less well defined, effect. Indeed, in the field of cation complexation[6] the vast majority of all results reported can be broadly categorized either as binding studies or transport studies. In each, the entropic component of the binding is intimately involved, but it often (although certainly not always) goes unrecognized.

A difficulty inherent to all extremely rigid structures is that while they might possess beautiful organization, they generally possess very poor dynamics. Nature, in pursuit of reactivity, requires not only excellent structural organization but also a certain degree of flexibility and dynamics. In an enzyme, for example, the ground state starting materials must be bound, the transition state must be stabilized, and then the products must be released. The human design can be optimized for only one of these components if a rigid structure is prepared.

1.2 Valinomycin

1.2.1 The Structure of Valinomycin

Valinomycin (Figure 1.1) is a cyclododecadepsipeptide.[7] It is composed of six amino and six hydroxy acids that alternate throughout the structure to form a 36-membered ring. Each of the amino or hydroxy acids possesses a sidechain. Thus nine isopropyl and three methyl groups are present on valinomycin's periphery. The presence of these hydrophobic chains is obviously important for any molecule that must serve as a transport agent (a

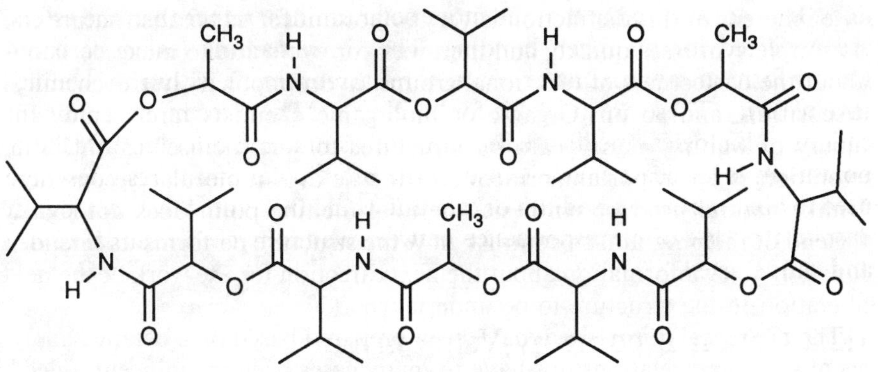

Figure 1.1. Molecular structure of valinomycin.

carrier) because the molecule must reside inside a hydrophobic membrane much of the time.

Because valinomycin possesses six ester and six amide linkages in its structural framework, the 12 carbonyl groups are of two distinct types. The ester groups are inherently less rigid (because of differences in resonance) and less polar (with lower dipole moment) than are the amide carbonyl groups. It is interesting that this molecule, a 36-membered ring, is apparently too large to bind the cation potassium but does so with remarkable selectivity.[8] It is also of considerable interest to note that although the amide carbonyl groups are more polar than the ester groups, only the latter are used in the binding.

Another interesting and important property of the valinomycin structure is its chirality. The amino and hydroxy acids alternate throughout the structure, but the stereochemistry is (D, D, L, L) repeated three times. D-Amino acids are unknown in proteins although they do occur in peptides, especially when the latter arise from a bacterial source. Even so, the stereochemistry is noticeable and has important consequences for the structure and function of valinomycin.

The structures of valinomycin in the presence and absence of a bound cation have been obtained and reported.[9] We can gain some useful insight about the biological function by considering these two structures. The unbound ligand folds into what Truter[10] has called a "tennis ball seam arrangement." Thus the molecule, which appears to be too large in its two-dimensional arrangement, folds to accomplish two things. First, the folding permits the ligand to present a cation with a three-dimensional array of ester donor groups. Second, the folding makes the ligand's hole size (internal cavity) appropriate for the potassium cation. Concurrent with this, the isopropyl and methyl groups turn outward and interact with membrane lipids during the transport process.

The three-dimensional structure of valinomycin is held in the prebinding arrangement by the formation of six hydrogen bonds between amide residues. It is not surprising that the more polar amides, rather than the esters, are preferred for hydrogen bonding. The six remaining (ester) carbonyl groups form an approximately octahedral arrangement within the valinomycin cavity. These are available for binding K^+. The three most prominent cations in biological media are sodium, potassium, and calcium.[11] Of these, potassium is the least charge-dense. If the more polar amide residues were used to bind the cation, sodium or calcium would be favored over potassium despite the lack of correspondence between valinomycin's internal cavity and cation size.[12]

1.2.2 Cation Binding by Valinomycin

In order to more fully appreciate the remarkable intricacy of valinomycin's structure, let us consider briefly the question of cation binding dynamics. In any given solvent, for any given host–guest pair, an equilibrium constant describes the extent of binding at that temperature. The equilibrium constant for the reaction

$$\text{Ligand} + \text{Cation} \underset{}{\overset{K_S}{\rightleftharpoons}} \text{Complex}$$

$$K_S = \frac{k_1 \text{ or } k_{complex}}{k_{-1} \text{ or } k_{release}}$$

is often called the stability constant and is termed K_S. It is sometimes overlooked that the equilibrium constant is also the ratio of the binding and release rates, $K_S = k_{complex}/k_{release}$ (or k_1/k_{-1}). The rate issue is not critical in cases where only the extent of binding (K_S) is of interest but is, however, very important for cation transport.

Transport through a membrane involves three important steps (see Figure 1.2). First, the cation must be bound rapidly (large k_1) at the source phase

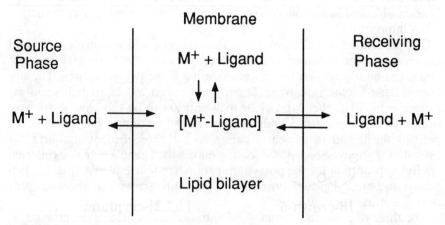

Figure 1.2. Schematic of the membrane transport process.

and bound strongly (large K_S) so that a reasonable concentration of cations can be bound by the ligands. Second, the cation must be bound strongly (large K_S) within the membrane's lipid phase. Cation binding constants are generally higher in lipids than in polar media[13], so the second step is usually not a problem. The problem and paradox arise in the third stage of cation transport. At the receiving phase (the exit side of the membrane), the cation must be released in order for transport to be successful. Thus a rapid decomplexation rate constant (large k_{-1}) and a weak stability constant (small K_S) are desired. It is obviously impossible for a system to simultaneously exhibit strong and weak binding. Since nature has no greater power than humanity to accomplish the impossible, it usually operates instead on the basis of a compromise. The rate constants for K$^+$ binding by valinomycin are shown in Table 1.1 along with the corresponding rate and stability constants for 18-crown-6 and [2.2.2]-cryptand (Figure 1.3).

It is apparent from the data shown in Table 1.1 that valinomycin exhibits intermediate cation binding strength and dynamics. 18-Crown-6 is a fast (dynamic) binder that has acceptable binding and release rate constants, but its overall binding constant in aqueous solution (115) is poor. Further, 18-crown-6 has high dynamics but has the disadvantage of lacking the three-dimensional structure of valinomycin. This latter difficulty may be overcome by adding a third strand to a crown ether to form the cryptands.[14]

Table 1.1. Potassium cation binding rates and strengths for valinomycin, 18-crown-6, and [2.2.2]-cryptand in water at 25°C.

Ligand	K_K	$k_1{}^b$	$k_{-1}{}^c$
18-crown-6	115	4.3×10^8	3.7×10^6
[2.2.2]-cryptand	200,000	7.5×10^6	3.8×10^1
valinomycin^a	31,000	4.0×10^7	1.3×10^3

^{a}Value determined in anhydrous methanol
^{b}In $M^{-1}s^{-1}$
^{c}In s^{-1}

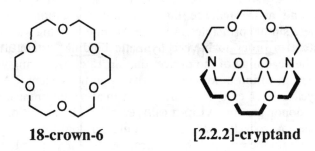

18-crown-6   **[2.2.2]-cryptand**

Figure 1.3. Molecular structure of 18-crown-6 and [2.2.2]-cryptand.

The cryptands offer to a bound cation a three-dimensional array of donor groups, which is very similar to the situation in valinomycin. Notice, however, that the binding dynamics for the cryptand are relatively poor. The cation binding rates (k_1) are generally acceptable, being in the 10^5 range. Decomplexation rates are very slow, however. Even given the exceptional binding strengths of cryptands, these compounds are generally unacceptable for transport simply because the cation release rate is too slow.

The difference between valinomycin and a cryptand is that the three-dimensional structure of the former is held together rather loosely by hydrogen bonds, whereas the cryptands' structure results from covalent binding. Here we see a very important example of the difference between nature's flexible framework approach and the attempts of clever organic chemists to duplicate these structural properties by using covalent skeletons.

Let us now consider the question of chirality in valinomycin. We noted that the chirality alternated between a pair of D-isomers and a pair of L-isomers. It would be counterproductive for nature to build a flexible framework, such as that found in valinomycin, and then to provide the bends in rigid form. Bends can be placed in proteins quite easily by the incorporation of the secondary amino acid proline. This, however, confers considerable rigidity on the structure. The fact that valinomycin needs several bends to attain the tennis ball seam arrangement means that it would be a far less flexible structure than it is. In this case, nature has evolved the extremely clever expedient of changing chirality so that the spatial relationships alternate but the overall rigidity of the binder does not increase.

In light of all these properties, we decided to undertake a program to build valinomycin mimics that might encompass flexibility and three-dimensionality.[15] We also thought that as the program developed,[16] we might be able to make greater use of the notions of feeble forces to help enforce structure on the materials we prepared and perhaps to organize materials within biological membranes.

1.3 The Lariat Ethers as Valinomycin Mimics

The lariat ethers,[15] so called because their molecular models resemble a looped rope and, as we shall see, they rope and tie a cation, were prepared based on these simple principles. A single crown-ether ring was used as the basic structure in order to have a dynamic binding system. It was also thought at the beginning of the project that an 18-crown-6 ring would help confer potassium cation binding selectivity upon the system.[17] Attached to the macrocycle in each case would be a sidearm also containing one or more Lewis basic donor groups. As originally envisioned, the cation would be bound by the macrocyclic ring. There would no doubt be productive encounters with both the ring and the sidearm, but the larger number of donors present in the ring suggested that the first productive encounter would be with it. Because the lariat ether was basically a crown ether, the encounter ought to be rapid and the system should therefore retain its dynamics.

Three-dimensionality would be obtained by a conformational change in which the sidearm swings over the macroring, adding a third dimension of solvation (Figure 1.4). Such a structure might have the cation binding selectivity properties of valinomycin and also its three-dimensionality and dynamics. Whether it would have the same selectivity exhibited by valinomycin was not known at the outset; neither was how to gauge selectivity known.

As the lariat ether program developed, novel methods for synthesis were developed and quite a broad range of structures was prepared.[18] Cation binding constants were ascertained for them.[19] NMR studies (both relaxation time[20] and shift reagent studies[21]) revealed differences in dynamics for the carbon- and nitrogen-pivot structures. We have recently reviewed our efforts in the lariat ether area; that discussion will not be duplicated here.[22] Suffice it to say that the information obtained from these studies suggested that the nitrogen-pivot molecules were inherently more flexible structures than the carbon-pivot compounds and also offered greater versatility and ease of synthesis. An overview of the structures prepared is shown in Figure 1.5. The two crowns shown at the top are carbon-pivot lariat ethers. Those at the bottom are lariat ether, bibrachial lariat ether (BiBLE), and a tribracchial lariat ether (TriBLE) compounds.

The two-armed lariat ethers pose some interesting questions.[23] In the carbon-pivot series, there are isomer difficulties in preparation. If one begins with a substituted derivative of glycerol and attempts to conduct cyclization in the "2+2" fashion used by Pedersen to prepare dibenzo-18-crown-6, an isomer mixture results. The isomerism is of two kinds. By considering the 18-crown-6 residue as if it were a benzene ring, we can imagine isomer substitution of sidearms as being either pseudo-*para* or pseudo-*meta* (Figure 1.6). In addition, in each pair of isomers there is the possibility of the sidearms being disposed *syn* or *anti*. Because of the inherently lower flexibilities of the carbon-pivot compound and because of the possibilities of

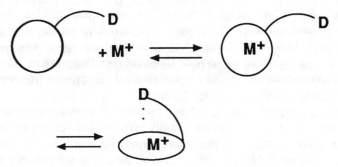

Figure 1.4. Schematic complexation of a metal ion by a lariat ether.

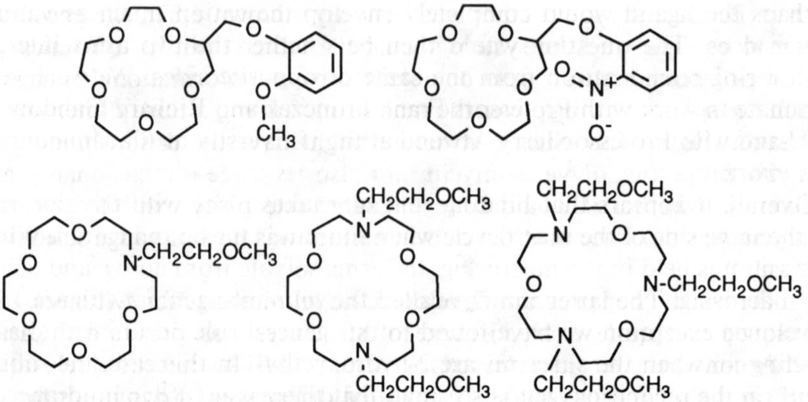

Figure 1.5. Molecular structures of selected lariat ethers.

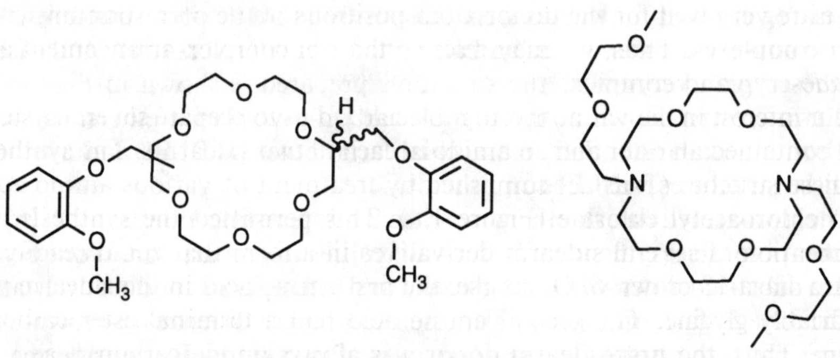

Figure 1.6. C- and N-pivot two-armed lariat ethers.

isomerism, relatively little experimental work was conducted on these structures.

In contrast, none of these problems are posed by the nitrogen-pivot systems. If the sidearms in the latter are attached in the 4,13-positions of an 18-membered ring as shown, the arms are always pseudo-*para*. Other substitution patterns are possible, but the uncertainty of the isomer pattern is not a problem. Moreover, because nitrogen readily inverts, there is no *syn, anti*-isomer problem either. It should be noted that a number of other groups have studied nitrogen-pivot bibracchial lariat ethers. These compounds have proved to be both interesting and useful.[24]

Our principal interest in these compounds was to discover if additional versatility was provided by the second sidearm. In particular, we noted from crystal structure studies that there was a ring sidearm cooperativity in the single-arm systems.[25] Nevertheless, the counteranion was often present in the solvation sphere. We felt that if a second sidearm was present, then perhaps the ligand would completely envelop the cation much as valinomycin does. The question would then be whether the two arms interact with a ring-bound cation from the same or opposite sides. We were very fortunate to work with Professors Frank Fronczek and Richard Gandour at LSU and with Professor Jerry Atwood at the University of Alabama during this effort.

Overall, it appears that binding generally takes place with the sidearms on the same side of the macrocycle when sodium is the bound cation. When potassium is held in the macroring, the arms solvate from above and below the macroring. The latter may be called the *anti*-arrangement (Figure 1.7). The single exception we have found to this general rule occurs with diaza-18-crown-6 when the sidearms are 2-hydroxyethyl. In this case, the substituent on the second oxygen is so small that there is no steric hindrance to binding potassium from the same side. It appears, therefore, that the *syn*-arrangement is preferred to the *anti*-arrangement primarily for steric reasons. It should be noted that the donor atom positions for the *syn*-complexes correlate very well for the donor atom positions of the corresponding cryptand complexes. Thus, we may refer to the *syn* complex arrangement as a *pseudo*-cryptand complex.[26]

Our interest in the valinomycin molecule led us to prepare several systems that contained an ester and an amide in each of two sidearms. The synthesis of these structures was accomplished by treatment of various amino acids with chloroacetyl chloride (Figure 1.8). This permitted the synthesis and purification of several sidearm derivatives in a form that could readily alkylate diaza-18-crown-6. Of course, the first amino acid in the sidearm was invariably glycine. The second amino acid had a terminal ester carbonyl group. Thus, the first sidearm donor was always amide carbonyl, and the second sidearm donor was always ester carbonyl.[27]

Complexation studies were undertaken to determine the selectivity of these systems and also to see if evidence could be obtained for ester sidearm participation by these structures. In fact, crystal structures were obtained by Fronczek for several complexes and also for several of the parent ligands.[28] In all of the sodium complexes we obtained, sidearm interactions occurred from the same rather than the opposite sides. This is in keeping with our previous observations. In addition to this, however, we noted with considerable interest that only the amide carbonyl groups were involved in

Figure 1.7. *Anti* and *syn* arrangements in bibracchial lariat ethers.

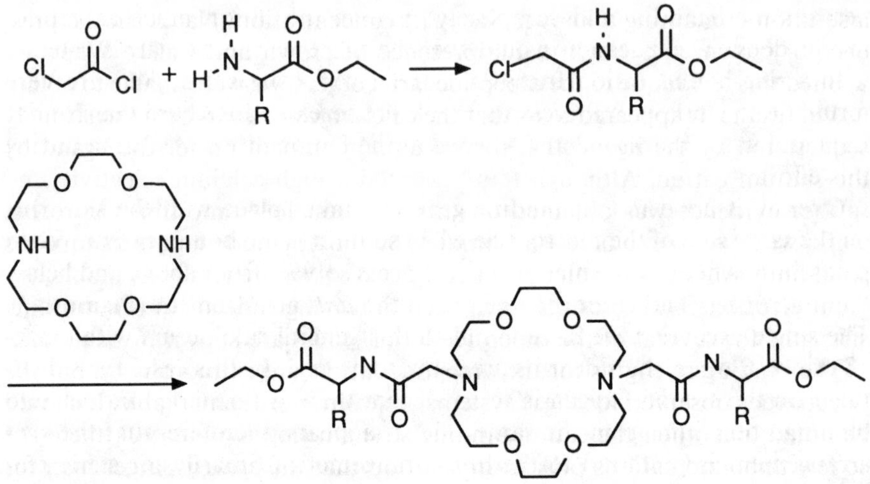

Figure 1.8. Synthesis of lariat ethers bearing two amide–ester binding arms.

cation binding. This was also true of the potassium complexes that formed in the *anti*-sidearm arrangement.

Binding selectivity was determined in the fashion that we have developed extensively in our group.[19] The reader may well be aware that ion selective electrodes (ISEs) offer convenience and versatility but in certain cases are unsuitable to the solvent of interest. An especially serious limitation occurs with commercially available calcium selective electrodes. These electrodes are constructed using a polycarbonate body that tends to crack when thrust into methanol rather than water. The polycarbonate body can be carefully conditioned so that no damage occurs to the electrode but during the conditioning process, the ionophore, a hydrophobic phosphorus compound, is completely dissolved and thus lost.

In order to circumvent these difficulties, we developed a method for measuring calcium stability constants in methanol solution that involved the use of glass electrodes.[17] There are two common alkali ion selective electrodes made of glass. The sodium electrode uses a sodium selective glass. The second type is the monovalent cation electrode that can be used for Li$^+$, Na$^+$, K$^+$ or even NH$_4^+$ cations, but there is interference unless the cations are measured individually. Because these electrodes both depend upon the glasses used in their manufacture, they may be used in alcohol rather than water. Our experience is that they do not perform well in aprotic solvents but work reasonably well in both methanol and water. Using methanol, however, requires conditioning in the solvent prior to use.[17]

The glass electrodes may be used in the presence of both calcium and a second cation. The ligand will bind both sodium and calcium, usually to different extents. The competitive method involves measuring the EMF of a solution containing a known Na^+ (Cl^-) concentration, Na^+Cl^- in the presence of crown, and NaCl in the presence of crown and $CaCl_2$. When we applied this technique to the dipeptide lariat ethers, we were unable to obtain useful results. It appeared to us that the calcium cation was being so strongly sequestered by the ligand that there was no competition for the ligand by the sodium cation. Although this suggested a high calcium selectivity, no further evidence was obtained on this question for some time. In recent work, we have re-explored Ca^{2+}-binding by the dipeptide lariats in aqueous solution.[29] Working in water permits the use of commercially available calcium electrodes. The results have been gratifying indeed. Cation binding strengths for several of the dipeptide lariat ethers are shown in Table 1.2.

The binding strengths for the dipeptide lariats are extraordinary, but the selectivities observed for these systems are even more remarkable. It should be noted that when the cation binding strength (log K_S) is $>10^7$, there are so few unbound cations present in solution that the error is larger than for measurements in the log K_S range of 10^2–10^6. Furthermore, it should be appreciated that the "selectivities" are the ratios $K_{S(Ca)}/K_{S(Na)}$ and not a single value determined when both cations were present.

It is especially interesting that the calcium cation binding strengths in aqueous solution are so large. Alkali metal cation binding in aqueous so-

Table 1.2. Cation binding and selectivity values for various macrocyclic ligands.

Ligand Ref.	Solvent	log K_S Ca^{2+}	log K_S Na$^+$	Selectivity Ca^2/Na$^+$
15-crown-5	water	NR^a	0.79	—
18-crown-6	water	0.5–1.8^b	0.5–1.8^b	ca. 1
[2.2.1]-cryptand	water	6.9^b	5.4^b	31.6
[2.2.2]-cryptand	methanol	8.0^b	8.14	1.3
[2.2.2]-cryptand	water	4.5^b	3.9^b	4.0
DA-12-c-4(gly-gly-OMe)$_2$^{d}	methanol	3.78	2.84	8.7
DA-12-c-4(gly-gly-OMe)$_2$	water	4.32	ND^e	—
DA-18-c-6(CH$_2$CH$_2$OMe)$_2$	methanol	4.48	4.75	0.5
DA-18-c-6(CH$_2$COOEt)$_2$	methanol	6.8	5.51	19
DA-18-c-6(CH$_2$COOEt)$_2$	water	4.26	2.0	182
DA-18-c-6(gly-gly-OMe)$_2$	water	6.7^g	2.2	$>10^4$
DA-18-c-6(gly-gly-OEt)$_2$	water	6.6^g	2.2	$>10^4$
DA-18-c-6(gly-ala-OMe)$_2$	methanol	CNBD^f	4.12	—
DA-18-c-6(gly-ala-OEt)$_2$	water	$\geqslant 7^g$	2.2	$\sim 10^5$
DA-18-c-6(gly-val-OEt)$_2$	water	$\geqslant 7^g$	2.2	$\sim 10^5$
DA-18-c-6(gly-leu-OMe)$_2$	water	$\geqslant 7^g$	2.3	$\sim 10^5$

^{a}NR = not reported
^{b}Average of two or more reported values
^{c}Average of three or more reported values
^{d}DA = 4,10-diaza
^{e}ND = not determined
^{f}CNBD = could not be determined, see text
^{g}The observed value of 7.7–7.8 in water was 7.1 in 40 mM aq. NaCl solution.

lution is usually much poorer than in nonpolar solvent systems.[13] Monovalent sodium and divalent calcium cations have approximately the same ionic radius. Because calcium has two positive charges, it is twice as charge-dense as sodium. Selectivity in this case appears to result from electronic factors rather than steric factors such as the so-called hole–size relationship.[17]

The binding strengths and selectivities of these didpeptide bibracchial lariat ethers are interesting and useful, but they do not mimic those of valinomycin. This is because valinomycin uses ester carbonyl groups, and the amide carbonyl groups are predominant in our own structures. Thus, we and nature began with the same goal but wound up with different results. On the other hand, we have succeeded in preparing completely synthetic structures that have useful cation binding abilities and unusually high cation binding selectivities.

1.4 Feeble Forces, Flexible Frameworks, and Cation Channels

Crown ethers and their relatives have been valued from their invention[1] as cation binders and extraction or transport agents. Valinomycin is a model for this application. Nature's alternative to active transport is to form a cation-conducting channel; the latter is far more common in natural systems. Of course, the challenges of design and synthesis are far greater for channel formers than for cation carriers.

Several efforts have been made during the past decade to prepare cation conducting channels.[30-36] For the most part, the structures used resemble a tunnel. This is entirely reasonable considering that nature's cation conducting channel, gramicidin, has a tunnel shape. In several cases, the compounds used were materials that resulted from other studies[32] or were naturally occurring materials joined together in an effort to emulate the structure of gramicidin.[31,37] The Schreiber channel[37] constitutes a special case. An interesting example is the "chundle" (Figure 1.9) reported by Jullien and Lehn. Its name, from *ch*annel + b*undle* (of fibers), appears to be inspired by "chunnel," the name given to the channel tunnel between England and France.

No data for function have been reported by Jullien and Lehn for the chundle.[34] The most successful example of the attempts to prepare channel molecules has been that reported by Fyles and co-workers[35], which is based upon the bola-amphiphile work of Fuhrhop.[33,38]

Based on all of the notions discussed previously in this chapter, it is probably obvious that our attempt would be to make a flexible system containing all of what we believed to be the appropriate functional groups in the hope that nature would organize the system into the appropriate channel. Such hopes were not ill founded, we thought, because gramicidin, a pentadecapeptide (containing 15 amino acid units), is a structure assembled for this purpose by nature. Although the compound itself is probably

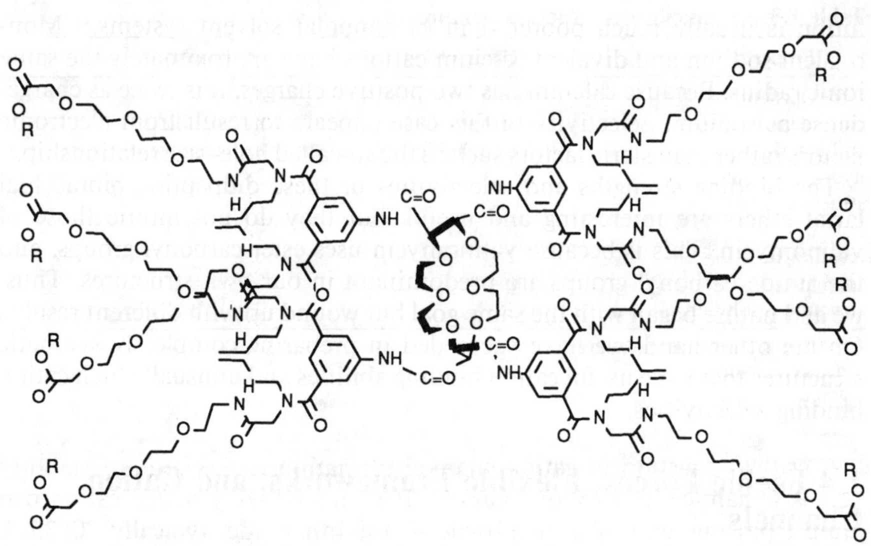

Figure 1.9. Molecular structure of the chundle.

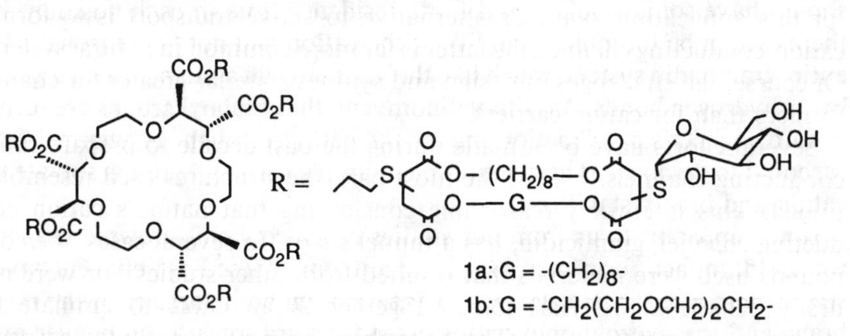

1a: G = -(CH₂)₈-
1b: G = -CH₂(CH₂OCH₂)₂CH₂-

Figure 1.10. Two earlier approaches to channel-forming molecules: left, that of Fyles et al. (ref. 35); right, that of Führhop et al. (refs. 33, 38).

long enough to span most membranes in a fully extended form, the molecule actually coils. The coiling occurs because there is an alternation of D,L-amino acids just as observed in the valinomycin system. Gramicidin bends to a helical shape determined by the changes in chirality. This shortens the molecule to about half the distance required to span a membrane, but the molecule dimerizes to achieve the required membrane-spanning distance. X-ray crystal structures were recently reported for the gramicidin A system by Wallace and Ravikumar[39] and Langs[40] and the data reported are shown in Table 1.3. The structure of gramicidin A is O=CH-L-val-gly-L-ala-D-leu-L-ala-D-val-L-val-D-val-L-trp-D-leu-L-trp-D-leu-L-trp-D-leu-L-trp-ethanolamine.

Table 1.3. Gramicidin A structural features.

Property	Channel[a]	Pore[b]
Gross structure	dimer	dimer
coil interior	hydrophilic	hydrophilic
coil exterior	hydrophobic	hydrophobic
hydrogen bonds	28	28
structure length	31 Å	26 Å
hole size	3.85–5.47 Å	4.9 Å
residues/turn	7.2	6.4
solvent	none	38 sites
structure resolution	0.86 Å	2.0 Å

[a]Wallace & Ravikumar (ref. 39)
[b]Langs (ref. 40)

A critical question in cation channel formation is the issue of stability versus dynamics in a cation channel. It seems unlikely that a cation can jump from one end of a membrane to the other side, typically 30–35 Å away. Gramicidin provides sites (carbonyl groups) that can temporarily solvate the cation as it passes through the channel. Preferably the cation should have some stability and finite residence time in each position, but there cannot be too much stability or the cation will not flow through. The entire gramicidin system, much like the valinomycin system, is held together by 28 hydrogen bonds. As with valinomycin, the carbonyl groups are turned inward. A "peristaltic" action moves the cation through the channel from carbonyl to carbonyl. The electron pairs on the carbonyl groups solvate the cation and briefly stabilize it.

It was apparent to us from our studies of hydrophobic crown ethers that we could, in fact, form membranes from them. These lipid bilayers have a structure in which there is a macrocyclic ring in both surfaces of the membrane and the hydrophobic chains extend inward toward the bilayer midplane. The question then became: Could the crown ethers capture cations and conduct them into the membrane? Furthermore, what sort of relay could be used within the membrane while not concentrating too much polarity at its center? Indeed, relays might be required at other positions throughout the channel to form in the membrane bilayer. It is well known how far a cation must jump from carbonyl to carbonyl in the gramicidin channel,[41] but it is not known how far a cation can jump in the absence of other donor groups. We assumed that lower energies are required for smaller increments of distance or any temporary instability and that this would be the case here as well.

Even though we were designing a system based on the flexible framework concept, we knew that we would have to impose a certain amount of order on the system for it to be successful. One way to provide such order is to introduce a crown ether ring. In a hydrophobic crown ether, the macroring must be at the surface of the membrane as it is very polar compared to the hydrophobic tails.[42] Another way to impose order is to use steroidal linkages in the system. Steroids are one way that nature imposes order and rigidity on its bilayers. Rat liver plasma membrane, for example, contains 24% cholesterol in its bilayer. One plan was to incorporate the steroidal subunit as a spacer within a three-macrocycle system. We hoped this would make the system reasonably rigid yet still insertable in the membrane. We had already demonstrated that the steroidal lariat ethers could, in fact, form bilayers themselves.[43] Two compounds that do so are illustrated in Figure 1.11.

As we conceived it, the concept of the membrane channel based on crown ethers was as follows. There would be a crown ether at each surface of the membrane, and these crown ethers would be spanned by a covalent link. Attached to the opposite side of each crown ether would be a flexible tail, which would generally add hydrophobicity to the structure. We felt that some sort of internal relay was required, and as a first attempt we thought this could be satisfied by a crown ether molecule. Thus the overall structure envisioned as a first attempt was composed of three diaza-18-crown-6 systems held together by 12 carbon chains and terminated by 12 carbon chains. It was thought that if this system was sufficiently hydrophobic to enter the membrane, it would fold up in a fashion schematically as illustrated in Figure 1.12.

Obviously, there are innumerable structural and conformational possibilities in such a complex structure. Even so, a number of specific questions can be asked:

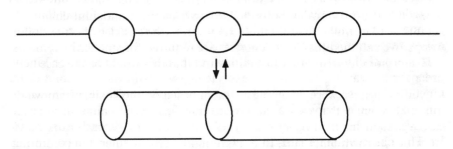

Figure 1.11. Macrocycles bearing steroidal spacer units.

Figure 1.12. Schematic depiction of channel formation by linked crown ethers.

- Can a macrocyclic (crown) ether ring function as a polar head group in a lipid bilayer membrane?
- If so, can a third macroring be used as the central relay, or will it be too polar?
- How far apart must be the macrorings if they are to span a membrane such as egg lecithin (phosphatidyl choline)?
- How many "residence" points for the cation will be required, and what spacing of them is appropriate?
- What role will water play in this complex system?

For the first attempt at a channel-forming molecule, we simply made some guesses. We noted that gramicidin had a span of 26–31 Å, so we chose 12-carbon spacers to separate the crown ether rings. The linear distance from carbon to carbon in an extended hydrocarbon chain is about 1 Å. Thus, 12 carbon chains would give us a span of ~24 Å. We chose 4,13-diaza-18-crown-6 as the basic macrocyclic subunit because it is a versatile and readily accessible molecule with which we had considerable experience. We did not know how to gauge the importance or location of water in the system but assumed that considerable water would permeate the channel if it formed.

The synthesis of our first channel-forming molecule (Figure 1.13) was accomplished in a fairly straightforward fashion, although with considerable difficulty.[44] Diaza-18-crown-6 was alkylated with a C_{12} sidechain that was then alkylated with a dibromide to permit the formation of the two exterior units. These were then connected again by an alkylation reaction to give the final structure. The structure was fully characterized by all of the normal methods, but no crystal structure could be obtained on the oily material.

Once the molecule was in hand, the question of how to assess its cation conduction ability arose. This phase of the project was done in collaboration with Professor Luis Echegoyen and his co-worker, Mr. Qianshan Xie. Echegoyen and Xie decided to use a procedure reported previously for the assessment of cation transport in a gramicidin system.[45] This is an NMR method in which ^{23}Na linewidths are measured. The membrane is constructed in the usual fashion, and then dysprosium is added to the system. External dysprosium serves as a shift reagent so that sodium inside and outside the membrane can be observed directly by NMR. The linewidths may then be correlated to cation transport rates. It was found that the channel former illustrated in Figure 1.13,[44] was poorer in cation conduction ability than gramicidin by only about 100-fold.[46]

It is especially interesting to note that there is a difference in the kinetic order for the various systems. As expected, gramicidin shows second-order kinetics. This is expected because the gramicidin molecule is known to dimerize when serving as a cation channel former. It was a surprise to discover that the diaza-18-crown-6 molecule exhibited second-order kinetics. This observation remains to be explained and is the subject of continuing work in our laboratories.

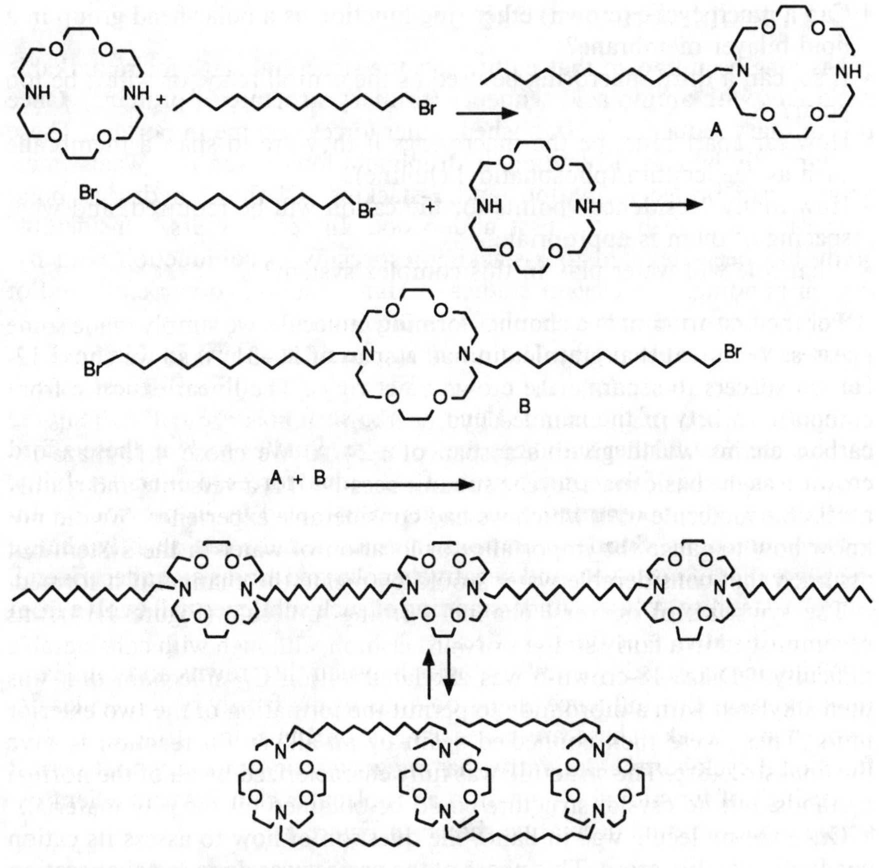

Figure 1.13. Synthesis of the channel-forming crown ethers.

At the present time, we imagine that the channel inserts into the membrane with the central ring extended rather than forming a "tunnel." In this way, the overall length of the channel is slightly greater and the central "relay" points are extended. Of course, this is still speculation.

1.5 Future Directions

The previous paragraph concluded with the admission of speculation. This entire section will be nothing but speculative lucubration. It may be easier to hypothesize than to write a fully justified proposal. On the other hand, it may be yet easier to suggest a direction for the next 36 or 48 months of effort rather than to peer a decade into the future. What follows is a combination of shorter-term estimates and longer-term guesses.

1.5.1 Flexible Frameworks

It has long been known that nature sets the structural stage for remarkable molecules with amino acid sequence (primary structure of proteins). Once this primary sequence is established, other forces become important. These are typically hydrogen bonding, hydrophobic forces, van der Waals interactions, salt bridge formation, and π-stacking. Of these, hydrogen bond formation is probably the best understood. In recent years, considerable study has been devoted to π-stacking, especially in conjunction with hydrogen bonding. The elegant studies of Hamilton and co-workers[47] and of Rebek and co-workers[48] exemplify this direction.

An early example of the Hamilton approach is shown in Figure 1.14. Here, hydrogen bond formation is the basic element of host–guest association. The ability of the naphthalene unit to shift position and overlay the barbiturate host further stabilizes the complex. Studies such as these afford an understanding of the contribution of individual forces to overall stability. Further, any disappointments encountered will lead us in previously undefined directions. Subtle alterations in chain lengths can slightly change the alignment of positively and negatively polarized atoms and affect overall complex stability. A better understanding of such subtleties will evolve from work in this area over the next decade.

While one can readily draw analogies between the crowns and the Hamilton receptor, this influence is less obvious in the family of molecules developed by Rebek. These are based largely on the Kemp triacid, a polyfunctional cyclohexane derivative having geometric properties that permit "focusing" of functional groups into a "molecular cleft." An excellent example is the acridine derivative shown in Figure 1.15.

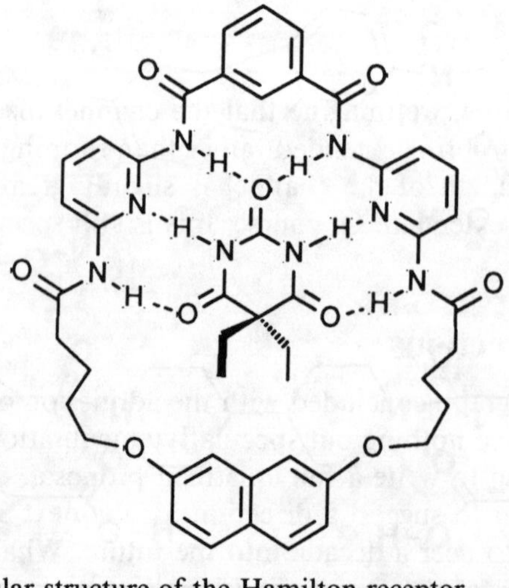

Figure 1.14. Molecular structure of the Hamilton receptor.

Although the effort from Rebek's laboratory has been both elegant and extensive, related examples have been slow to develop. In recent work from our own group,[49] we have found that the ferrocene subunit can be utilized as a molecular cleft complexing agent. Two structures representative of this group are shown in Figure 1.16.

Overall, these structures and families of structures yet to be born will permit the study of weak forces as they relate to molecule complexation. Selectivities will be tailored by adjusting distances, orientation, type of primary interaction (so far usually hydrogen bonding), and secondary interaction (steric forces, π-stacking, etc.). This area promises to be unusually fertile because the structures are of relatively low molecular weight but are

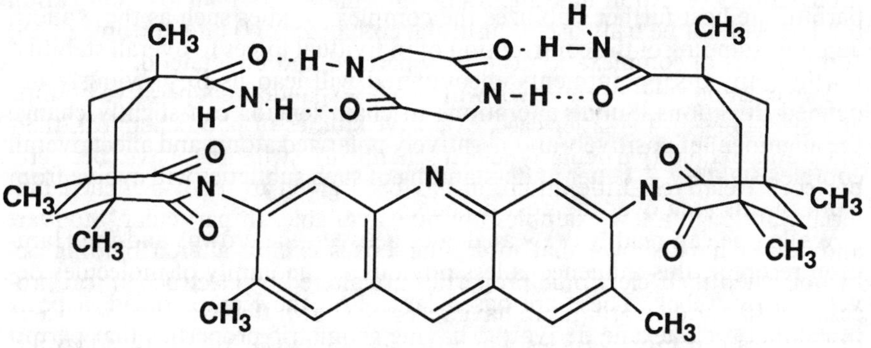

Figure 1.15. A molecular "cleft" based upon Kemp's triacid.

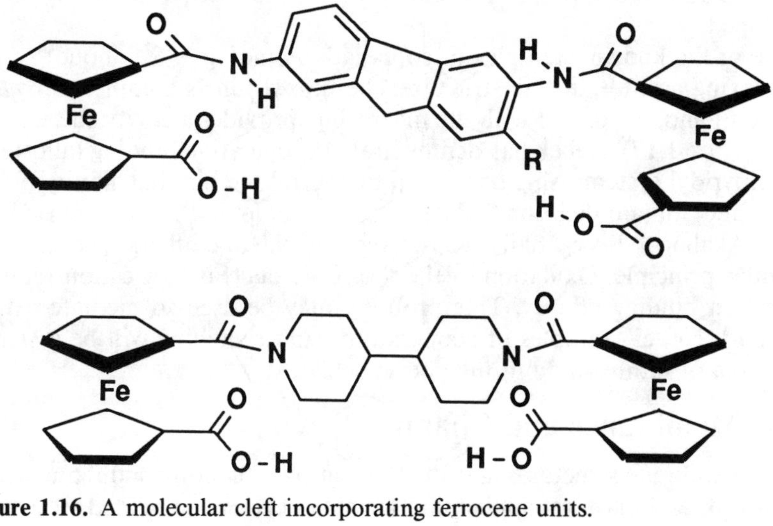

Figure 1.16. A molecular cleft incorporating ferrocene units.

rich in functional groups and have defined binding pockets and predictable orientation or auxiliary groups.

1.5.2 Complexation

Extensive collections of data on cation binding now exist,[50,51] a good bit of which still involves the two-phase picrate extraction process. Such methodology presents problems for developing a comprehensive view of cation binding strength and selectivity. The picrate extraction data are not really comparable with binding constants determined in homogeneous solution. One hopes that there will be a trend toward using techniques that will produce true homogeneous cation binding constants. Only by doing so will it be possible to truly understand cation binding strength and selectivity.

Additional attention will be focused in the foreseeable future on the thermodynamics of cation binding. Some information is available, but cation binders will not be truly useful until the enthalpic (ΔH) and entropic ($T\Delta S$) components of complexation can be separated and understood.

Two existing areas of endeavor will also be expanded: cooperative binding and switching. In the latter case, extensive studies of photochemical, thermal, and redox (particularly electrochemical) switching have been reported.[52] Relatively little attention has been given to redox-switched materials, although a few examples are now available. In particular, Stoddart and Kaifer have shown that molecular boxes exhibit altered binding behavior when their electronic properties are altered by electrochemical processes.[53] In our own group, we have recently shown that steroidal ferrocene derivatives can form vesicles when the ferrocene (Fe^{II}) is oxidized to ferricinium (Fe^{III}).[54] This process is reversible, which means that a guest may be trapped inside the vesicle and that reduction will cause vesicle rupture and guest release. Shinkai and co-workers[55] have demonstrated thermally altered permeability properties, and more such systems are likely to be developed.

One of the known examples of cooperative binding is Reinhoudt's urea complexing agent.[56] In this structure, the uranyl ion is complexed by the salen unit and, in concert with the macroring, provides a cavity suitable for urea (Figure 1.17). Rebek has demonstrated cooperative binding (allostery) in a bipyridyl system using one metal to control the binding affinity of the compound for another ion.[57] The ferrocene cryptand work reported by Hall,[58] Akabori,[59] Beer,[60] Saji,[61] Kaifer and Gokel,[62] and others is based upon a similar principle. Oxidation of the ferrocene nucleus to a cation reduces the cation binding affinity. This property may be used to mediate cation transport. Novel examples of cooperativity and switching will be a significant area of future study in macrocycle chemistry.

1.5.3 Membranes and Channels

Novel membrane structures and the fabrication of cation-conducting channels will be an important application for macrocycles and related structures

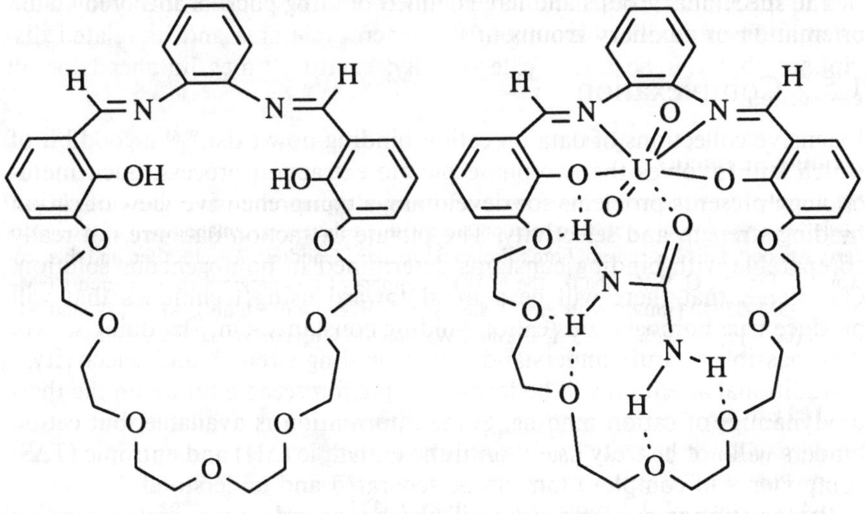

Figure 1.17. Cooperative binding of uranyl ion and urea by a macrocyclic salen derivative.

during the next decade. After many years of attempts, the first successful cation channels have been reported. New cation channel structures will undoubtedly follow, and new approaches such as the recently reported covalent linking of two gramicidin structures will also emerge.[37] The successful cation channels have thus far been inherently flexible structures, and these successes will likely reinforce the flexible framework approach outlined here.

1.5.4 Mixed Systems

It also seems certain that lines between disciplines will blur, as suggested by Beer's report of ferrocenyl calixarenes[63] and cavitands.[64] The observations of Atwood and co-workers that calixarenes form polyhydrated bilayers reminiscent of certain clays reinforces this notion.[65] Molecules are likely to get larger and more complex. Our ability to predict the properties of new structures will also improve, but this may well be a slow process.

Three recent reports reinforce this new direction. Saigo[66] and co-workers have merged a bisphenol-A box with a crown ether to make a ditopic receptor for a hydrophobic ammonium ion. Kuroda, Hiroshige, and Ogoshi have sandwiched a porphyrin between two cyclodextrins, resulting in a catalyst for epoxidation reminiscent of cytochrome P450.[67] Finally, calaix[4]arene has been converted into its *bis*(crown) derivatives by Pochini and co-workers.[68] This hybrid structure is a new type of cation complexing agent.

The speculation presented here is limited in scope and probably in vision as well. There is one thing about the macrocycle area and its related disciplines that can be stated categorically: exciting times lie ahead for all concerned.

Acknowledgments

We warmly thank the many co-workers whose names appear on the literature cited. In particular, this work would not have been possible without the expert collaboration of Professors Jerry Atwood, Luis Echegoyen, Frank Fronczek, Richard Gandour, Angel Kaifer, and their co-workers. Finally, G. W. G. thanks the NIH for support of this program through grants GM–36262, GM–33940 (jointly with Prof. Echegoyen, University of Miami), AI–27179, and AI–30188 (both jointly with Prof. J. I. Gordon, Washington University Medical School).

References

1. (a) Pedersen, C. J. *J. Am. Chem. Soc.* **1967,** *89,* 2495.
   (b) Pedersen, C. J. *J. Am. Chem. Soc.* **1967,** *89,* 7077.
   (c) Pedersen, C. J. *J. Inclusion Pheom.* **1988,** *6,* 337.

2. Gokel, G. W. *Crowns and Cryptands;* Royal Society of Chemistry; London, 1991.

3. *Host-Guest Complex Chemistry Macrocycles;* Vögtle, F., and E. Weber, Eds. Springer Verlag; Berlin, 1985.

4. Lehn, J.-M.; Potvin, P. G. *Progr. Macrocycle Chem.* **1987,** *3,* 1969.

5. Benner, S. A.; Ellington, A. D. *Bioorganic Chemistry Frontiers;* Springer Verlag; Heidelberg, 1990; Chapter 1, p 1.

6. Inoue, Y.; Gokel, G. W. *Cation Binding by Macrocycles;* Marcel Dekker; New York, 1990.

7. (a) Brockmann, H.; Schmidt-Kastner, G. *Chem. Ber.* **1955,** *88,* 57.
   (b) Grell, E.; Funck, T.; Eggers, F. *Membranes* **1975,** *5,* 1.
   (c) Pinkerton, M.; Steinrauf, L. K.; Dawkins, P. *Biochem. Biophys. Res. Commun.* **1969,** *35,* 512.

8. Ovchinnikov, Y. A.; Ivanov, V. T.; Shkrob, A. M., *Membrane-Active Complexones,* Elsevier, Amsterdam, 1974.

9. (a) Duax, W. L.; Hauptman, H.; Weeks, C. M.; Norton, D. A. *Science* **1972,** *176,* 911.
   (b) Smith, G. D.; Duax, W. L.; Langs, D. A.; DeTitta, G. T.; Edmonds, J. W.; Rohrer, D. C.; Weeks, C. M. *J. Am. Chem. Soc.* **1975,** *97,* 7242.

10. Truter, M. R. *Structure and Bonding* **1973,** *16,* 71.

11. Henry, J. B. *Clinical Diagnosis and Management,* 16th ed. W.B. Saunders; Philadelphia, 1979; p 144.

12. Morf, W. E.; Ammann, D.; Bissig, R.; Pretsch, E.; Simon, W. *Progr. Macrocyclic Chem.* **1979,** *1,* 1.

13. Dishong, D. M.; Gokel, G. W. *J. Org. Chem.* **1982,** *47,* 147.

14. Lehn, J.-M.; *Angew. Chem., Int. Ed. Engl.* **1988,** *27,* 89.

15. (a) Gokel, G. W.; Dishong, D. M.; Diamond, C. J. *J. Chem. Soc., Chem. Commun.* **1980,** 1053.
    (b) Gokel, G. W.; Dishong, D. M.; Diamond, C. J. *Tetrahedron Lett.* **1981,** 1663.

16. Gokel, G. W.; Echegoyen, L.; Kim, M. S.; Hernandez, J.; de Jesus, M. *J. Inclusion Phenom.* **1989,** *7,* 73.

17. Gokel, G. W.; Goli, D.M.; Minganti, C.; Echegoyen, L. *J. Am. Chem. Soc.* **1983,** *105,* 6786.

18. (a) Dishong, D. M.; Diamond, C. J.; Cinoman, M. I.; Gokel, G. W. *J. Am. Chem. Soc.* **1983**, *105*, 586.
(b) Schultz, R. A.; White, B. D.; Dishong, D. M.; Arnold, K. A.; Gokel, G. W. *J. Am. Chem. Soc.* **1985**, *107*, 6659.

19. Arnold, K. A.; Gokel, G. W. *J. Org. Chem.* **1986**, *51*, 5015.

20. (a) Kaifer, A.; H. D. Durst; Echegoyen, L.; Dishong, D. M.; Schultz, R. A.; Gokel, G. W. *J. Org. Chem.* **1982**, *47*, 3195.
(b) Echegoyen, L.; Kaifer, A.; Durst, H. D.; Gokel, G. W. *J. Org. Chem.* **1984**, *49*, 688.
(c) Kaifer, A.; Echegoyen, L.; Durst, H.; Schultz, R. A.; Dishong, D. M.; Goli, D. M.; Gokel, G. W. *J. Am. Chem. Soc.* **1984**, *106*, 5100.

21. Kaifer, A.; Echegoyen, L.; Gokel, G. W. *J. Org. Chem.* **1984**, *49*, 3029.

22. Gokel, G. W. In *Inclusion Phenomena;* Atwood, J. L., E. Davies, and D. D. MacNicol, Eds.; Oxford University; Oxford, 1990; Vol. 4, 287.

23. (a) Gatto, V. J.; Gokel, G. W. *J. Am. Chem. Soc* **1984**, *106*, 8240.
(b) Gatto, V. J.; Arnold, K. A.; Viscariello, A. M.; Miller, S. R.; Gokel, G. W. *J. Org. Chem.* **1986**, *51*, 5373.

24. (a) Fages, F.; Desvergne, J. P.; Bouas-Laurent, H.; Lehn, J.-M.; Konopelski, J. P.; Marsau, P.; Barrana, Y. *J. Chem. Soc., Chem. Commun.* **1990**, 655.
(b) Zabirov, N. G.; Scherbakova, V. A.; Cherkasov, R. A. *Zh. Obsch. Khim.* **1990**, *60*(4), 786.
(c) Ossowski, T.; Schneider, H. *Chem. Ber.* **1990**, *123*(8), 1673.
(d) Bradshaw, J. S.; Krakowiak, K. E.; An, H.; Izatt, R. M. *Tetrahedron* **1990**, *46*, 1163.
(e) Golchini, K.; Mackovic-Basic, M.; Gharib, S. A.; Masilamani, D.; Lucas, M. E.; Kurtz, I. *Am. J. Physiol.* **1990**, *258*, F538.
(f) McDaniel, C. W.; Bradshaw, J. S.; Tarbet, K. H.; Lindh, B. C.; Izatt, R. M. *J. Inclusion Phenom.* **1989**, *7*, 545.
(g) Akabori, S.; Kumagai, T.; Habata, Y.; Sato, S. *J. Chem. Soc., Perkin Trans. 1* **1989**, 1497.
(h) Tsukube, H.; Adachi, H.; Morosawa, S. *J. Chem. Soc., Perkin Trans. 1* **1989**, 89.
(i) Lukyanenko, N. G.; Reder, A. S. *Khim. Geterotsikl. Soedin.* **1989**, 1673.
(j) Wambeke, D. M.; Lippens, W.; Herman, G. G.; Goeminne, A. M. *Bull. Soc. Chim. Belg.* **1989**, *98*, 307.
(k) Chekhlov, A. N.; Zabirov, N. G.; Cherkasov, R. A.; Martynov, I. V. *Dokl. Akad. Nauk SSSR* **1989**, *307*, 129.
(l) Minta, A.; Tsien, R. Y. *J. Biol. Chem.* **1989**, *264*, 19449.
(m) Bradshaw, J. S.; Krakowiak, K. E.; Izatt, R. M. *J. Heterocycl. Chem.* **1989**, *26*, 565.
(n) Wang, D.; Jiang, L.; Gong, Y.; Hu, H. *Gaodeng Xuexiao Huaxue Xuebao* **1989**, *10*, 148.
(o) Jurczak, J.; Ostaszewski, R.; Salanski, P. *J. Chem. Soc., Chem. Commun.* **1989**, 184.
(p) Akabori, S.; Kumagai, T.; Habata, Y.; Sato, S. *J. Chem. Soc., Chem. Commun.* **1988**, 661.
(q) Kleinpeter, E.; Gaebler, M.; Schroth, W. *Magn. Reson. Chem.* **1988**, *26*, 380.
(r) Simonov, Y. A.; Fonar, M. S.; Popkov, Y. A.; Andronati, S. A.; Orfeev, V. S.; Dvorkin, A. A.; Malinovskii, T. I. *Dokl. Akad. Nauk SSSR* **1988**, *301*, 913.
(s) Bradshaw, J. S.; Krakowiak, K. E.; Bruening, R. L.; Tarbet, B. J.; Savage, P. B.; Izatt, R. M. *J. Org. Chem.* **1988**, *53*, 3190.
(t) Tsukube, H.; Yamashita, K.; Iwachido, T.; Zenki, M. *Tetrahedron Lett.* **1988**, *29*, 569.
(u) Bradshaw, J. S.; Krakowiak, K. E. *J. Org. Chem.* **1988**, *53*, 1808.
(v) Schultz, A. Z.; Pinto, D. J. P.; Welch, M.; Kullnig, R. K. *J. Org. Chem.* **1988**, *53*, 1372.
(w) Shinkai, S.; Nakamura, S.; Ohara, K.; Tachiki, S.; Manabe, O.; Kayijama, T. *Macromolecules* **1987**, *20*, 21.
(x) Lukyananko, N. G.; Kirichenko, T. I.; Shcherbakov, S. V.; Nazarova, N. Y.; Karpenko, L. P.; Bogatskii, A. V. *Zh. Org. Khim* **1986**, *22*, 1769.

(y) Babb, D. A.; Czech, B. P.; Bartsch, R. A. *J. Heterocycl. Chem.* **1986**, *23*, 609.

(z) Bogatskii, A. V., Gorodnyuk, V. P.; Kotlyar, S. A.; Bondarenko, N. N.; Kostyanovskii, R. G. *Otkrytiya, Izobret* **1986**, *98; Chem. Abstr. 106: 67362v.*

(aa) Ricard, A.; Capillon, J.; Quiveron, C. *Polymer* **1985**, *25*, 1136.

(bb) Tsukube, H. *Bull. Chem. Soc. Jpn.* **1984**, *57*, 2685.

(cc) Tsukube, J. *J. Chem. Soc., Chem. Commun.* **1984**, 315.

(dd) Tsukube, H. *J. Chem. Soc., Chem. Commun.* **1983**, 970.

(ee) Keana, J. F. W.; Cuomo, J.; Lex, L.; Seyedrezai, S. E. *J. Org. Chem.* **1983**, *48*, 2647.

(ff) DeJong, F. A.; Van Zon, A.; Reinhoudt, D. N.; Torny, G. J.; Tomassen, H. P. M. *J. R. Neth. Chem. Soc.* **1983**, *102*, 164.

(gg) Shinkai, S.; Kinda, H.; Araragi, Y.; Manabe, O. *Bull. Chem. Soc. Jpn.* **1983**, *56*, 559.

(hh) Kobayashi, H.; Okahara, M. *J. Chem. Soc., Chem. Commun.* **1983**, 800.

(ii) Bogatskii, A. V.; Lukyanenko, N. G.; Pastushok, V. N.; Kostyanovsky, R. G. *Synthesis* **1983**, 992.

(jj) Frère, Y.; Gramain, P. *Makromol. Chem.* **1982**, *183*, 2163.

(kk) Tazaki, M.; Nita, K.; Takagi, M.; Ueno, K. *Chem. Lett.* **1982**, 571.

(ll) Cho, I; Chang, S.-K. *Bull. Korean Chem. Soc.* **1980**, 145.

(mm) Gramain, P.; Kleiber, M.; Frère, Y. *Polymer* **1980**, *21*, 915.

(nn) Kulstad, S.; Malmsten, L. A. *Acta Chem. Scand. Ser. B* **1979**, *B33*, 469.

(oo) Wester, N.; Vögtle, F. *J. Chem. Res. (S)* **1978**, 400.

(pp) Takagi, M.; Tazaki, M.; Ueno, K. *Chem. Lett.* **1978**, 1179.

25. Gandour, R. D.; Fronczek, R. R.; Gatto, V. J.; Minganti, C.; Schultz, R. A.; White, B. D.; Arnold, K. A.; Mazocchi, D.; Miller, S. R.; Gokel, G. W. *J. Am. Chem. Soc.* **1986**, *108*, 4078.

26. Arnold, K. A.; Echegoyen, L.; Fronczek, F. R.; Gandour, R. D.; Gatto, V. J.; White, B. D.; Gokel, G. W. *J. Am. Chem. Soc.* **1987**, *109*, 3716.

27. (a) White, B. D.; Arnold, K. A.; Gokel, G. W. *Tetrahedron Lett.* **1987**, 1749.

(b) White, B. D.; Fronczek, F. R.; Gandour, R. D.; Gokel, G. W. *Tetrahedron Lett.* **1987**, 1753.

28. White, B. D.; Mallen, J.; Arnold, K. A.; Fronczek, F. R.; Gandour, R. D.; Gehrig, L. M. B.; Gokel, G. W. *J. Org. Chem.* **1989**, *54*, 937.

29. Trafton, J. E.; Li, C.; Mallen, J.; Miller, S. R.; Nakano, A.; Schall, O. F.; Gokel, G. W. *J. Chem. Soc., Chem. Commun.* **1990**, 1266.

30. Behr, J. P.; Lehn, J. M.; Dock. A. C.; Moras, D. *Nature* **1982**, *295*, 526.

31. Tabushi, I.; Kuroda, Y.; Yokoto, K. *Tetrahedron Lett.* **1982**, 4601–4604.

32. (a) Neevel, J. G.; Nolte, R. J. M. *Tetrahedron Lett.* **1984**, 2263–2266.

(b) Kragton, U. F.; Roks, M. F. M.; Nolte, R. J. M. *J. Chem. Soc., Chem. Commun.* **1985**, 1275–1276.

33. Führhop, J.-H.; Liman, U.; David, H. H. *Angew. Chem. Int. Ed. Engl.* **1985**, *24*, 339–340.

34. Julien, L.; Lehn, J.-M. *Tetrahedron Lett.* **1988**, 3803–3806.

35. Carmichael, V. E.; Dutton, P. J.; Fyles, T. M.; James, T. D.; Swan, J. A.; Zojaji, M. *J. Am. Chem. Soc.* **1989**, *111*, 767–769.

36. Menger, F. M.; Davis, D. S.; Persichetti, R. A.; Lee, J.-J. *J. Am. Chem. Soc.* **1990**, *112*, 2451.

37. Stankovic, C. J.; Heinemann, S. H.; Schreiber, S. L. *J. Am. Chem. Soc.* **1990**, *112*, 3702.

37. (a) Fuhrhop, J.-H.; Mathieu, J. *J. Chem. Soc., Chem. Commun.* **1983**, 144.

(b) Fuhrhop, J.-H.; Liman, U. *J. Am. Chem. Soc.* **1984**, *106*, 4643.

(c) Fuhrhop, J.-H.; Fritsch, D.; Tesche, B.; Schmiady, H. *J. Am. Chem. Soc.* **1984**, *106*, 1998.

(d) Fuhrhop, J.-H.; David, H.-H.; Mathieu, J.; Liman, U.; Winter, H.-J.; Boekema, E. *J. Am. Chem. Soc.* **1986,** *108,* 1785.

(e) Fuhrhop, J.-H.; Fritsch, D. *Acct. Chem. Res.* **1986,** *19,* 130.

(f) Fuhrhop, J.-H.; Liman, U.; Koesling, V. *J. Am. Chem. Soc.* **1988,** *110,* 6840.

39. Wallace, B. A.; Ravikumar, K. *Science* **1989,** *241,* 182.

40. Langs, D. A. *Science* **1989,** *241,* 188.

41. (a) Schmidt, G. M. J.; Hodgkin, D. C.; Oughton, B. M. *Biochem. J.* **1957,** *65,* 744.

(b) Liquori, A. M.; DeSantis, P.; Kovacs, A. L.; Mazzarella, L. *Nature* **1966,** *211,* 1039.

(c) Urry, D. W. *Proc. Natl. Acad. Sci. USA* **1971,** *68,* 672.

(d) Urry, D. W.; Goodall, M. C.; Glickson, J. D.; Mayers, D. F. *Proc. Natl. Acad. Sci. USA* **1971,** *68,* 1907.

(e) Hull, S. E.; Karlsson, R.; Main, P.; Woolfson, M. M.; Dodson, E. J. *Nature* **1978,** *275,* 207.

(f) Urry, D. W.; Venkatachalam, C. M.; Spisni, A.; Bradley, R. J.; Trapane, T. L.; Prasad, K. U. *J. Membrane Biol.* **1980,** *55,* 29.

(g) Urry, D. W.; Venkatachalam, C. M.; Spisni, A.; Laüger, P. *Proc. Natl. Acad. Sci. USA* **1980,** *77,* 2028.

(h) Urry, D. W.; Prasad, K. U.; Trapane, T. L. *Proc. Natl. Acad. Sci.* **1982,** *79,* 390.

(i) Urry, D. W.; Walker, J. T.; Trapane, T. L. *J. Membrane Biol.* **1982,** *69,* 225.

(j) Urry, D. W.; Trapane, T. L.; Prasad, K. U. *Science* **1983,** *221,* 1064.

(k) Urry, D. W.; Alonso-Romanowski, C. M.; Venkatachalam, C. M.; Bradley, R. J.; Harris, R. D. *J. Membrane Biol.* **1984,** *81,* 205.

(l) Urry, D. W.; Trapane, T. L.; Venkatachalam, C. M.; Prasad, K. U. *J. Am. Chem. Soc.* **1986,** *108,* 1448

(m) Urry, D. W.; Trapane, T. L.; Venkatachalam, C. M. *J. Membrane Biol.* **1986,** *89,* 107.

42. (a) Kuwamura, T.; Kawachi, T. *Yakugaku* **1979,** *28,* 195; *Chem. Abstr.* 90:206248d.

(b) Kuwamura, T.; Akimuru, M.; Takahashi, H.; Arai, M. *Kenkyu Hokoku-Asahi Garasu Kogyo Gijutsu Shorekai* **1979,** *35,* 45; *Chem. Abstr.* 95:61394q.

(c) Kuwamura, T.; Yoshida, S. *Nippon Kagaku Kaishi* **1980,** 427; *Chem. Abstr.* 93:28168e.

(d) Ikeda, I.; Yamamura, S.; Nakatsuji, Y.; Okahara, M. *J. Org. Chem.* **1980,** *45,* 5355.

(e) Okahara, M.; Kuo, P. L.; Ikeda, I. *J. Chem. Soc., Chem. Commun.* **1980,** 586.

(f) Kuo, P.; Tsuchiya, K.; Ikeda, I.; Okahara, M. *J. Colloid Interfac. Sci.* **1983,** *92,* 463.

43. Gokel, G. W., Echegoyen, L. E. In *Bioorganic Chemistry Frontiers;* Dugas, H., Ed., **1990,** *1,* 115 and references cited therein.

44. Nakano, A.; Xie, Q.; Mallen, J. V.; Echegoyen, L.; Gokel, G. W. *J. Am. Chem. Soc.* **1990,** *112,* 1287.

45. Riddell, F. G.; Hayer, M. K. *Biochim. Biophys. Acta.* **1985,** *817,* 313.

46. Buster, D. C.; Hinton, J. F.; Millett, F. S.; Shungu, D. C. *Biohys. J.* **1988,** *53,* 145.

47. Hamilton, A. D. In *Advances in Supramolecular Chemistry;* Gokel, G. W., Ed.; JAI; Greenwich, CT, 1990; Vol. 1, p 1.

48. (a) Rebek, Jr., J. *Acct. Chem. Res.* **1984,** *17,* 258.

(b) Rebek, Jr., J. *Science* **1987,** *235,* 1478.

49. Medina, J. C.; Li, C.; Bott, S. G.; Atwood, J. L.; Gokel, G. W. *J. Am. Chem. Soc.* **1991,** *113,* 366.

50. (a) Izatt, R. M.; Bradshaw, J. S.; Nielsen, S. A.; Lamb, J. D.; Christensen, J. J., Sen, D. *Chem. Rev.* **1985,** *85,* 271.

(b) *Cation Binding by Macrocycles;* Inoue, Y., and G. W. Gokel, Eds.; Marcel Dekker; New York, 1990.

51. Bruening, R. L.; Izatt, R. M.; Bradshaw, J. S. In *Cation Binding by Macrocycles;* Inoue, Y., and G. W. Gokel, Eds.; Marcel Dekker; New York, 1990; Chapter 2, p 111.

52. (a) Shinkai, S.; Manabe, O. *Top. Curr. Chem.* **1984**, *121*, 67.
    (b) Kaifer, A. K.; Echegoyen, L. E. In *Cation Binding by Macrocycles;* Inoue, Y., and G. W. Gokel, Eds.; Marcel Dekker; New York, 1990; Chapter 8, p 363.
    (c) Shinkai, S. In *Bioorganic Chemistry Frontiers;* Dugas, H. Ed.; **1990**, *1*, 161.

53. Ashton, P. R.; Goodnow, T. T.; Kaifer, A. E.; Reddington, M. V.; Slawin, A. M. Z.; Spencer, N.; Stoddart, J. F.; Vincent, C.; Williams, D. J. *Angew. Chem., Int. Ed. Engl.* **1989**, *28*, 1369.

54. Medina, J. C.; Gay, I., Chen, Z., Echegoyen, L.; Gokel, G. W. *J. Am. Chem. Soc.* **1991**, *113*, 365.

55. (a) Shinkai, S.; Nakaura, S.; Tachiki, S.; Manabe, O.; Kajiyama, T. *J. Am. Chem. Soc.* **1985**, *107*, 3363.
    (b) Shinkai, S.; Nakamura, S.; Ohara, K.; Tachiki, S.; Manabe, O.; Kajiyama, T. *Macromolecules* **1987**, *20*, 21.

56. Van Staveren, C. J.; van Erden, J.; van Veggel, C. J. M.; Harkema, S.; Reinhoudt, D. N. *J. Am. Chem. Soc.* **1988**, *110*, 4994.

57. Rebek, Jr., J.; Costello, T.; Marshall, J.; Wattley, R.; Gadwood, R. C.; Onan, K. *J. Am. Chem. Soc.* **1985**, *107*, 7481.

58. (a) Bell, A. P.; Hall, C. D. *J. Chem. Soc., Chem. Commun.* **1980**, 163.
    (b) Hammond, P. J.; Beer, P. D.; Hall, C. D. *J. Chem. Soc., Chem. Commun.* **1983**, 1161.
    (c) Beer, P. D.; Elliot, J.; Hammond, P. J.; Dudman, C.; Hall, C. D. *J. Organomet. Chem.* **1984**, *263*, C37.

59. (a) Sato, M.; Kubo, M.; Ebine, S.; Akabori, S. *Tetrahedron Lett.* **1982**, 185.
    (b) Akabori, A.; Ohtomi, M.; Sato, M.; Ebine, S. *Bull. Chem. Soc. Jpn.* **1983**, *56*, 1455.

60. (a) Beer, P. D. *J. Organomet. Chem.* **1985**, *297*, 313.
    (b) Beer, P. D. *Chem. Soc. Rev.* **1989**, *18*, 409.

61. Saji, T.; Kinoshita, I. *J. Chem. Soc., Chem. Commun.* **1986**, 716.

62. Medina, J. C.; Goodnow, T. T.; Bott, S.; Atwood, J. L.; Kaifer, A. E.; Gokel, G. W. *J. Chem. Soc., Chem. Commun.* **1991**, 290.

63. Beer, P. D.; Keefe, A. D.; Drew, M. G. B. *J. Organomet. Chem.* **1988**, *353*, C10.

64. Beer, P. D.; Smythe, A. C.; Tite, E. L.; Ibbotson, A. *J. Organomet. Chem.* **1989**, *376*, C11.

65. Bott, S. G.; Coleman, A. W.; Atwood, J. L. *J. Am. Chem. Soc.* **1988**, *110*, 610.

66. Saigo, K.; Kihara, N.; Hashimoto, Y.; Lin, R. J.; Fujimura, H.; Suzuki, Y.; Hasegawa, M. *J. Am. Chem. Soc.* **1990**, *112*, 1144.

67. Kuroda, Y.; Hiroshige, T.; Ogoshi, H. *J. Am. Chem. Soc.* **1990**, *112*, 1594.

68. Arduini, A.; Casnati, A.; Dodi, L.; Pochini, A.; Ungaro, R. *J. Chem. Soc., Chem. Commun.* **1990**, 1597.

CHAPTER 2

Redox-Active Polyether Ligands: Toward Metal Ion Isotopic Separations

Zhihong Chen and Luis Echegoyen
Department of Chemistry
University of Miami
Coral Gables, FL 33124
U.S.A.

2.1 Introduction

One of the initial interests following the first syntheses of macrocyclic poly-ethers was to prepare analogous systems with higher stability constants and selectivities for the various cations. In general, crown ethers have been shown to be relatively strong cation binders, yet they lack selectivity and are not particularly efficient as cation-transporting agents.[1] On the other hand, cryptands have three-dimensional cavities that can accommodate a metal ion of suitable size. Cryptands also form very stable complexes with alkali and alkaline-earth metal ions while exhibiting high selectivity.[2] However, their binding constants are too high (typically in the order of 10^7–10^8 M^{-1}) for optimum cation transport properties due to ligand saturation in the membrane phase.[3,4]

In order for a cation carrier to exhibit optimum transporting abilities across a membrane phase, it should have a high cation-binding constant at the membrane–donor interface and possess a relatively low cation-binding constant at the membrane-receiving interface. The only way to meet this apparently contradictory requirement is to build into the molecular structure a so-called switching mechanism, based on the fact that some molecules have two different binding states. These two states can be easily and reversibly interconverted by external forces such as pH gradients,[5] light,[6] temperature,[7] and redox gradients.[8-25]

2.2 Enhanced Cation Binding Using Redox-Switched Ligands

The most efficient redox-switched ligands should be those having one or more negative charges on the redox-active center, a similar situation to that resulting upon deprotonation of pH-switched ligands.[5] In these ligands, an electroactive group is close to and in the proper geometrical orientation with respect to the cation-binding center. Electrochemical (or chemical) reduction of the ligand then leads to an excess negative charge that in turn enhances the binding of the cation. The binding of the cation and the redox process are thus coupled. Since the redox process is reversible, a switch is present that allows control between low and high cation-binding states. An advantage offered by these redox-active ligands results from the development of an additional negative charge on the ligand, which enhances the electrostatic interaction with a cation and increases the binding ability of the ligand. Some of the redox-active groups that have been used for this switching mechanism are nitrobenzene,[10–14] azo,[15] and quinone.[16–27]

The first series of redox-switched ligands to be studied were lariat ethers with nitrobenzene as the redox center.[10–14] Nitrobenzene can undergo a reversible, one-electron reduction. The anion radical of the o-nitrobenzyl substituent is able to form an intramolecular ion-pair with the cation bound by the macroring.[16] Although the nitrobenzene-substituted lariat ethers worked well in terms of cation-binding enhancement upon electrochemical reduction, they were not suitable as carriers due to the rapid decomposition of their anion radicals in the presence of water. On the other hand, the anion radical of anthraquinone was known to be very stable, even in aqueous and neutral pH media, as long as oxygen was excluded.[28] Therefore, efforts were made to replace the nitrobenzene groups with anthraquinone groups as the sidearms of the lariat ethers and also to synthesize a series of podands, crown ethers, and cryptands connected with the anthraquinone groups. (See structures 1–15 in Figure 2.1.)[16–26] The electrochemical behavior of anthraquinone differs from that of nitrobenzene in the ability of the former to undergo a second, quasi-reversible reduction step to form the corresponding dianion. As a result, the formation of the dianion leads to an additional complexation equilibrium in the redox process in the presence of the cation. The equilibria occurring in solution can be represented by Scheme I as shown in Figure 2.2. K_1, K_2, and K_3 are binding constants for the neutral, the monoanion, and the dianion states of the ligand, respectively. The binding enhancements K_2/K_1 and K_3/K_2 can be obtained from the corresponding formal potentials, as long as these are well resolved. Binding enhancement values between 10^2–10^6 have been reported for anthraquinone-substituted ligands.[16–26] The general trend of binding enhancement for anthraquinone compounds is the same as that for the nitrobenzene compounds. The binding enhancement order is $Li^+ > Na^+ > K^+$ for the formation of the complexes by both the monoanion and the dianion.

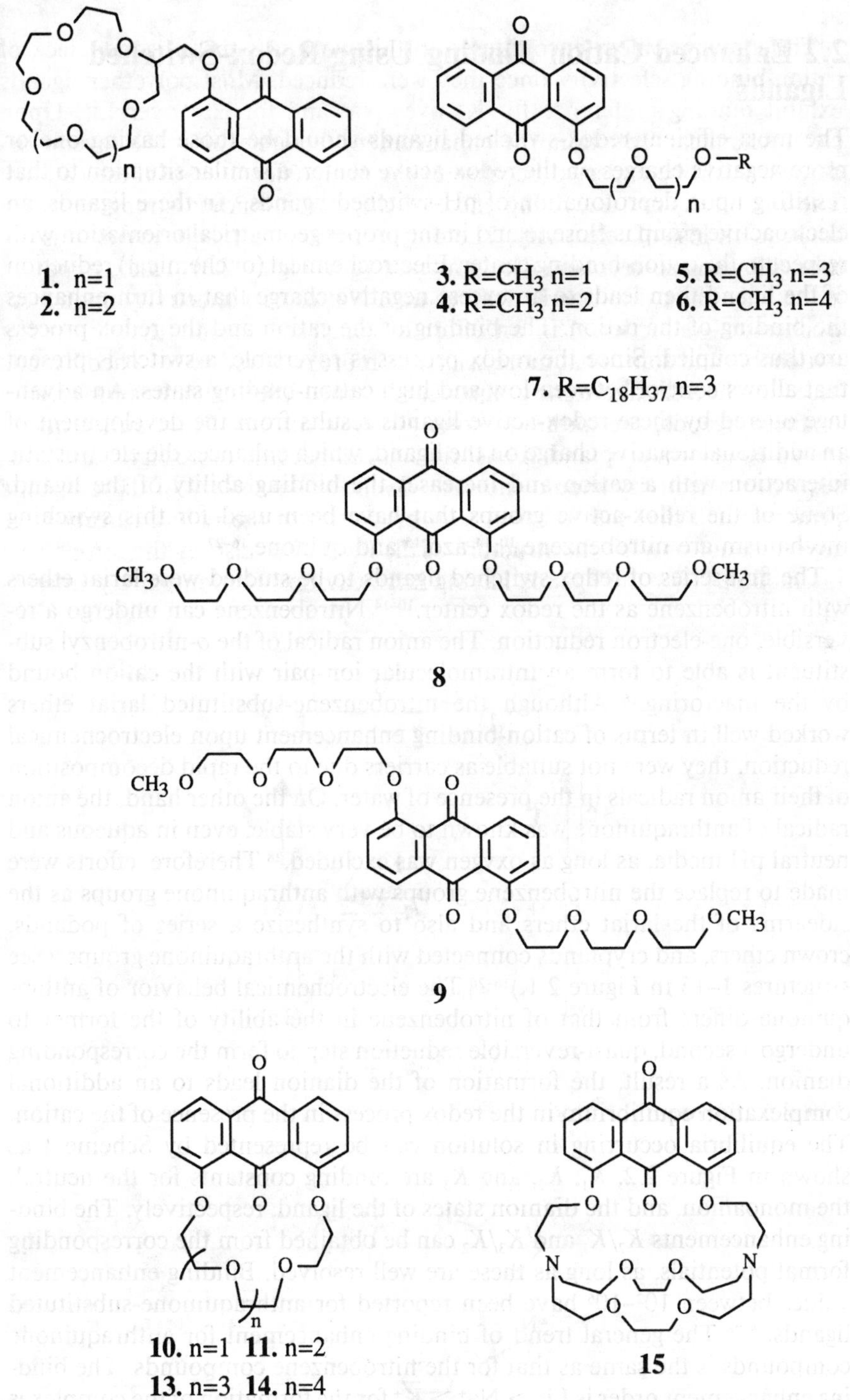

1. n=1
2. n=2

3. R=CH₃ n=1
4. R=CH₃ n=2

5. R=CH₃ n=3
6. R=CH₃ n=4

7. R=C₁₈H₃₇ n=3

8

9

10. n=1 **11.** n=2
13. n=3 **14.** n=4

15

Figure 2.1. Redox-switched polyethers bearing quinone groups.

The major problem encountered in this work was these ligands' lack of cation-binding selectivity once they were reduced. Most polyether ligands exhibit binding preference for K^+ over Na^+ and for Na^+ over Li^+. Upon electrochemical reduction, the enhancement factors exhibit the reverse order, as would be expected because of the charge densities of these cations. The net results are a leveling effect of the cation-binding strengths and a consequent loss of selectivity. The only exception to this behavior was exhibited by an azo-cryptand (See structure 16 in Figure 2.3). This compound is the only one exhibiting K^+ (rather than Na^+ or Li^+) selectivity upon electrochemical reduction.[15] In an effort to construct systems that are redox switchable and somewhat more selective, more rigid redox-switched ligands based on the anthraquinone group were designed. (See structures 10–15.)[24-26] In these ligands, the anthraquinone group is directly fused either into a crown ether or a cryptand. Thus, the ligands become more rigid, and consequently selectivity should increase. Cyclic voltammetry of these ligands demonstrated that the complexation properties of the reduced ligands are governed not only by the structure of the ligand but also by the ion-pairing interaction.[26] For the ligands having large cavities, the ion-pairing interac-

$$L + M^+ \underset{}{\overset{K_1}{\rightleftarrows}} LM^+$$

$$+e \Big\updownarrow 1 \qquad\qquad +e \Big\updownarrow 1'$$

$$L^- + M^+ \underset{}{\overset{K_2}{\rightleftarrows}} LM^-$$

$$+e \Big\updownarrow 2 \qquad\qquad +e \Big\updownarrow 2'$$

$$L^= + M^+ \underset{}{\overset{K_3}{\rightleftarrows}} LM^-$$

Figure 2.2. Scheme I

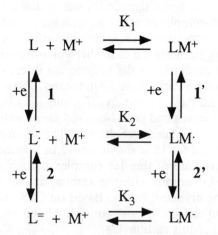

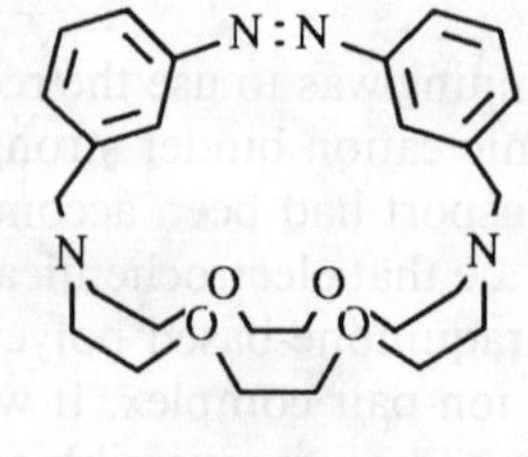

Figure 2.3. Structure *16*

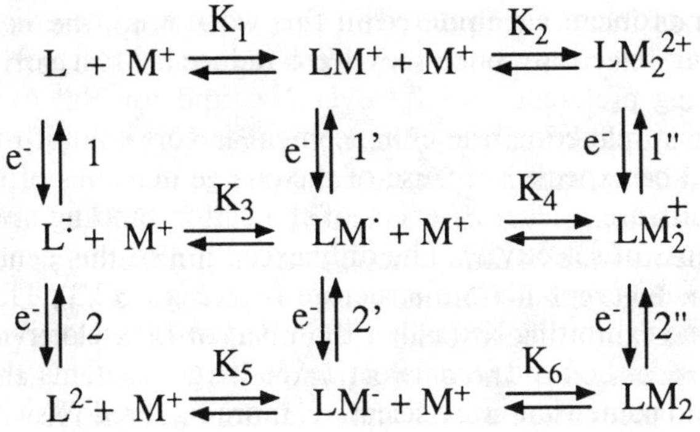

$$L + M^+ \underset{}{\overset{K_1}{\rightleftharpoons}} LM^+ + M^+ \underset{}{\overset{K_2}{\rightleftharpoons}} LM_2^{2+}$$

$$e^- \updownarrow 1 \qquad\qquad e^- \updownarrow 1' \qquad\qquad e^- \updownarrow 1''$$

$$L^{\cdot-} + M^+ \underset{}{\overset{K_3}{\rightleftharpoons}} LM^{\cdot} + M^+ \underset{}{\overset{K_4}{\rightleftharpoons}} LM_2^{\cdot+}$$

$$e^- \updownarrow 2 \qquad\qquad e^- \updownarrow 2' \qquad\qquad e^- \updownarrow 2''$$

$$L^{2-} + M^+ \underset{}{\overset{K_5}{\rightleftharpoons}} LM^- + M^+ \underset{}{\overset{K_6}{\rightleftharpoons}} LM_2$$

Figure 2.4. Scheme II

tion controls the binding strength. Thus, the selectivity of the ligand upon reduction is leveled off. However, the cation-binding affinity of smaller anthraquinone-based and reduced crown ethers is determined by a combination of the binding ability of the neutral ligand and the charge/size ratio of the cation.

Most anthraquinone-based ligands (as well as the nitrobenzene-based ligands) form cation complexes of 1:1 stoichiometry. An interesting case in exception to this rule was ligand 15, for which further additions of the cations beyond the 1:1 stoichiometry led to further voltammetric changes.[25] For the first time, the voltammetric results showed the direct resolution of six reduction waves—two for each of the ligand states. These are the free ligand, the 1:1 complex, and the 1:2 complex. Previously, a maximum of only four resolved reduction waves had been observed for an anthraquinone-podand or a crown ether, as stipulated by Scheme I. The scheme that best describes the behavior observed for 15 is shown above in Scheme II as illustrated in Figure 2.4. The structure of this 1:2 complex probably involves an interaction of the two Li+ cations with the anthraquinone carbonyl that points to the inside of the cryptand cavity. Based on CPK models, it is possible for each of the two cations to "sit" comfortably on two "pockets" created by the crown around this carbonyl.[25,26]

2.3 Enhanced Cation Transport Using Redox-Switching Ligands

The strategy from the beginning was to use the redox-switching mechanism to make a weak but dynamic cation binder stronger and then to deactivate the binder after cation transport had been accomplished. The first stage of this effort was to demonstrate that electrochemical reduction of the sidearm in nitrobenzene- and anthraquinone-based polyethers led to the formation of a strong intramolecular ion-pair complex. It was also demonstrated that the electrochemical redox process was reversible and that the binding affinity of these ligands increased upon reduction. Therefore, the next stage was obviously to utilize those redox-switchable ligands as the carriers in cation transport processes.

The first example of cation–electron coupled transport in which both activation and deactivation of the carriers were conducted electrochemically in a liquid membrane was demonstrated in 1989.[29] The carriers used in these experiments were anthraquinone-derived lariat ethers and podands 1, 2, 5, and 7 in Figure 2.1. Cation transport through a CH_2Cl_2 bulk liquid membrane was controlled by electrochemical means. The procedure involved both reduction of the anthraquinone-based carrier near the source interface to enhance cation uptake and oxidation near the receiving interface to increase the rate of cation release.[29]

Three distinct transport stages were monitored. The *neutral stage,* which simply involves cation transport by the parent (and thus neutral) ligand, is conducted in the absence of any applied potential. The *reduced stage* involves only controlled potential reduction of the carrier at the source interface. Transport is thus effected by the reduced ligand during this stage of the experiment. In addition to reduction, the *pumped stage* involves oxidation of the reduced carrier complexes at the receiving interface, which is performed by electrochemical means. The rates of cation transport increased from the neutral to the reduced to the pumped state in most cases.

The first electrochemically switched cation transport experiments necessitated the use of bulk liquid membrane models due to experimental demands. However, bulk liquid membranes are rather poor models of real lipid bilayers, and they are plagued with many other limitations such as interfacial disruption caused by stirring. Furthermore, redox switching of carriers was initially achieved via electrochemical reduction and oxidation using electrodes. Although separate potentiostats were used, voltage application and subsequent currents through the solutions were not completely independent. Since the goal was to mimic electron–cation coupled transport in a lipid bilayer system, it was impossible to use an electrochemical switching scheme coupled with a real vesicle system. Therefore, it was necessary to use chemical means instead of electrochemical means to obtain a redox potential gradient. The well-defined and little-studied solid-supported liquid membrane system was the first choice for examining the mechanism of enhanced electron–cation coupled transport by chemically redox-controlled lariat ethers and podands.

It was shown that crown-ether- and podand-based redox-switchable ligands (1 and 7 in Figure 2.1, respectively) can be used to enhance cation transport via chemical-redox control.[26,30] The transport studies were conducted using a modified H-cell design.[26,30] In this design, the membrane used was orthonitrophenyl octyl ether, which is supported on a polypropylene film.[26,30] Na^+ transport was monitored by measuring the cation concentration (using atomic absorption) in the receiving phase as a function of time. The transport rates of Na^+ through a solid-supported, o-nitrophenyl octyl ether membrane depend on the charge state of the ligand and the cooperation between reduction at the source phase and oxidation at the receiving phase. Sodium borohydride in the aqueous source phase served as the reducing agent, and potassium ferricyanide in the receiving phase served as the oxidizing agent. The solid-supported system permits better control of experimental parameters than the bulk liquid membrane model. The relative rates of neutral, reduced, and pumped Na^+ transport were 1:4:16 for carrier 1.[30]

Preliminary results for transport experiments using liposomes have shown that electron transfer through a lipid bilayer membrane is facilitated by the redox-active lipophilic podands, such as 7, when a chemical redox gradient is imposed across the bilayer membranes. Enhanced coupled cation–electron transport has also been observed using this system.[26,31] Much future work is needed in this area, especially regarding liposomes and other bilayer systems, to further understand and quantitate the processes responsible for cation transport across membranes. Efforts are underway in many laboratories to expand on these initial results.[32]

2.4 Isotope Separation by Redox-Switched Ligands—Back to the Future

The study of isotopic separation is very important not only from a fundamental but also a practical point of view. Most methods employed in metal isotopic separations nowadays involve physical separations such as diffusion cells. Since chemical separation systems should be much cheaper than physical separation methods, the search for effective and practical chemical systems is an important endeavor.

The idea of using cyclic polyethers and cryptands to effect metal ion isotope separations is definitely not new.[33] The first report of calcium isotope separation using a crown ether appeared in 1976.[34] Since this publication appeared, a rather small number of researchers have attempted similar experiments using a wide variety of crown ethers and cryptands. The most comprehensive account on the subject of crown ethers and cryptands as applied to isotopic separations was published in 1985, and it offers a thorough review of the field at that time.[33] As a matter of fact, little progress seems to have been made since this review article was published, which may be a reflection of the rather small isotope fractionation factors that are available using these cyclic polyethers.

Most of the experiments that have been reported involve the use of liquid–liquid extraction systems.[33] In a typical experiment, a relatively lipophilic crown ether or cryptand is dissolved in an appropriate organic solvent, such as CH_2Cl_2 or $CHCl_3$, and the metal cation isotope mixture is dissolved in water. Both solutions are then put in contact, vigorously shaken, allowed to separate into their corresponding phases, and physically separated. Analysis of the cation isotopes remaining in both phases, which is usually performed via mass spectrometry, allows the determination of the isotope-partitioning power of the particular ligand. It should be noted that all isotopic separations involving crown ethers and cryptands are based on equilibrium, not kinetic, considerations.[35] A detailed theoretical description of equilibrium isotope fractionations is beyond the scope of this monograph. The reader is referred to several authoritative reviews for details.[33,36]

Some general trends and observations are in order concerning isotope effects using crown ethers and cryptands, and these are presented here. In the interest of brevity, only general results are discussed, rather than supplying detailed tables, and are as follows:

1. *All* crown ethers and cryptands exhibit some degree of isotopic recognition.
2. The isotopic fractionation ability seems to be somewhat related to the ligand cavity size–cation radius ratio[37,38]
3. Typical maximum α values observed are 1.027 for Ca^{2+} and 1.057 for Li^{+}[38]

Although these separation factors may appear too low to be of interest, they are the most promising chemical systems for practical enrichment of some metal isotopes.[39] Up until these studies were made with crown ethers and cryptands, the separation of most metal isotopes was always conducted using expensive physical methods. No practical chemical separation method for calcium isotopes had existed until these ligands were investigated, and some were successfully immobilized on solid supports, such as $[2_B.2.2]$.[40] The latter is a polymer resin that incorporates a benzo-[2.2.2] cryptand moiety and is commercially available from Merck-Schuchardt as Kryptofix 222B polymer.

Somewhat related to the topic of metal ion separations using synthetic crown and cryptand compounds is other work involving natural ionophores[41] and the observation of lithium isotope effects in vivo.[42–46] Monactin, a naturally occurring neutral ionophore, is able to separate $^{24}Na^{+}/$ $^{22}Na^{+}$ with a separation factor $\alpha = 1.038$.[41] To our knowledge, this is the only reported investigation of isotopic fractionation using naturally occurring ionophores. In vivo effects have been reported by Lieberman et al.[42–45] They have observed that when equal doses of ^6Li or ^7Li are administered to rats, the animals receiving ^6Li are initially less active than those receiving ^7Li.[42] They have also observed that the ^6Li isotope was translocated 5–8% faster than ^7Li across the membrane of human erythrocytes.[43]

The general interest in redox-switchable ligands discussed at the beginning of this monograph, together with the observations that have just been discussed, has led to the consideration of these ligands as alternatives for metal isotopic separations. One of the strongest inspirations leading to this decision came from the research work done by Stevenson and co-workers.[47–53] They had shown that very pronounced equilibrium isotope effects were observed for electron transfer reactions.[47-53] The general reaction describing the electron exchange equilibrium is presented schematically in Figure 2.5.

When $x=y=0$, $K_{eq}=1$ by definition. When $y=0$, and $x=1, 2, 3$, or 6, however, the corresponding K_{eq} values are 0.86, 0.55, 0.37, and 0.26, respectively. The monotonically decreasing values for K_{eq} clearly show that successive increases in deuteration disfavor the formation of the correspondingly heavier anion radical ion-pair. Further introduction of neutrons into the system by introducing ^{13}C atoms in addition to the six deuteriums results in an even more drastically reduced value of K_{eq}, down to 0.10. Other experiments have been performed by the same authors, including physical separation experiments that have yielded α values of 1.7 for the $x=6, y=0$ system and an α of 3.85 for the $x=y=6$ system. Although these controversial results have found many critics[54] and theoretical explanations are still lacking, these unusual isotopic effects have now been confirmed by various methods such as ESR,[47,49,51,53] NMR,[52] mass spectral, scintillation counting[48], and cyclotron resonance[50] in Stevenson's group and also by others using cyclic voltammetry.[55]

As shown in the reaction in Figure 2.5, a one-electron reduced group (benzene) is able to interact with a cation bound by a crown ether. Both of these functional groups are totally independent in this reaction. On the other hand, the redox-switchable ligands described in this monograph typically possess both of these structural and functional moieties covalently bound to each other in such a way as to provide optimal interaction with the cation. It is thus clear that these systems should be more effective isotope fractionating agents than the simple neutral crown ethers. In addition to the ion–dipole interactions, which are the only ones present in the crown–cation cases, these molecules exhibit strong intramolecular ion-pairing interactions due to both the presence of the negative charge upon reduction of the ligands and the unpaired electron effect, as described in Stevenson's work.

Prompted by these expectations, an investigation of isotopic recognition using a series of redox-switched lariat ethers, crown ethers, and cryptands was recently initiated. Only very preliminary data are presently available. The first experiment performed was to assess the degree of recognition (if any) between the naturally occurring lithium isotopes, ^{6}Li and ^{7}Li, by ligand

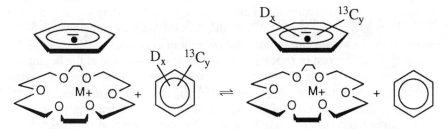

Figure 2.5. Stevenson's isotopic equilibrium indicated by electron exchange.

1.[56] The extent of isotopic recognition was determined from the corresponding reduction potentials measured from the cyclic voltammograms of the isotopically pure complexes in nonpolar media.

As described before, there are two additional redox waves corresponding to processes 1' and 2' in Scheme I after the addition of cation to the solution containing anthraquinone-substituted ligands. However, only one redox wave corresponding to 1' was observed after the addition of lithium ions due to an electron transfer reaction between the anion radical species.[16] Therefore, the redox potential for the 1:1 (ligand: Li$^+$) complex is readily determined when an excess of Li$^+$ is present in the solution. The isotopic recognition factor, α, which is equivalent to the equilibrium constant for the reaction shown below, can be calculated

$$^7Li^+ - 1^- \cdot + \ ^6Li^+ - 1 \ \overset{K}{\rightleftharpoons} \ ^7Li^+ - 1 + \ ^6Li^+ - 1^- \cdot$$

from the difference of the formal potentials for the complexes of 1 with $^6Li^+$ and $^7Li^+$. Experiments were performed in nonpolar solvents with different ratios of acetonitrile and dichloromethane. Ferrocene was used as an internal potential standard. The results are presented in Table 2.1.

As shown in this table, ligand *1* is unable to discriminate between the two lithium isotopes in a relatively polar medium, CH_3CN in this case. As the percentage of CH_2Cl_2 is increased, $\Delta\Delta E$ increases up to 12 mV, and the lithium isotopic recognition factor increases to 1.6. The fact that the redox potential of the $^6Li^+$-ligand complex appeared at a more positive potential than that of the $^7Li^+$ complex indicates that the reduced ligand can form a stronger complex with $^6Li^+$ than with $^7Li^+$.

Recently, the actual separation of lithium isotopes using these reducible ligands has been attempted.[57] The redox-switchable ligands were electrochemically reduced to the corresponding anion radicals or dianions in CH_2Cl_2. The resulting solutions were contacted with a solid lithium perchlorate sample of known $^6Li^+/^7Li^+$ composition. After separation of the solution from the solid lithium salts, the mixture of different isotopic lithium ions was extracted from the CH_2Cl_2 with water. Then the compositions of the lithium ions were analyzed by atomic absorption using ^6Li and ^7Li lamps following well-established analytical procedures.[58] Preliminary results in-

Table 2.1. Formal potentials and isotopic recognition factor for the complexes of *1* with $^6Li^+$ and $^7Li^+$

Solvent (volume percent)		$^6Li^+$	$^7Li^+$		
CH_3CN	CH_2Cl_2	ΔE (mV)	ΔE (mV)	$\Delta\Delta E$ (mV)	K
100	0	−1173	−1172	−1 ± 3	1.0 ± 0.1
50	50	−1195	−1202	7 ± 3	1.4 ± 0.1
25	75	−1162	−1174	12 ± 3	1.6 ± 0.2

dicate that the isotopic separation factors are between 1.13–1.18 for 1, 10, 11, and 15. It must be stressed at this point that these values correspond to real isotopic separation factors and not to indirectly measured values inferred from physical parameters.

An important note regarding these results with the separation of Li isotopes is now in order. The largest $^6Li/^7Li$ separation factor reported for a simple crown ether is 1.057 at 0°C.[38] Industrial processes presently employed for the separation of these isotopes have α values barely above 1.05. Based on the evidence obtained so far, it seems reasonable to anticipate that these reduced systems will be far more efficient than the simple crown ethers and cryptands in effecting isotopic separations. These are potentially useful commercial systems that could be exploited in the future.

Acknowledgments

The authors wish to express their gratitude to the Chemistry Division of the National Science Foundation (Grant CHE 9011901) and to the National Institute of Health (Grant GM 33940) for financial support of the work described in this monograph.

References

1. (a) Pedersen, C. J. *J. Am. Chem. Soc.* **1967**, *89*, 2495.
   (b) Pedersen, C. J. *J. Am. Chem. Soc.* **1967**, *89*, 7017.

2. (a) Dietrich, B.; Lehn, J.-M.; Sauvage, J. P. *Tetrahedron Lett.* **1969**, 2885.
   (b) Dietrich, B.; Lehn, J.-M.; Sauvage, J. P. *Tetrahedron Lett.* **1969**, 2889.

3. Lamb, J. D.; Christensen, J. J.; Oscarson, J. L.; Nielsen, B. L.; Asay, B. W.; Izatt, R. M. *J. Am. Chem. Soc.* **1980**, *102*, 6820.

4. Behr, J.-P.; Kirch, M.; Lehn, J. M. *J. Am. Chem. Soc.* **1985**, *107*, 241.

5. (a) Izatt, R. M.; Lamb, J. D.; Hawkins, R. T.; Brown, P. R.; Izatt, S. R.; Christensen, J. J. *J. Am.Chem. Soc.* **1983**, *105*, 1782.
   (b) Izatt, S. R.; Hawkins, R. T.; Christensen, J. J.; Izatt, R. M. *J. Am. Chem. Soc.* **1985**, *107*, 63.
   (c) Izatt, R. M.; Lindh, G. C.; Clark, G. A.; Bradshaw, J. S.; Nakatsuji, Y.; Lamb, J. D.; Christensen, J. J. *J. Chem. Soc., Chem. Commun.* **1985**, 1676.

6. Shinkai, S. *Pure & Appl. Chem.* **1987**, *59*, 425, and references therein.

7. (a) Shinkai, S.; Nakamura, S.; Tachiki, S.; Manabe, O.; Kajiyama, T. *J. Am. Chem. Soc.* **1985**, *107*, 3363.
   (b) Shinkai, S.; Nakamura, S.; Ohara, K.; Tashiki, S.; Manabe, O.; Kajiyama, T. *Macromolecules* **1987**, *20*, 21.
   (c) Shinkai, S.; Kazufumi, T.; Manabe, O.; Kajiyama, T. *J. Am. Chem. Soc.* **1987**, *109*, 4458.

8. Shinkai, S.; Inuzuka, K.; Hara, K.; Sone, T.; Manabe, O. *Bull. Chem. Soc. Jpn.* **1984**, *57*, 2150.

9. (a) Saji, T.; Kinoshita, I. *J. Chem. Soc., Chem. Commun.* **1986**, 716.
   (b) Saji, T. *Chem. Lett.* **1986**, 275.

10. Kaifer, A.; Echegoyen, L.; Gustowski, D. A; Goli, D. M.; Gokel, G. W. *J. Am. Chem. Soc.* **1983**, *105*, 7168.

11. Gustowski, D. A.; Echegoyen, L.; Goli, D. M.; Kaifer, A.; Schultz, R. A.; Gokel, G. W. *J. Am. Chem. Soc.* **1984**, *106*, 1633.

12. Kaifer, A.; Gustowski, D. A.; Echegoyen, L.; Gatto, V. J.; Schultz, R. A.; Cleary, T. P.; Morgan, C. R.; Goli, D. M.; Rios, A. M.; Gokel, G. W. *J. Am. Chem. Soc.* **1985,** *107,* 1958.

13. Delgado, M. Ph.D. Dissertation, University of Miami, Dec. 1987.

14. Miller, S. R.; Gustowski, D. A.; Chen, Z.; Gokel, G. W.; Echegoyen, L.; Kaifer, A. E. *Anal. Chem.* **1988,** *60,* 2021.

15. Gustowski, D. A.; Gatto, V. J.; Kaifer, A.; Echegoyen, L.; Godt, R. E.; Gokel, G. W. *J. Chem. Soc., Chem. Commun.* **1984,** 923.

16. Kaifer, A.; Echegoyen, L. In *Cation Binding by Macrocycles: Complexation of Cationic Species by Crown Ethers;* Inoue, Y., and G. W. Gokel, Eds.; Marcel Dekker; New York, 1990; Chapter 8, p 363.

17. Echegoyen, L.; Gokel, G. W.; Echegoyen, L. E.; Chen, Z.; Yoo, H. K. *J. Incl. Phen.* **1989,** *7,* 257.

18. Morgan, C. R.; Gustowski, D. A.; Cleary, T. P.; Echegoyen, L.; Gokel, G. W. *J. Org. Chem.* **1984,** *49,* 5008.

19. Echegoyen, L.; Gustowski, D. A.; Gatto, V. J.; Gokel, G. W. *J. Chem. Soc., Chem. Commun.* **1986,** 220.

20. Gustowski, D. A.; Delgado, M.; Getto, V. J.; Echegoyen, L.; Gokel, G. W. *Tetrahedron Lett.* **1986,** *27,* 3487.

21. Gustowski, D. A.; Delgado, M.; Gatto, V. J.; Echegoyen, L.; Gokel, G. W. *J. Am. Chem. Soc.* **1986,** *108,* 7553.

22. Gustowski, D. A. Ph.D. Dissertation, University of Miami, May 1987.

23. Delgado, M.; Echegoyen L.; Gatto, V. J.; Gustowski, D. A.; Gokel, G. W. *J. Am. Chem. Soc.* **1986,** *108,* 4135.

24. Delgado, M.; Gustowski, D. A.; Yoo, H. K.; Gokel, G. W.; Echegoyen, L. *J. Am. Chem. Soc.* **1988,** *110,* 119.

25. Chen, Z.; Schall, O. F.; Alcalá, M.; Li, Y.; Gokel, G. W.; Echegoyen, L. *J. Am. Chem. Soc.* **1991,** *113,* 365.

26. Chen, Z. Ph.D. Dissertation, University of Miami, Dec. 1991.

27. (a) Wolf, R. E.; Cooper, S. R. *J. Am. Chem. Soc.* **1984,** *106,* 4646.
    (b) Bock, H.; Hierholzer, B.; Schmalz, P. *Angew. Chem. Int. Ed. Engl.* **1987,** *26,* 791, and references therein.
    (c) Marayama, K.; Sohmiya, H.; Tsukube, H. *Tetrahedron Lett.* **1985,** 3583.
    (d) Toga, H.; Hashimoto, K.; Morihashi, K.; Kikuchi, O. *Bull. Chem. Soc. Jpn.* **1988,** *61,* 3026.
    (e) Ozeki, E.; Kimura, S.; Imanishi, Y. *J. Chem. Soc., Chem. Commun.* **1988,** 1353.

28. *The Chemistry of Quinonoid Compounds;* Patai, S., Ed.; Wiley; New York, 1974; vols. 1 and 2.

29. (a) Echeverria, L.; Delgado, M.; Gatto, V. J.; Gokel, G. W.; Echegoyen, L. *J. Am. Chem. Soc.* **1986,** *108,* 6825.
    (b) Echegoyen, L. E.; Yoo, H. K.; Gatto, V. J.; Gokel, G. W.; Echegoyen, L. *J. Am. Chem. Soc.* **1989,** *111,* 2440.

30. Chen, Z.; Gokel, G. W.; Echegoyen, L. *J. Org. Chem.* **1991,** *56,* 3369.

31. Chen, Z.; Xie, Q.; Echegoyen, L. unpublished results.

32. (a) Patterson, B. C.; Thompson, D. H.; Hurst, J. K. *J. Am. Chem. Soc.* **1988,** *110,* 3656.
    (b) Kobuke, Y.; Hamachi, I. *J. Chem. Soc., Chem. Commun.* **1989,** 1300.
    (c) Liu, M. D.; Patterson, D. H.; Jones, C. R.; Leidner, C. R. *J. Phys. Chem.* **1991,** *95,* 1858.

33. Heumann, K. G. *Top Curr. Chem.* **1985,** *127,* 77.

34. Jepson, B. E.; De Witt, R. *J. Inorg. Nucl. Chem.* **1976,** *38,* 1175.

35. Cox, B. G.; Schneider, H.; Stroka, J. *J. Am. Chem. Sc.* **1978,** *100,* 4746.

36. Bigeleisen, J.; Lee, M. W.; Mandel, F. *Ann. Rev. Phys. Chem.* **1973,** *24,* 407.

37. Knöchel, A.; Wilken, R.-D. *J. Am. Chem. Soc.* **1981,** *103,* 5707.

38. Nishizawa, K.; Takano, T.; Ikeda, I.; Okahara, M. *Sep. Sci. and Tech.* **1988,** *23,* 333.

39. Nishizawa, K. *J. Nucl. Sci. Technol.* **1984,** *21,* 133.

40. Heumann, K. G.; Schiefer, H.-P. *Angew. Chem., Int. Ed. Engl.* **1980,** *19,* 406.

41. Räde, H.-S.; Wagener, K. *Radiochim. Acta* **1972,** *18,* 141.

42. Lieberman, K. W.; Alexander, G. J.; Stokes, P. E. *Pharm. Biochem. Behav.* **1979,** *10,* 933.

43. Lieberman, K. W.; Stokes, P. E.; Kocsis, J. *Biol. Psychiatry* **1979,** *14,* 854.

44. Alexander, G. J.; Lieberman, K. W.; Stokes, P. E. *Biol. Psychiatry* **1979,** *15,* 469.

45. Stokes, P. E.; Okamoto, M.; Lieberman, K. W.; Alexander, G.; Triana, E. *Biol. Psychiatry* **1982,** *17,* 413.

46. Sherman, W. R.; Munsell, L. Y.; Wong, Y. H. *J. Neurochem.* **1984,** *42,* 880.

47. Stevenson, G. R.; Espe, M. P.; Reiter, R. C.; Lovett, D. J. *Nature* **1986,** *323,* 522.

48. Stevenson, G. R.; Lauricella, T. L. *J. Am. Chem. Soc.* **1986,** *108,* 5366.

49. Stevenson, G. R.; Espe, M. P.; Reiter, R. C. *J. Am. Chem. Soc.* **1986,** *108,* 5760.

50. Stevenson, G. R.; Reiter, R. C.; Espe, M. E.; Bartmess, J. E. *J. Am. Chem. Soc.* **1987,** *109,* 3847.

51. Lauricella, T. L.; Pescatore, J. A.; Reiter, R. C.; Stevenson, R. D.; Stevenson, G. R. *J. Phys. Chem.* **1988,** *92,* 3687.

52. Stevenson, G. R.; Sturgeon, B. E; Vines, K. S.; Peters, S. J. *J. Phys. Chem.* **1988,** *92,* 6850.

53. Stevenson, G. R.; Reidy, K. A.; Peters, S. J. *J. Am. Chem. Soc.* **1989,** *111,* 6578.

54. Marx, D.; Kleinhesselink, D.; Wolfsberg, M. *J. Am. Chem. Soc.* **1989,** *111,* 1493.

55. Goodnow, T. T.; Kaifer, A. E. *J. Phys. Chem.* **1990,** *94,* 7682.

56. Muñoz, S.; Echegoyen, L. *J. Chem. Soc., Perkin Trans. 2* **1991,** 1735.

57. Chen, Z.; Echegoyen, L. unpublished results.

58. Kushma, K. *Anal. Chim. Acta* **1986,** *133,* 225.

The Ins and Outs of Macromolecules

George R. Newkome

Center for Molecular Design and Recognition
Department of Chemistry
University of South Florida
Tampa, FL 33620
U.S.A.

3.1 Introduction

In 1968, we initiated a long-term research project that, simply stated, was to replace the oxygen atoms in the crown ether backbone with suitable N-heterocycles so that transition metal ions could be incorporated within the cavity. Nearly 15 years later, despite the little-known synthetic difficulties associated with this simple concept, the preparation of sexipyridine was completed. Herein is an encapsulated overview of our research in this area, as well as a projection of our approach to "the ins and outs of macromolecules."

With the reported "crown ethers" by Pedersen in the mid-1960s, researchers were instantly captivated by their synthetic, mechanistic, and structural possibilities. The potential use of this electron-rich cavity to encapsulate alkali and alkaline ether metals was obvious; a quarter century later studies continue—with marginal, truly new usable products resulting. In 1970, we approached "crown ethers" by addressing the inclusion of pyridine(s) in the framework, via sequential replacement of $-CH_2OCH_2-$ with $-C=N-C-$, to introduce rigidity into the macroring, to probe the interaction of directed N-electrons within the ring, to create a suitable locus for transition metals, and to devise a polyfunctionalized cavity possessing both a metal ion with redox capabilities and juxtaposed binding site for specific organic guests.

3.2 Heteromacrocycles

Introduction of a pyridino moiety to generate structure I was initially accomplished by direct nucleophilic substitution of dihalo- (di- and

The Ins and Outs of Macromolecules

George R. Newkome
Center for Molecular Design and Recognition
Department of Chemistry
University of South Florida
Tampa, FL 33620
U.S.A.

3.1 Introduction

In 1968, we initiated a long-term research project that, simply stated, was to replace the oxygen atoms in the crown ether backbone with suitable N-heterocycles so that transition metal ions could be incorporated within the cavity. Nearly 15 years later, despite the little-known synthetic difficulties associated with this simple concept, the preparation of sexipyridine was completed. Herein is an encapsulated overview of our research in this area, as well as a projection of our approach to "the ins and outs of macromolecules."

With the report of "crown ethers" by Pedersen[1] in the mid-1960s, researchers were instantly captivated by their synthetic, mechanistic, and structural possibilities. The potential use of this electron-rich cavity to encapsulate alkali and alkaline ether metals was obvious; a quarter century later studies continue—with marginal, truly new useable products resulting! In 1970, we approached "crown ethers" by addressing the inclusion of pyridine(s) in the framework, via sequential replacement of $-CH_2OCH_2-$ with $-C=N-C-$, to introduce rigidity into the macroring, to probe the interaction of directed N-electrons within the ring, to create a suitable locus for transition metals, and to devise a polyfunctionalized cavity possessing both a metal ion with redox capabilities and juxtaposed binding site for specific organic guests.

3.2 Heteromacrocycles

Introduction of a pyridino moiety to generate structure 1 was initially accomplished by direct nucleophilic substitution of dihalo- (di- and poly-)pyridines with dialkoxides generated from polyethylene glycols.[2] Such displacements generated pyridyl "crowns" possessing an imidate moiety, thus imposing an unfavorable bonding constraint into the resultant macrocycle.[3] Diverse nucleophiles[4] and electron-poor heterocycles[5,6] were readily adapted to this simple procedure. Simple homologation[7] of the starting haloheterocycles easily circumvented this problem and gave rise to the second series of macrocycles [e.g., structure 2; ($m = 0$)], which possess a suitable locus for transition metal complexation. Such simple technology was easily adopted to the synthesis of polypyridine macrocycles [structure 3; ($m \neq 0$)].[8] (See Figure 3.1 for structures 1–3.)

Our quest toward synthesizing sexipyridine (structure 3) was initiated by repeating the early work of others, in which dihalopyridines were treated with metals at elevated temperatures; needless to say, our studies were no more than successful than the dozens before us. This approach was doomed from the beginning due to the unfavorable metal ion complexation directing the termini away from coupling range; helical complexes have recently been reported to support this conjecture.[9] Thus a new approach was devised to introduce the rigidity in the last step via preparation of at least one pyridine moiety, as shown in Figure 3.2, ($n=3$).[10,11] Application ($n \neq 3$) of this *bis*-dithiane procedure to other related yet unknown macrocycles needs to be attempted.

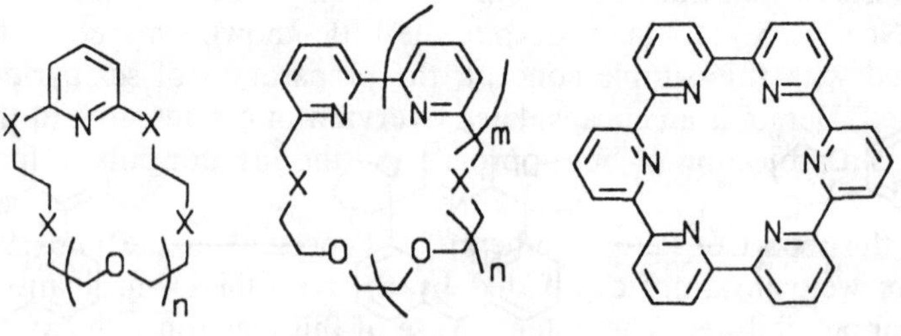

Figure 3.1. Pyridino and polypyridino macrocycles (see text).

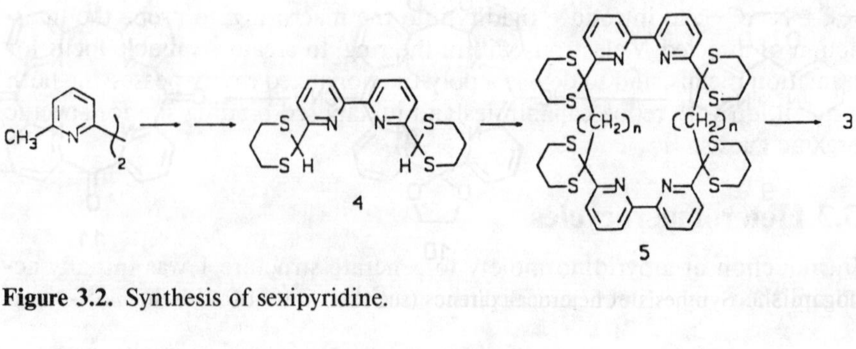

Figure 3.2. Synthesis of sexipyridine.

Simultaneous with these "crown ether" studies, preparation of hetero-macrocycles that possess directed *N*-electrons within the cavity gave rise to the area of heterocalixarenes.[12] In view of the marvelous works of Gutsche and others in the design and preparation of calixarenes,[13] the dearth of heteroderivatives was overpoweringly obvious. The key factor was that al-most all synthetic routes to the calixarenes utilized aromatic electrophilic substitution procedures—which are totally unfavorable for the incorporation of electron-withdrawing heterocycles, such as pyridine. We thus devised a novel nucleophilic procedure[14] using lithioacetonitrile; application of this procedure to structure 6 or 9 gave rise to cyclized intermediates, which were hydrolyzed to generate the desired trione 8[15] and the pyridyl porphyrin 11,[16] respectively. (See Figure 3.3 for structures 6–11.) The chemistry and inherent directivity of the inner *N*-electron pairs of the macrocycles possessing con-tiguous electron-deficient moieties give rise to novel radical processes,[17] resulting in a facile coupling procedure generating a lattice of polypyridines.

3.3 Cascade polymers

Over the past several decades, diverse cavities within a macrocyclic frame-work have been created to probe all sorts of host–guest interactions with the promise of generating futuristic catalytic properties. As time has pro-gressed, the complexity of the molecular designs has increased to include or incorporate appendages, inner design, specific binding sites, and so on;

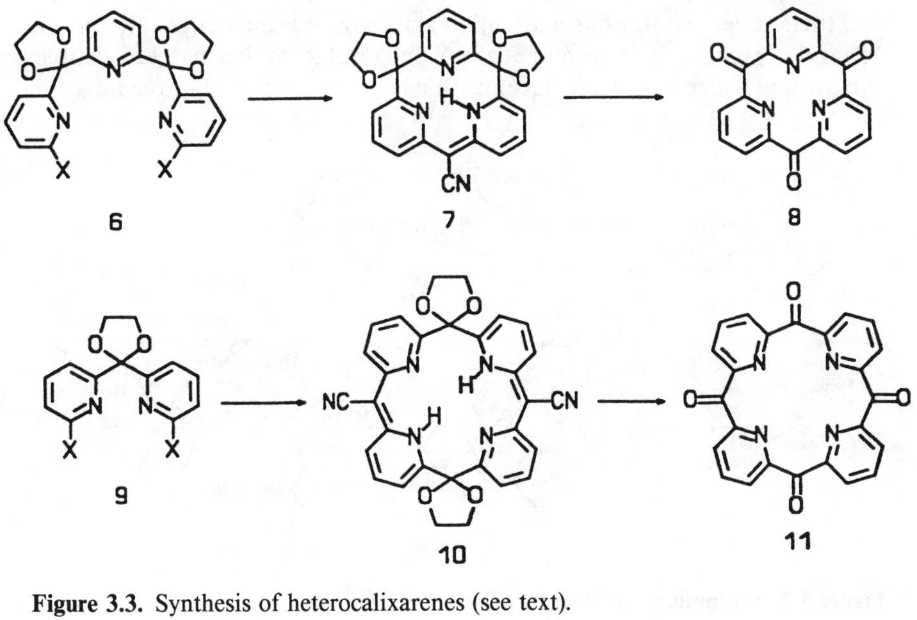

Figure 3.3. Synthesis of heterocalixarenes (see text).

however, cavity design may not necessarily be the best approach, especially because nature generally utilizes a catalytic surface within a canyon (or crevice), where reagents can easily ingress and products can egress. Thus, this concept leads in fact to an anti-"crown ether" model. A hole or cavity may not be the important factor, rather, the essential component may be a macrosurface or canyon. The resultant molecules must be large enough to engulf the "guest," and must possess a predetermined inner surface for docking with an outer hydrophilic surface for application within an aqueous environment. We created "arborols" to address this concept; these are molecules possessing a general lipophilic inner region with a hydrophilic surface that are based on a mathematical mode of synthesis related to fractal construction. Such macromolecules have specific size, shape, and inner–outer functionalized surfaces capable of host–guest interactions similar to those present in nature. Such species are truly superior in dimensions, and so unimolecular micelles, new materials, and catalytic surfaces within an aqueous environment are now forthcoming.

3.3.1. Arborols

Model arborol 12 best exemplifies the foundation for the construction of cascade polymers, which are micellar systems derived from the architectural model of trees.[18] This arborol is a "one-directional" cascade polymer, based on an alkyl tail with a spherical polyfunctional head comprising three tiers. The application of this technology to create two-[19] and three-[20] directional arborols (silvanols) has afforded 13, 14, and 15, respectively. (See Figure 3.4.) Most of the arborol series are slightly hygroscopic, water-soluble solids, and their aqueous solutions foam upon agitation. It is amazing that arborols 13 (and 14) with $n = 10$ or more ($n = 8$ or more) form thermally reversible, thixotropic aqueous gels at concentrations as low as 1.0 wt %; transmission

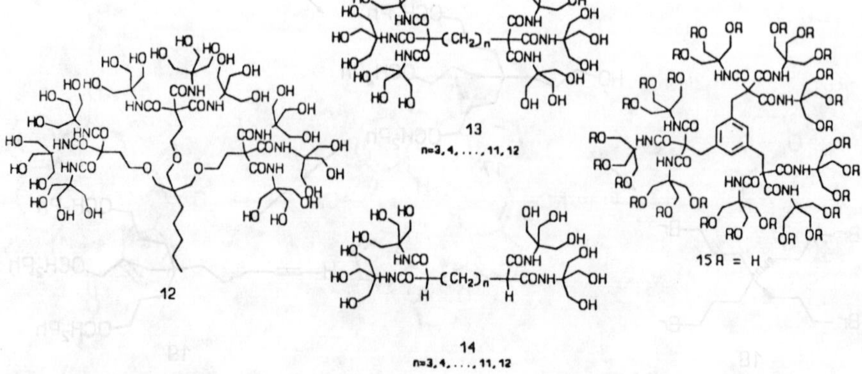

Figure 3.4. Molecular structures of arborols and silvanols.

electron microscopy confirms a specific aggregation composed of long fibrous rod structures, capable of molecular intercalation of lipophilic probes.[21] Thus, attachment of the cascade polymer moieties grants novel properties to the core molecule.

3.3.2 Micellanes™

A cascade spherical polymer, which best mimicked micellar topology and possessed a monodispersed molecular weight, was demonstrated by preparation of a series of four-directional saturated hydrocarbon cores with a multiple functionalized surface. The term *Micellane*™ was chosen to denote the family, with prefix numbers to describe the carbon building core and traditional terminology to characterize the surface functionality.[22]

Nitrotriol 16 was converted in several steps (see Figure 3.5) to novel tetra(*bis*-homologated) analogues of pentaerythritol, thus circumventing the need for "spacer" technology. The desired tetrabromide core 18 was obtained directly from the key intermediate 17, which also served to give the desired alkyne building block 19 in several steps. Alkylation of the four-directional core 18 with excess 19 using HMPA and LDA gave C{(CH$_2$)$_3$C≡C(CH$_2$)$_3$C[(CH$_2$)$_3$OCH$_2$Ph]$_3$}$_4$ (20), which was simultaneously reduced and deprotected, affording dodecaol 21. Facile conversion of 21 to the corresponding dodecabromide and repetition of the alkylation–reduc-

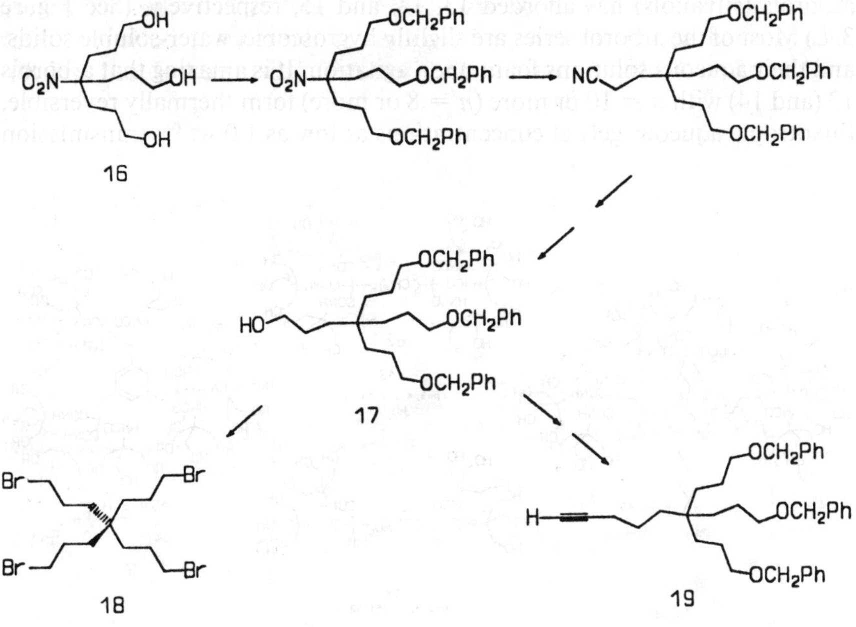

Figure 3.5. Synthesis of micellane building blocks.

tion sequence generated the hexatricontaol [$8^{2\cdot}3$]Micellanol™. Subsequent oxidation with RuO_4 gave [$8^{2\cdot}3$]Micellanoic Acid™ (22), whose tetramethylammonium salt exhibited micellar properties.[23] (See Figure 3.6.)

Other polymers possessing a quaternary carbon branching point, maximized density of terminal functionality, equal bond distances between the surface terminal groups and cascade center, and the same chemical connection ($-CH_2OCH_2CH_2CONH-$) among all adjacent quaternary carbons have been prepared and possess a symmetrical, three-dimensional ball-shape strucure.[24] Polyhomologation of pentaerythritol with acrylonitrile (See Figure 3.7), using Bruson's procedure,[25] gave the tetranitrile 23, which was hydrolyzed and converted smoothly into the corresponding acyl chloride 24 (core). Similarly, "tris" was alkylated to give the triethyl ester 25 (building block). Treatment of core 24 with excess amine 25 gave [12]-ester 26, which was hydrolyzed to [12]-acid 27. (See Figure 3.8.) Subsequent cascade generation was conducted via peptide formation using the [12]-acid and excess amine 25 in the presence of 1-hydroxybenzotriazole and dicyclohexylcarbodiimide, affording [36]-acid in reasonable yields. Repetitive reactions were conducted to generate additional tiers; the corresponding [108]- and [324]-acids have been prepared.

Construction of supramolecular molecules that are homogeneous spherical polymers possessing a predetermined porosity, molecular inclusion characteristics, and precise dense-packing limits is possible. Modification of the surface functionality can be readily accomplished, such as the attachment of polypyridines[26] and diverse metal ions. Chirality has been introduced into the core[27], and diverse dyes and molecular probes have been incorporated within the lyophilic region of the cascade polymers. Such control of molecular shapes and architectures utilizing covalent bond for-

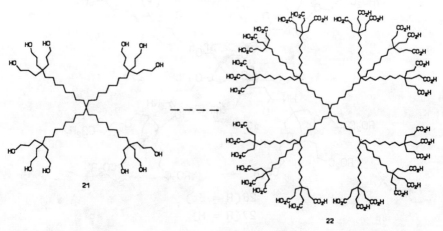

Figure 3.6. Conversion of dodecaol 21 to [$8^{2\cdot}3$] Micellanoic Acid™ (22).

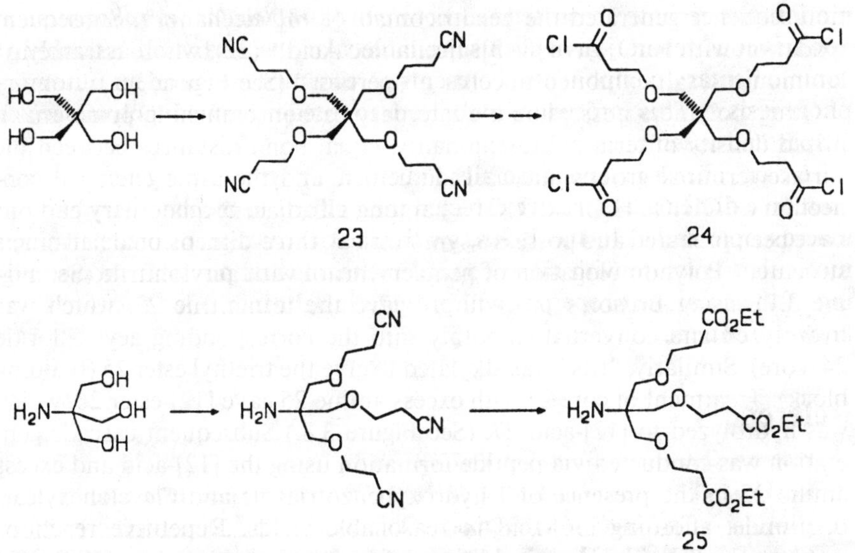

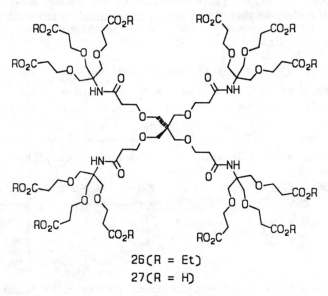

Figure 3.7. Homologation of pentaerythritol and 1,1,1-tris (hydroxymethyl) aminomethane to give Micellane™ building blocks.

26(R = Et)
27(R = H)

Figure 3.8. Structure of [12]-ester (R=Et) and [12]-acid (R=H).

mation has recently been termed *covalent* or *molecular morphogenesis.*[28] Spontaneous generation of such supramolecular species by self-assembly of complementary components represents a recognition-dependent automorphogenesis.[28] This procedure will lead to predetermined polymolecular shapes and assemblies.

To date, chemists have generally limited thier synthetic ingenuity to microscopic dimensions or have directed their efforts to biochemistry and the macroscopic arena. In the 1990s, synthetic construction of predetermined molecular structure with mesoscopic dimensions and physical understanding of their unique properties will provide insight into one of nature's unknown frontiers.

References

1. Pedersen, C. J. *J. Am. Chem. Soc.* **1967**, *89*, 7017.

2. Newkome, G. R.; Robinson, J. M. *J. Chem. Soc., Chem. Commun.* **1973**, 831.

3. Newkome, G. R.; McClure, G. L.; Broussard-Simpson, J.; Danesh-Khoshboo, F. *J. Am. Chem. Soc.* **1975**, *97*, 3232.

4. Newkome, G. R.; Danesh-Khoshboo, F.; Nayak, A.; Benton, W. H. *J. Org. Chem.* **1978**, *43*, 2685.

5. Newkome, G. R.; Nayak, A.; Sorci, M. G.; Benton, W. H. *J. Org. Chem.* **1979**, *44*, 3812.

6. Newkome, G. R.; Garbis, S. J.; Majestic, V. K.; Fronczek, F. R. and Chiari, G. *J. Org. Chem.* **1981**, *46*, 833.

7. Newkome, G. R.; Kiefer, G. E.; Xia, Y.-J.; Gupta, V. K. *Synthesis* **1984**, 676.

8. Newkome, G. R.; Sauer, J. D.; Roper, J. M.; Hager, D. C. *Chem. Rev.* **1977**, *77*, 313.

9. (a) Garrett, T. M.; Koert, U.; Lehn, J.-M.; Rigault, A.; Meyer, D.; Fischer, J. *J. Chem. Soc., Chem. Commun.* **1990**, 557.
   (b) Lehn, J.-M.; Rigault, A. *Angew. Chem., Int. Ed. Engl.* **1988**, *27*, 1095.
   (c) Lehn, J.-M.; Rigault, A.; Siegel, J.; Harrowfield, J.; Chevrier, B.; Moras, D. *Proc. Natl. Acad. Sci.* **1987**, *84*, 2565.

10. Newkome, G. R.; Lee, H.-W. *J. Am. Chem. Soc.* **1983**, *105*, 5956.

11. Newkome, G. R., Lee, H.-W.; Fronczek, F. R. *Isr. J. Chem.* **1986**, *27*, 87.

12. Newkome, G. R.; Joo, Y. J.; Fronczek, F. R. *J. Chem. Soc., Chem. Commun.* **1987**, 854.

13. Gutsche, C. D. In *Monographs in Supramolecular Chemistry;* Stoddart, J. F., Ed.; Royal Society Chemistry; London, 1989.

14. Newkome, G. R.; Joo, Y. J.; Evans, D.W.; Pappalardo, S.; Fronczek, F. R. *J. Org. Chem.* **1988**, *53*, 786.

15. Newkome, G. R.; Joo, Y. J.; Fronczek, F. R. *J. Am. Chem. Soc.* **1986**, *108*, 6074.

16. Newkome, G. R.; Joo, Y. J.; Fronczek, F. R. *J. Chem. Soc., Chem. Commun.* **1987**, 854.

17. Newkome, G. R.; Joo, Y. J.; Evans, D. W.; Fronczek, F. R. *J. Org. Chem.* **1990**, *55*, 5714.

18. Newkome, G. R.; Yao, Z.-Q.; Baker, G. R.; Gupta, V. K. *J. Org. Chem.* **1985**, *50*, 2003.

19. Newkome, G. R.; Baker, G. R.; Saunders, M. J.; Russo, P. S.; Gupta, V. K., Yao, Z.-Q.; Miller, J. E.; Bouillion, K. *J. Chem. Soc., Chem. Commun.* **1986**, 752.

20. Newkome, G. R.; Yao, Z.-Q.; Baker, G. R.; Gupta, V. K.; Russo, P. S.; Saunders, M. J. *J. Am. Chem. Soc.* **1986**, *108*, 849.

21. Newkome, G. R.; Baker, G. R.; Arai, S.; Saunders, M. J.; Russo, P. S.; Theriot, K. J.; Moorefield, C. N.; Rogers, L. E.; Miller, J. E.; Lieux, T. R.; Murray, M. E.; Phillips, B.; Pascal, L. *J. Am. Chem. Soc.* **1990,** *112,* 8458.

22. Newkome, G. R.; Moorefield, C. N.; Baker, G. R.; Johnson, A. L.; Behera, R. K. submitted for publication in *J. Am. Chem. Soc.* and presented at the Symposium on Self-assembling Structures at the 199th National Meeting of the American Chemical Society, Boston, MA, April, 1990.

23. Newkome, G. R.; Moorefield, C. N.; Baker, G. R.; Saunders, M. J.; Grossman, S. H. submitted for publication in *J. Am. Chem. Soc.*

24. Newkome, G. R.; Lin, X. *Macromolecules* in press.

25. Bruson, H. A. U.S. Patent 2,401,607; *Chem. Abstr.* **1946,** *40,* 5450.

26. Behera, R. K. unpublished results.

27. Weis, C. D.; Lin, X. unpublished results.

28. Lehn, J.-M. *Angew. Chem., Int. Ed. Engl.* **1990,** *29,* 1304.

CHAPTER 4

Tailoring Macrocycles for Medical Applications

David Parker
Department of Chemistry
University of Durham
South Road
Durham DH1 3LE
U. K.

4.1 Introduction

Crown chemistry has now come of age. The chemical framework and physical principles that underpin this exciting area of research have been defined. Two of the properties of macrocylic complexes that set them apart from related acyclic complexes are selectivity in binding and kinetic stability with respect to dissociation. The pronounced tendency of crowns to selectively bind a given cation in the presence of other competing ions is exemplified par excellence in the binding of sodium by spherand-6 and of potassium by cryptand [2.2.2]. Of course, the selectivity is tempered by the slow kinetics of association, particularly for spherands. The tendency of macrocyclic complexes to undergo slow kinetics of dissociation (whether direct, acid, or metal catalyzed) has been well documented.[1]

In seeking to harness these attractive features in useful applications, there are practical considerations to bear in mind. For any sensor, selectivity in binding one ion in the presence of competing ions is only one of several requirements that must be fulfilled. In an ion-selective electrode, for example, the essential feature is a thin polymeric membrane (containing the ionophore) that transports the ion under consideration. Binding of the ion should be fast and reasonably strong, equilibrium should be reached quickly, and the ionophore should be sufficiently lipophilic to inhibit leaching out to the aqueous phase.[2] These considerations—determining the response time and lifetime of a sensor—are equally important to the achievement of good selectivity.

When kinetically inert radiolabeled complexes of macrocyclic ligands are used in vivo, either for diagnostic or therapeutic applications, once again other considerations must be borne in mind.[3] For example, the *formation* of the radiolabeled complex must also occur rapidly and preferably under ambient conditions of pH and temperature. For a protein conjugated to a macrocyclic ligand, direct radiolabeling imposes even more exacting constraints. Nonspecific binding of the radioisotope to the protein must be minimized while retaining a good and rapid radiolabeling yield. Finally, many radioisotopes are available in a highly impure chemical form, notwithstanding their usually good radioisotopic purity. Contaminant metal cations (e.g., Ca^{2+}, Zn^{2+}, Cu^{2+}, Fe^{3+}) are often in 100-fold excess over the radioisotope itself, so that high selectivity in binding is also required.

The following examples illustrate these features when medical applications of macrocyclic ligands are sought in analysis (using a lithium-selective, ion-selective electrode[4,5]), diagnosis (in developing γ- or β^+-labeled antibodies for radioimmunoscintigraphy),[6,7] and in therapy (for β^--labeled tumor-localizing antibodies).[8,9,10]

4.2 Clinical Analysis

Ion-selective electrodes (ISEs) and related ISFETs are being developed for the continuous monitoring of blood cations and pH.[11] Most ISEs comprise a thin membrane composed of an inert polymer matrix (e.g., PVC) that holds an organic solution of an electrically neutral, ion-specific complexing agent. The membrane separates a reference solution (containing a fixed concentration of this ion) from the sample. By selectively transferring the ion to be measured from the sample solution to the membrane phase, a potential difference is set up that is logarithmically related to the activity of the sample ion in solution. The electrochemical cell is completed by an external reference electrode. With the aid of the Nicolsky–Eisenman equation (Eq. 4.1), the measured EMF may be related to the activity of the ion in solution.

$$\text{EMF} = E_0 + s \log [a_i + \Sigma_{ji} K_{ij}^{POT} (a_j)^{z_i/z_j}] \qquad (4.1)$$

(where E_0 is composed of two constant internal potential differences and a variable liquid junction potential generated between the reference electrolyte and the sample solution.)

$s = \dfrac{2.303\ RT}{Z_i F}$

Z_i^{ai} = charge and activity of the ith ion

$K_{ij}^{POT} = a_i/a_j$ is a selectivity factor that measures the preference of the sensor for the interfering ion j relative to the ion i to be detected. Ideally, all $K_{ij}^{POT} \to 0$.

Using calibration solutions, ISEs can therefore directly measure the activity of ions in whole blood. Representative concentrations of selected ionic species in different body fluids are listed in Table 4.1. Inspection of this table reveals that for the clinical analysis of K^+, Ca^{2+}, and Li^+, good selectivity over Na^+ is required.

Table 4.1. Representative Concentrations of Selected Ionic Species in Body Fluids.[a]

Ion	Serum			Intracellular Fluid			Urine		
	max.	mean	min.	max.	mean	min.	max.	mean	min.
H^+	5.6×10^{-5}	5×10^{-5}	4.3×10^{-5}	6×10^{-5}	—	4×10^{-5}	2×10^{-2}	—	4.1×10^{-5}
Li^+	—	$(<0.01)^b$	—	—	—	—	6.3×10^{-3}	—	4.3×10^{-4}
Na^+	150	140	135	~18	~10	~6	220	—	120
K^+	5	4	3.5	—	~200	—	80	—	35
Mg^{2+}	0.8	0.6	0.45	~3.2	—	~1.6	8.3	—	2.5
Ca^{2+}	1.2	1.1	1.0	—	~0.05	—	3.6	—	0.7
Cl^-	110	103	95	76.9	—	58.9	240	—	120
Br^-	~0.17	~0.03	$~9 \times 10^{-3}$	—	—	—	0.11	—	0.04
HCO_3^-	26.5	24	21.3	11.9	—	10.9	—	—	—
HPO_4^{2-}	~1	~0.5	~0.3	—	—	—	—	~18	—
SO_4^{2-}	~0.5	~0.3	~0.15	—	—	—	—	~25	—

[a] Concentrations in mmol dm^{-3}

[b] For normal healthy patients (For those being treated with Li_2CO_3, the mean value is 1.0)

Using the neutral carrier antibiotic valinomycin, excellent selectivity over sodium is found (log $K^{POT}_{K,Na} = -5.5$),[12] whereas excellent acyclic ionophores for Ca^{2+} have been developed by Simon.[13] No such well-behaved ISE for Li^+ analysis has been developed. At the outset of our own work, the most useful Li^+ sensors employed acyclic oxa-amide ligands *1* and *2* for which log $K^{POT}_{Li,Na} = -1.9$ and -2.4, respectively. The more selective ionophore *2* does not perform adequately in serum analysis, owing to substantial protein interference. The target figure for Li^+/Na^+ selectivity is log $K^{POT}_{Li,Na} = -4.3$ when less then 1% interference is found. The search for a good lithium ISE is a keen one, as Li_2CO_3 is administered in daily doses of ca. 1 g to patients suffering from manic depression. Close monitoring of lithium concentration is required during treatment in order to secure a therapeutic effect while avoiding overdose, which could lead to fatal poisoning.

Prior work has established that a 14-crown-4 skeleton has an optimum "cavity size" to incorporate a lithium ion. In order to enhance Li^+/Na^+ discrimination in binding, two additional ligating substituents were introduced so that the Li^+ ion (sitting in the plane of the four oxygens of the 14-C-4 ring) became 6-coordinate. The favors 1:1 complexation and inhibits competitive 2:1 (ligand to metal) complexation that is particularly important for sodium. In addition, the preferred additional donors were amide carbonyl oxygens, causing the ion/ion–dipole (and induced dipole) interactions with the more highly polarizing lithium cation to further promote Li^+/Na^+ selectivity. Thus ligand *3* (Figure 4.1) was designed,[4,5] and both the *n*-butyl and iso-butyl amides were examined and compared to both the dibenzyl-substituted 14-crown-4 derivative *4* and to ligand *1*, which forms the basis of lithium analysis in the commercial Corning ISE. The diamide-based electrodes behaved very well, showing log $K^{POT}_{ij} = -2.92$ for *3a* and -3.25 for *3b* in the presence of a simulated clinical background (i.e., 150 mmol NaCl, 4.3 mmol KCl, and 1.26 mM CaCl$_2$). The slope of the EMF versus lithium activity response was 61 mV at 37 °C (i.e., ideal Nernstian behavior almost up to the clinical range) (Figure 4.2). The sensor based on *3a* is stable in plasma, has a short response time (ca. 12 seconds), and a lifetime of at least 50 days, features that augur well for its clinical use.[5]

4.2.1 Future Developments

When considering which of the other ionic species present in body fluids (see Table 4.1) require improved methods of analysis, two ions are immediately apparent: Mg^{2+} and HCO_3^-. In clinical practice, most methods for estimating blood bicarbonate use values of pH and pCO_2 and calculate bicarbonate concentration from the Henderson–Hasselbalch equation (Eq. 4.2). Not only is the method slow, but it is somewhat inaccurate.

$$pH = pK + \log\frac{[HCO_3^-]}{[CO_2]_{dissolved}} \text{ (with } pK = 6.35\text{)} \qquad (4.2)$$

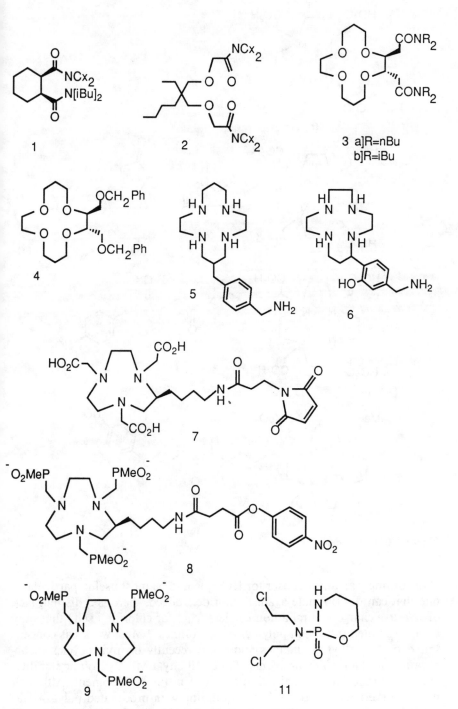

Figure 4.1. Ligands discussed in this chapter (see text).

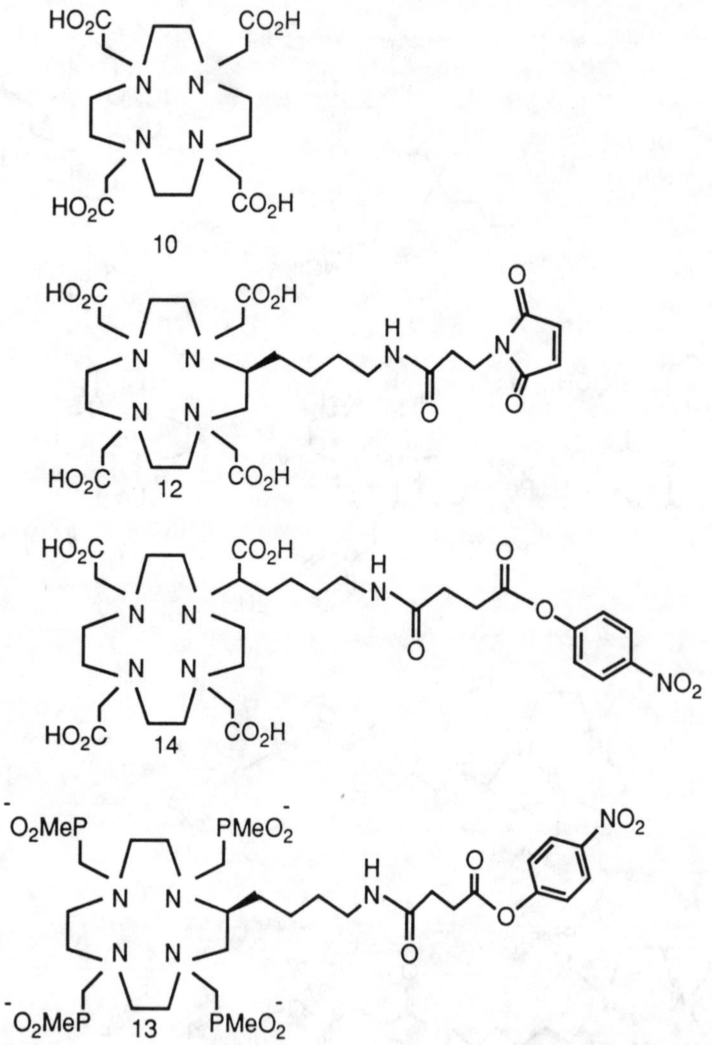

Figure 4.1. *Continued*

Accordingly an anion sensor for HCO_3^- would be most useful, particularly one that can discriminate against chloride. Indeed, there is a distinct lack of *selective* charge neutral anion carriers (e.g., for chloride itself) that may be incorporated in solvent polymeric membranes.[14] Most work has concentrated on polyalkyl tin halides and more recently on macrocycles incorporating tin,[15] for which problems of lipophilicity, EMF, hydrolytic stability, and membrane lifetime have limited any fruitful development. Although many workers have described anion binding with protonated polyaza-macrocycles,[16] these are inappropriate for sensor application because of their

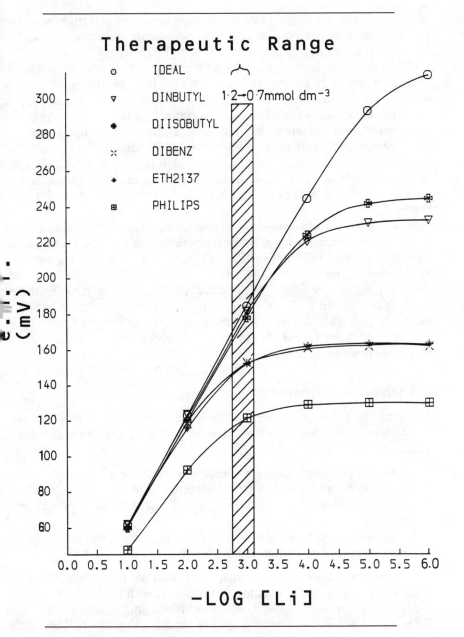

Figure 4.2. EMF response of various lithium electrodes (37°C) as a function of lithium Concentration in the presence of 150 mM NaCl, 4.3 mM KCl and 1.26 mM CaCl$_2$.

pH dependence. Certainly there is considerable scope in further studying the anion-binding behavior of suitable Lewis acidic elements (e.g., B, Zn, Sn, and Ti) either incorporated into or bound by a lipophilic macrocycle.

Of the common cations, Mg^{2+} ion-selective electrodes that show reasonable selectivity over sodium and calcium have been described,[17] but there is scope for improvement. In designing a suitable ligand, note should be taken of magnesium's preference for 6-coordination (and no more) and for binding to "hard" donor atoms with a large ground-state dipole moment (e.g., amides and phosphorus–oxygen bonds). In passing, it is remarkable that none of the synthetic crown ethers exhibits good selectivity for the ammonium cation; a mixture of the macrotetrolide antibiotics nonactin and monactin remains pre-eminent as membrane compounds for NH_4^+ electrodes.

Finally, there is a challenge to be met in the development of chiral sensors based on enantiomerically pure macrocyclic receptors. The best-reported selectivity with an ISE involves Lehn's [18]-O_6 tetramide binding α-phenyl ethylammonium ions with rather modest enantioselectivity.[18] Clearly, additional binding interaction to the ether and/or amide oxygen (hydrogen bonding interaction with the ammonium ion) are required in order to enhance the difference in free energy for binding the enantiomeric ions. Furthermore, a chiral sensor should also be chemoselective, which may be equally difficult to address.

4.3 Diagnostic Agents

There are three obvious classes of clinical diagnostic agents in which synthetic macrocyclic chemistry has made or will make a substantial impact. These are

1. Radioisotopic agents (either as simple complexes of a γ or β^+ emitter or as a complex conjugated to a tissue-specific molecule)
2. Luminescent complexes[19] (where enhanced or delayed fluorescence is observed for a suitable complex)
3. Paramagnetic contrast agents (for use in magnetic resonance imaging where improved resolution is often needed)

When using radioisotopes for imaging, a minimal interaction of the radiation with tissue and a maximum interaction with an external detector is desired. A selection of some suitable radioimaging isotopes is given in Table 4.2. If radioimmunoscintigraphy is required, then the half-life of the radionuclide needs to be sufficiently long to allow its transportation to, say, a tumor site. This is determined by the nature of the targeting antibody and limits the physical half-life to between six hours and eight days. Similar arguments apply for other targeting vehicles but with different time windows. In our own work with radiolabeled tumor-localizing monoclonal antibodies, initial efforts were concentrated on irreversibly labeling ^{64}Cu, ^{111}In

Table 4.2. Selected radioisotopes for imaging.

Radionuclide	$t_{1/2}$	E_{photon} (keV, %)	Source
^{99m}Tc	6.02 h	141 (89)	generator
^{67}Ga	3.25 d	184 (24)	cyclotron
^{111}In	2.83 d	171 (88)	cyclotron
^{131}I[a]	8.05 d	364 (82)	reactor
^{68}Ga	1.20 h	511 (178)	generator
^{82m}Rb[b]	6.4 h	511	generator
^{64}Cu[a]	12.8 h	511 (120)	reactor

[a] ^{64}Cu has an accompanying β-emission (37%, $E_{max} = 0.57$ MeV); ^{131}I also has an accompanying β-emission (0.188 MeV)

[b] ^{81}Rb decays by positron emission and electron capture ($t_{1/2} = 4.7$ h)

and ^{67}Ga to the protein such that the conjugate was stable with respect to isotope loss for the in vivo time period. Prior work with acyclic complexing agents, which were covalently bound to an antibody and then radiolabeled, had demonstrated the potential efficacy of this method.[21] The complexes of the chelates used (based on EDTA or DTPA) were insufficiently stable with respect to acid and cation-promoted dissociation, leading to an unacceptable build up of the radioisotope in the liver. These early studies also highlighted another practical problem of this approach: a fast forward rate of association is essential in order to achieve a decent radiolabeling yield. Furthermore, when radiolabeling an antibody–ligand conjugate, nonspecific binding of the metal ion to the protein must be minimized, the reaction should be fast ($\leq$ 40 minutes), and the labeling conditions are further constrained by the need to avoid protein denaturation. The approach adopted in Durham was to use macrocyclic complexing agents that were appropriately functionalized to permit antibody linkage. Where possible, it was intended to use charge neutral (or cationic) complexes (over pH range 3 to 8) in order to minimize the risk of proton- or cation-catalyzed dissociation. In addition, the ligand (in particular the ring size) was designed to be as "pre-organized" as possible[22] in order to limit unfavorable contributions to the enthalpy and entropy of complexation.

In selecting the parent macrocyclic system to be functionalized, the primary considerations are the nature of the radioisotope to be bound and the desire to form a complex that is kinetically inert in vivo. For copper, complexes with cyclam (14[N$_4$]) and the related [13]-N$_4$ tetra-aza coronand are well known to form thermodynamically[23] and kinetically stable complexes.[1] Accordingly, a set of C-functionalized [13]- and [14]-N$_4$ ligands 5 and 6 was prepared; these were conjugated to the antibody B72.3 (which binds to the TAG-72 antigen found in 80% of human colorectal and breast cancers) via their vinyl pyridine or maleimide derivatives. When radiolabeling the resultant antibody–macrocycle conjugate, care was taken to avoid the nonspecific binding of copper to the protein by working at low pH.[24,25] In the presence of a succinate or citrate buffer at pH 4 and 37 °C, good ^{64}Cu radiolabeling yields were obtained. A further kinetic analysis of this complexation step (involving ionic strength and concentration dependence) strongly suggested that anionic copper species (e.g., Cu [succinate]$_2^{2-}$) were reacting with the monoprotonated ligand in the rate-limiting step.[6]

The ultimate test of stability with respect to dissociation in vivo monitors the fate of the radiolabel in a given tissue as a function of time. Free copper, for example, tends to accumulate in the kidney and liver. Data for ^{64}Cu (and ^{67}Cu) bound to B72.3 revealed virtually no build up of radiolabel in any of these organs over 72 hours, which was consistent with the kinetic stability found in vitro.[1,6] The application of this ^{64}Cu work for diagnosis in humans involves two further steps—first, the development of "humanized" or "chimeric" antibody fragments[26,27] that can localize to a tumor site more rapidly, and second, the availability both of good quality ^{64}Cu (i.e., free from cold isotopes and other contaminant metal ions) as well as the positron tomographic equipment that is needed to obtain good images of the two colinear photons that result from positron annihilation.

The easily hydrolyzed ions Ga^{3+} and In^{3+} require a ligand that is tribasic to satisfy their charge and hexadentate in order to form coordinatively saturated complexes. Carboxylate and alkylphosphinic acid donors are preferred over other less acidic groups as they are less sensitive to protonation at low pH–inhibiting acid-catalyzed dissociation; they are ionized in the pH range 4.5 to 7, permitting rapid complexation. Ligands based on the [9]-N$_3$ ring 7 and 8 were accordingly prepared bearing malemide[7] or active ester functionality[26] to allow antibody linkage via a thiol or lysine ε-amino group. Prior work had established the stability at low pH of the parent ligand indium and gallium complexes—the latter, for example, could be observed unchanged by ^{71}Ga NMR in 6M HNO$_3$ over a period of six months.[26,27] In addition, the complexes of NOTA (NOTA = 1,4,7-triazacyclononane triacetate) with indium and gallium have been structurally characterized by X-ray crystallography (Figure 4.3.)[27,28] The syntheses of 7 and 8 used (2S)-lysine as the starting material.[29] Shorter syntheses of two N-functionalized variants (Figure 4.4) have also been effected. The compound bearing two malemides in Figure 4.4 may serve as a cross-linking agent and was isolated as a mixture of the two diastereoisomers. The malemide 7 and the monosubstituted N-linked analogue (Figure 4.4) have been conjugated to a chimeric B72.3 antibody, radiolabeled with ^{111}In and the biodistribution of the radiolabel examined as a function of time. After 24 hours, blood-to-liver ratios (for nontumor-bearing mice) of 3.6 and 4.2 for the C- and N-linked analogues conjugated respectively were obtained, which was consistent with good in vivo stability. As the liver is perfused with blood, a value for the liver of 30% of the blood value is expected, consistent with those obtained. For the N-linked conjugate, at 96 hours, the blood-to-liver ratio was still 3.0 to 1, notwithstanding the onset of antibody catabolism by this time. This work augurs well for trials in human patients using similar conjugates.

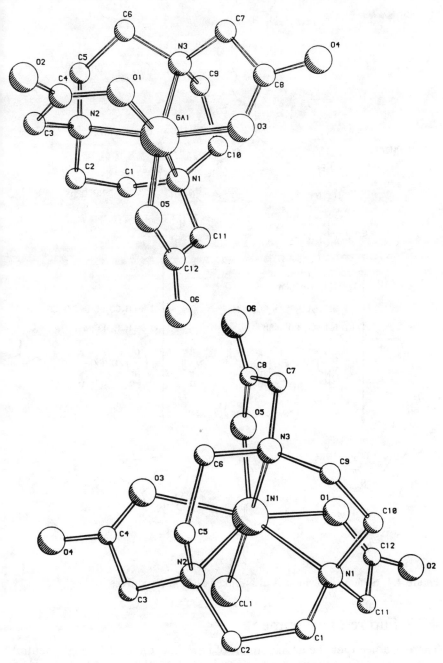

Figure 4.3. Crystal Structures of [Ga·NOTA] and [H In·NOTA] (as the chloro-adduct).

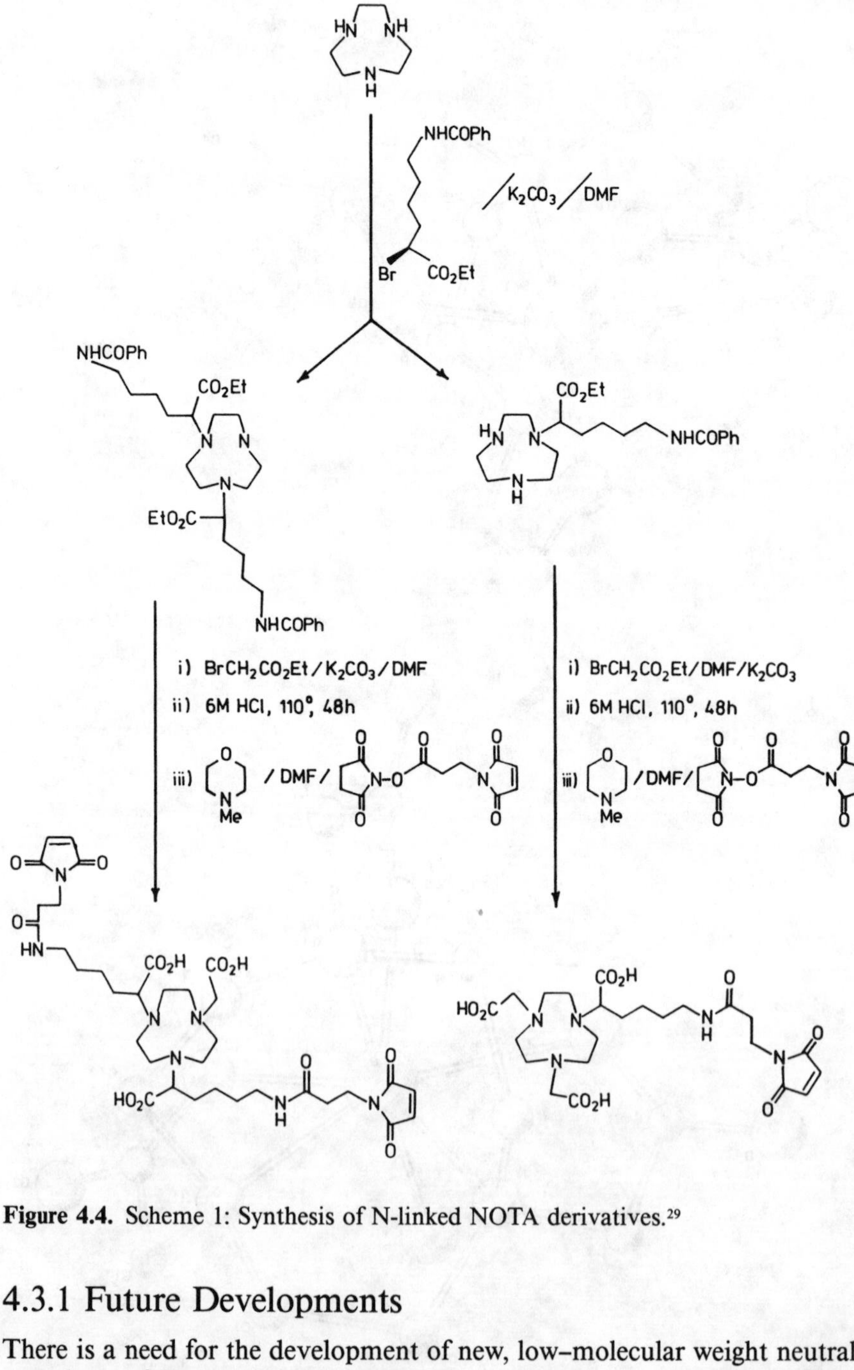

Figure 4.4. Scheme 1: Synthesis of N-linked NOTA derivatives.[29]

4.3.1 Future Developments

There is a need for the development of new, low–molecular weight neutral complexes of the γ-emitting isotopes ^{67}Ga, ^{99m}Tc, and [111]In. Such macrocyclic complexes may find use in diagnostic nuclear medicine, for example, in monitoring myocardial blood flow or cerebral perfusion as renal scanning agents or perhaps even directly as tumor-imaging agents. A particular problem remains in the γ-ray tomographic imaging of blood flow (and less so for tumor imaging) in the brain, where only ^{99m}Tc-HMPAO[30] has had success because it can penetrate the blood–brain barrier effectively. Ligands related to 9[26] are promising in this respect, because the alkyl group may be varied to control complex lipophilicity.

As positron emission tomography becomes established, more research will be needed in the development of suitable, kinetically stable complexes of ^{64}Cu, ^{68}Ga, and perhaps $^{81/82}$Rb. Stable calixspherand complexes of rubidium have been studied by Reinhoudt and others,[31] although stabilities in vivo are probably insufficient. It is much easier, of course, to prepare kinetically stable complexes of copper and gallium.

In magnetic resonance imaging, complexes of gadolinium with DOTA, *10,* and its variants (e.g., a tribasic macrocycle with an additional amide or alcohol) are being used as paramagnetic contrast agents[20] to enhance the tissue images. This is an attractive area of research, where high kinetic stability, charge neutrality, and lipophilicity are needed in a single complex and one needs to be able to prepare tens of grams easily, as a single human dose for one scan may be 2 g of complex!

4.4 Therapeutic Agents

There are two main types of therapeutic agent: those that are active when administered (e.g., β- or α-emitting radioisotopes) and those that may be rendered active in vivo by subjecting them to a suitable stimulus. In the latter case, much attention has been focused on "pro-drugs," which could be activated by a particular enzyme at the site of a tumor. An early example of this approach involved the drug cyclophosphamide *11,* which was introduced[32] in the belief that it would be inactive in the body until its ring was cleaved by an enzyme more common in cancer cells than in normal cells, leaving a potent alkylating agent. Although it is mainly "activated" in the liver, this drug (and its more recent analogues, e.g., ifosfamide) is still used in treating Hodgkin's disease and lymphosarcomas. More recent attention has been focused on boron neutron capture therapy in which a large number of boron-10 atoms are required ($\geq$ 1,000) to be localized at a tumor site and are then activated to the potent α-emitter boron-12 by external neutron irradiation. More promising are phototherapeutic methods, wherein a suitable porphyrin (attached, for example, to a tumor-localizing antibody) is irradiated at a specific wavelength at the tumor site, leading to local formation of destructive hydroxyl radicals of sufficient energy to cleave cellular DNA. It is clear that this approach is promising for skin melanoma at least.

In radioimmunotherapy, where the aim is to selectively target a sterilizing dose of radiation to cancer cells without harming normal cells, a tumor-seeking antibody is radiolabeled with a suitable β- or perhaps α-emitter. A selection of such isotopes is given in Table 4.3. Yttrium-90 is the preferred therapeutic isotope. It is a pure β-emitter of relatively high energy, enabling it to penetrate tumors that may express low levels of surface antigen or are poorly vascularized (inhibiting antibody penetration). It is relatively cheap, available from a strontium generator, and decays to stable zirconium. The key feature of its behavior in vivo is that the free ion is a bone-localizing cation. Any yttrium dissociation in vivo is particularly dangerous because the buildup of ^{90}Y in the bone may lead to a depletion of the immune cell population through irradiation of the proximate marrow. This heightens the risk of infection and can be fatal. Very clearly, complexes of ^{90}Y (or any other therapeutic isotope) should be as stable as possible for in vivo use.

Macrocyclic ligands that bind yttrium quickly (37 °C, 30 mins, pH 6, 10 μmol dm^{-3}) yet form a complex that is resistant to acid- and/or cation-promoted dissociation in the pH range 2.5 to 7.5 are not easy to find. Octadentate ligands based on DOTA, *10,* are most appropriate[9] and had been studied previously for their ability to bind Gd^{3+} and the other lanthanides.[33] Two C-functionalized derivatives related to DOTA have been prepared,[8,29,34] *12* and *13,* again from the precursor (2S)-lysine methyl ester. In addition, a shorter synthesis of the N-functionalized derivative has been carried out[29] (Figure 4.5). Other workers have described p-nitrophenyl–substituted derivatives of DOTA using a synthetic route from the tetrapeptide (2S)-NO$_2$-Phe-Gly-Gly-Gly.[35]

Antibody–macrocycle conjugates, comprising *12, 13,* or *14* linked to chimeric B72.3 antibody, have been evaluated in vivo.[10] All show that the deposition of ^{90}Y in the bone, which comprised earlier studies using DTPA–

Table 4.3. Therapeutic radioisotopes.

Radionuclide	Half-life (hours)	Mean range in tissue (mm)	Gamma (keV, %)	β max (MeV, %)
^{90}Y	64	3.9	none	2.25 (100)
^{67}Cu	62	0.2	93 (17)	0.40 (45)
			184 (47)	0.48 (3)
				0.58 (20)
^{111}Ag	179	1.1	342 (6)	1.04 (93)
			247 (1)	0.69 (6)
				0.79 (1)
^{199}Au	75	0.1	158 (76)	0.25 (22)
				0.30 (72)
^{188}Re	17	3.3	155 (9)	1.96 (18)
				2.12 (80)
^{131}I	193	0.4	364 (79)	0.61 (90)
^{161}Tb	166	0.3	75	0.45 (26)
			57 (21)	0.57 (64)
				0.58 (10)

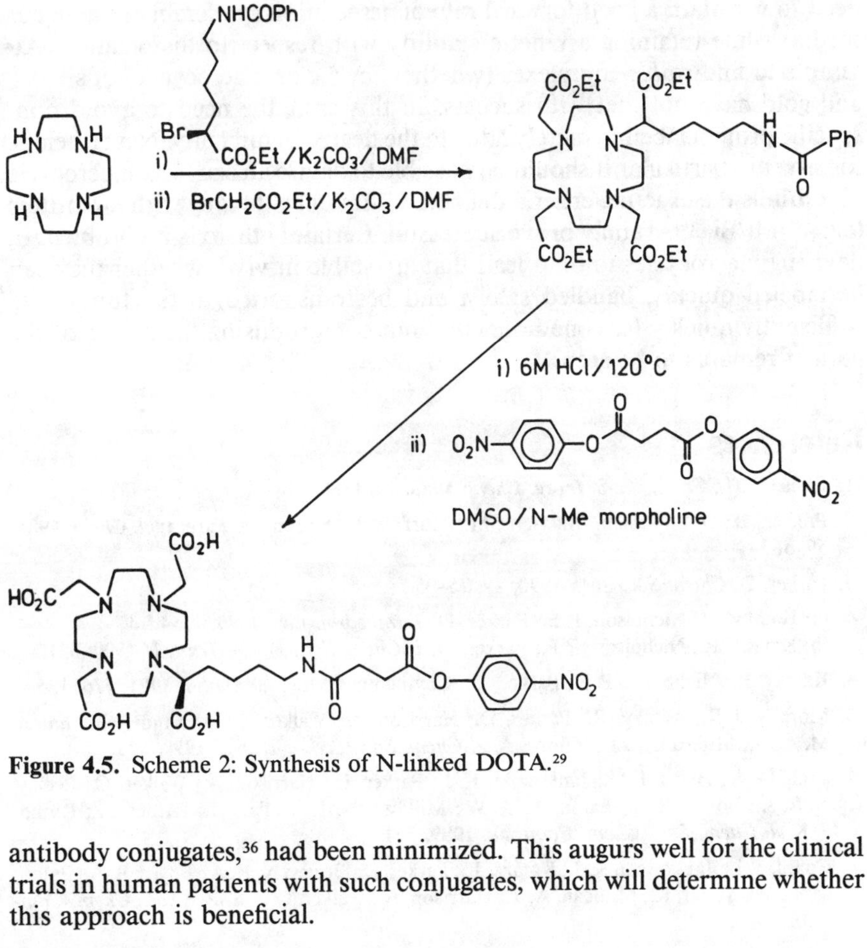

Figure 4.5. Scheme 2: Synthesis of N-linked DOTA.[29]

antibody conjugates,[36] had been minimized. This augurs well for the clinical trials in human patients with such conjugates, which will determine whether this approach is beneficial.

4.4.1 Future Developments

Although considerable effort has been devoted to radioimmunotherapy with ^{90}Y-labelled antibodies, this is not the only way forward in tumor targeting. The use of other isotopes and of other tumor-localizing molecules should be considered. In the latter case, antigen-binding antibody fragments (e.g., F(ab)s or F(ab)$_2$s) localize more rapidly, penetrate more deeply, and clear from the blood more rapidly than a whole antibody. Indeed, even a single complementarity determining region (i.e., a short peptide) may mimic the avidity of the parent antibody and be used as the targeting vehicle. Other tumor-localizing small molecules can also be considered for this purpose, such as *meta*-iodobenzylguanidine (and its derivatives) which localizes in many neuroectodermal tumors.[37]

There are a large number of radioisotopes which could be used in principle. The major limitations—as defined in detail earlier—are related to the need to maintain a good forward rate of association (preferably in aqueous media) while retaining a kinetic stability with respect to dissociation. Attempts to find *stable* complexes (whether acyclic or macrocyclic) of silver[38] and gold have not met with success. In this case, the need to avoid nonspecific protein labeling merely adds to the degree of difficulty. Nevertheless, for silver in particular it should be possible to devise an aza–thia macrocycle that fulfills the exacting criteria defined. Ultimately, therapy with α-emitters (e.g., $^{212}Pb/Bi$, ^{211}At) may prove successful. Certainly there is no problem in devising macrocycles to bind lead that are stable in vivo; whether they can be labeled quickly, handled safely, and be transported to the tumor site sufficiently quickly (ca. one hour) without compromising the health of the patient remains to be seen.

References

1. Chen, L.-H.; Chung, C.-S. *Inorg. Chem.* **1988,** *27,* 1880.

2. Pretsch, E.; Badintschem, M.; Welti, M.; Morf, W. E.; Simon, W. *Pure Appl. Chem.* **1988,** *60,* 567.

3. Parker, D. *Chem. Soc. Revs.* **1990,** *19,* 75–95.

4. (a) Kataky, R.; Nicholson, P. E.; Parker, D. *Tetrahedron Lett.* **1989,** 4554.
   (b) Kataky, R.; Nicholson, P. E.; Parker, D. *J. Chem. Soc., Perkin Trans II* **1990,** 321.

5. Kataky, R.; Nicholson, P. E.; Parker, D.; Covington, A. K. *The Analyst* **1991,** *116,* 135.

6. Morphy, J. R.; Kataky, R.; Parker, D.; Harrison, A.; Walker, C.; Alexander, R.; Eaton, M. A. W.; Millican, A. T.; Phipps, A. *J. Chem. Soc., Perkin Trans I* **1990,** 573.

7. Craig, A. S.; Helps, I. M.; Jankowski, K. J.; Parker, D.; Harrison, A.; Walker, C.; Beeley, N. R. A.; Boyce, B. A.; Eaton, M. A. W.; Millican, A. T.; Millar, K.; Phipps, A.; Rhind, S. K. *J. Chem. Soc., Chem. Commun.* **1989,** 794.

8. Cox, J. P. L.; Jankowski, K. J.; Kataky, R.; Parker, D.; Beeley, N. R. A.; Boyce, B. A.; Eaton, M. A. W.; Millar, K.; Millican, A. T.; Harrison, A.; Walker, C. *J. Chem. Soc., Chem. Commun.* **1989,** 797.

9. Broan, C. J.; Cox, J. P. L.; Craig, A. S.; Kataky, R.; Parker, D.; Randall, A. M.; Harrison, A.; Ferguson, G. *J. Chem. Soc., Perkin Trans. II* **1991,** 87.

10. Harrison, A.; Sansom, J.; Walker, C.; Cox, J. P.; Jankowski, K. J.; Parker, D.; Beeley, N. R. A.; Eaton, M. A. W.; Millican, A. T.; Secher, D.; Millar, K.; Phipps, A.; Farnsworth, A. *Int. J. Nucl. Med. Biol.* **1991,** *18,* 469.

11. (a) Sibbald, A.; Covington, A. K.; Carter, R. F. *Clin. Chem.* **1984,** *30,* 135.
    (b) Drake, H. F.; Treasure, T. *Intensive Care Med.* **1986,** *12,* 104.

12. Ammann, D.; Bissig, R.; Cimerman, Z.; Friedler, U.; Güggi, M.; Morf, M. E.; Oehme, M.; Osswald, H.; Pretsch, E.; Simon, W. In *Ion and Enzyme Electrodes in Biology and Medicine;* Kessler, M., L. C. Clark, D. W. Lubbers, I. A. Silver, and W. Simon, Eds.; Urban and Schwarzenberg; Vienna, 1976; p 22.

13. Ammann, D.; Güggi, M.; Pretsch, E.; Simon, W. *Anal. Lett.* **1975,** *8,* 709.

14. Oesch, U.; Ammann, D.; Pham, H. V.; Wuthier, U.; Zund, R.; Simon, W. *J. Chem. Soc., Faraday Trans I* **1986,** 1179.

15. Newcomb, M.; Horner, J. H.; Blanda, M. T.; Squattrito, P. J. *J. Am. Chem. Soc.* **1989,** *111,* 6294.

16. (a) Dietrich, B.; Guilhem, J.; Lehn, J.-M.; Pascard, C.; Sonveaux, E. *Helv. Chim. Acta* **1984,** *67,* 91.
    (b) Simmons, H. E.; Park, C. H. *J. Am. Chem. Soc.* **1968,** *90,* 2428.

17. Erne, D.; Stojanoc, N.; Ammann, D.; Hofstetter, P.; Pretsch, E., Simon, W. *Helv. Chim. Acta.* **1980,** *63,* 2271.

18. Bussmann, W.; Lehn, J. M.; Oesch, U.; Plumere, P.; Simon, W. *Helv. Chim. Acta.* **1981,** *64,* 657.

19. Alpha, B.; Balzani, V.; Lehn, J.-M.; Perathaner, S.; Sabbatini, N. *Angew. Chem. Int. Ed. Engl.* **1984,** *26,* 1266.

20. Lauffer, R. B. *Chem. Rev.* **1987,** *87,* 90.

21. Brechbiel, M. W.; Gansow, O. A.; Atcher, R. W.; Schlom, J.; Esteban, J.; Simpson, D. E.; Colcher, D. *Inorg. Chem.* **1986,** *25,* 2772.

22. Cram, D. J. *Science* **1983,** *219,* 1177.

23. Kodama, M.; Kimura, E. *J. Chem. Soc., Dalton Trans.* **1976,** 116 and 1720; **1977,** 1473 and 2269.

24. Morphy, J. R.; Parker, D.; Alexander, R.; Bains, A.; Eaton, M. A. W.; Harrison, A.; Millican, A. T.; Titmas, K.; Weatherby, D. *J. Chem. Soc., Chem. Commun.* **1988,** 156.

25. Morphy, J. R.; Parker, D.; Kataky, R.; Harrison, A.; Eaton, M. A. W.; Millican, A. T.; Phipps, A.; Walker, C. *J. Chem. Soc., Chem. Commun.* **1989,** 792.

26. Broan, C. J.; Jankowski, K. J.; Kataky, R.; Parker, D. *J. Chem. Soc., Chem. Commun.* **1990,** 1738.

27. Craig, A. S.; Parker, D.; Adams, H.; Bailey, N. *J. Chem. Soc., Chem. Commun.* **1989,** 1793.

28. Craig, A. S.; Helps, I. M.; Parker, D.; Adams, H.; Bailey, N.; Williams, M. G.; Ferguson, G. *Polyhedron* **1989,** *8,* 2841.

29. Cox, J. P. L.; Craig, A. S.; Helps, I. M.; Jankowski, K. J.; Parker, D.; Eaton, M. A. W.; Beeley, N. R. A.; Boyce, B. A.; Millican, A. T.; Millar, K. *J. Chem. Soc., Perkin Trans. I* **1990,** 2567.

30. Jurisson, S.; Schlemper, E. O.; Troutner, D. E.; Canning, L. R.; Nowotnick, D. P.; Neirinckx, R. D. *Inorg. Chem.* **1986,** *25,* 543.

31. Reinhoudt, D. N.; Dijkstra, P. J.; in't Veld, P. J. A.; Bugge, K. E.; Harbema, S.; Ungaro, R.; Ghidini, E. *J. Am. Chem. Soc.* **1987,** *109,* 4761.

32. Brock, N.; Wilmanns, H. *Dtsch. Med. Wochenschr.* **1958,** *83,* 453.

33. (a) Desreux, J. F.; Loncin, M. E.; Merciny, E. *Inorg. Chem.* **1986,** *20,* 987.
    (b) Desreux, J. F. *Inorg. Chem.* **1980,** *19,* 1319.

34. Broan, C. J., Jankowski, K. J.; Kataky, R.; Parker, D.; Randall, A. M.; Harrison, A. *J. Chem. Soc., Chem. Commun.* **1990,** 1739.

35. Moi, M. K.; Meares, C. F.; DeNardo, S. J. *J. Am Chem. Soc.* **1988,** *110,* 6266.

36. (a) Sharkey, R. M.; Kaltovich, F. A.; Shih, L. B.; Fand, I.; Govelitz, G.; Goldenberg, D. M. *Cancer Res.* **1988,** *48,* 3270.
    (b) Hnatowich, D. J.; Chinol, M.; Siebecker, D. A.; Gionet, M.; Griffin, T.; Doherty, P. W.; Hunter, R.; Kase, K. A. *J. Nucl. Med.* **1988,** *29,* 1428.
    (c) Roselli, M.; Schlom, J.; Gansow, O. A.; Raubitschek, S. M.; Brechbiel, M. W.; Colcher, D. *J. Nucl. Med.* **1989,** *30,* 672.

37. McEwan, A. J., Wyeth, P.; Ackery, D. *Int. J. Radiat. Appl. Instrum.* **1986,** *37,* 765.

38. Craig, A. S.; Parker, D.; Kataky, R.; Matthews, R. C.; Ferguson, G.; Schneider, H.; Adams, H.; Bailey, N. A. *J. Chem. Soc., Perkin Trans II* **1990,** 1523.

CHAPTER 5

Computer Modeling of Metal-Containing Macrocyclic Ligand Systems—the Present and the Future

Kenneth R. Adam and Leonard F. Lindoy

Department of Chemistry and Biochemistry
James Cook University
Townsville, Queensland 4811
Australia

5.1 Introduction

Over a number of years, we have been investigating macrocyclic ligand design for metal ion recognition and hence discrimination. The potential of cyclic ligands for achieving such discrimination has long been recognized,[1] especially with respect to the crown ether and cryptand complexes of the alkali and alkaline earth ions.[2] Our studies have focused on achieving recognition within a range of transition and post-transition ions and, in particular, between the following groups of industrially important metals: Co/Ni/Cu, Zn/Cd, and Ag/Pb.[3] Although our research in this area has been for the most part experimentally based, we have also had an interest in applying simulations of the molecular mechanics pertinent to particular investigations. To this end, a new molecular mechanics computational package (called MOLMEC) was developed at an early stage of the project.[4]

This chapter will be chiefly concerned with the present and future role of molecular modeling in studies such as those just mentioned. To put the discussion in context, however, it is appropriate that a description of a representative metal ion discrimination study, as performed in our laboratory, be outlined first.

5.2 A Strategy for Achieving Metal Ion Discrimination

A typical strategy employed in our discrimination studies has involved the initial synthesis of a cyclic ligand system that might exhibit favored coordination of a specific metal ion in the presence of others. The choice of the initial ring system has usually been either largely intuitive or based on analogy with the metal ion–binding properties of known related systems. Following its synthesis, a detailed investigation of the relevant coordination chemistry of the chosen macrocycle has been performed. Typically, this part of the study has involved some or all of the following: a kinetics study of complex formation and/or dissociation, X-ray structural determinations of selected solid complexes, spectrophotometric and NMR studies of the complexes in solution, and other (mainly spectroscopic) studies directed at obtaining a fuller understanding of the coordination behavior of the individual metal ion systems. A further important component of the overall investigation has been to determine (usually potentiometrically) the stability constants for each metal complex; in some cases, complementary calorimetric (ΔH) data has also been obtained.

Based on the collective results from these studies, the factors that appear to contribute to any observed discrimination by the chosen macrocycle are then assessed; namely, an attempt is made to predict which electronic and structural influences appear to be important in controlling the relative affinities of the ligand for the respective metal ions involved. Armed with this knowledge, the original ring system is then modified in an attempt to "tune up" any discrimination, and the entire sequence so far described is repeated. If necessary, this is followed by further macrocycle modification until the observed discrimination is maximized. For particular studies up to six experimental cycles have proved necessary before this latter goal was achieved, and in such cases a "matrix" of ligand structures is produced— each structure usually bearing a stepwise relationship to others in the series. It has been our experience that the matrix approach often leads to a much more complete understanding of the nature of any observed discrimination than is usually possible when a more restricted investigation is undertaken.

Details of individual studies of this type have been reported elsewhere and are not repeated here.[3,5] Rather, as mentioned previously, the focus of the remainder of this chapter is on both the use and future potential of molecular modeling for the simulation of metal complex structures (including those involved in metal ion discrimination studies). Initially, an outline of the nature of the molecular mechanics technique is presented; this is followed by a brief discussion of its use for modeling metal-containing species. Finally, the promise for the future of other (molecular orbital–based) modeling procedures is also discussed. Indeed, the latter methods are predicted to result in very significant advances in the modeling of systems of the type we have mentioned.

5.3 The Molecular Mechanics Technique

Molecular mechanics has proved to be an accurate and popular method for modeling the structure and thermodynamic properties of organic molecules, often yielding information that rivals in accuracy similar data obtained experimentally. This technique is based on obtaining the total strain energy for a molecule of interest. It uses a classical force field to describe the interactions between the atoms in the molecule, with different terms in the force field being used to describe the various interactions present, such as nonbonded interactions, bond stretch, angle bend and torsions. The total strain energy is then minimized such that the final structure generated corresponds to a minimum energy geometry. The MM2 force field of Allinger[6] and Kollman's AMBER[7] force field are examples of two widely used fields.

Molecular mechanics calculations involving metal-containing species are usually not as straightforward as those for organic species, especially when transition and other heavy metals are involved. One reason for this is the difficulty in developing suitable force fields for structures incorporating these elements. Thus the parameters in the standard MM2 force field have been chosen so that in most cases the method will accurately reproduce both structural data and enthalpy data for simple organic compounds. However, in the case of transition and post-transition metal compounds for example, the experimental data (and especially thermodynamic data) covering a wide range of compounds are generally not available. In addition, these metals exhibit a very wide range of bonding situations in different complexes, and this results in difficulties in developing a set of force field parameters that are transferable from one species to the next; that is, the method is generally successful only when applied to compounds whose structural and electronic features closely parallel those for which the force field parameters were evaluated. Nevertheless, a large number of molecular mechanics calculations have now been carried out on transition and other heavy metal ion complexes.[8,9] Many of these have been performed on the complexes of metals such as Co(III), with its relatively rigid octahedral coordination sphere; for well-defined metal systems such as this, it is usually easier to define a set of force field parameters that will apply over a range of complex types.

It needs to be noted that, as a consequence of the usual lack of thermodynamic data available for calibrating the parts of the force field involving the metal, strain energies calculated using the corresponding (extended) force field cannot be expected to have high quantitative precision. This appears not always to have been appreciated in the past.

Various macrocyclic complexes have also been investigated using molecular mechanics. For example, this technique has been used with reasonable success to investigate alkali metal complexes of crown ethers[10] and natural antibiotic ionophores.[11] Similarly, attempts by us[4,12,13] and others[9,14,15] to use molecular mechanics to model macrocyclic species containing heavier elements such as the transition metals have also often been successful—especially in the simulation of the structural features of individual complexes. Nevertheless, the studies so far reported have been restricted to the complexes of a limited number of ions and, even for these, widely tested force fields are not generally available [although recently we have published a set of parameter calibrated against X-ray data cited in the literature for 26 low-spin, square-planar complexes of Ni(II) incorporating N_4-donor macrocycles].[4] As mentioned previously, testing over a range of different structural and/or electronic environments is desirable in order to verify that a given force field is applicable across a range of structures containing a metal ion of interest.

5.4 Molecular Mechanics and Discrimination Studies

In the strategy for obtaining metal ion recognition discussed previously, a range of experimental data is used both to monitor and to act as a control of the organic synthesis program. However, the ligand modification steps are often far from facile, and this aspect usually constitutes the most time-consuming part of an entire investigation.

The majority of the (thermodynamic) discrimination observed by us so far is also manifested in terms of a variation of complex structure. In a number of cases we have been able to use the molecular mechanics technique to simulate such structural changes with good results.[12,16] More importantly, we have extended the calculations to the prediction of the effect of ligand modification on the resulting complex structures.[16] Provided that the ligand modification represents only a small structural variation from a previous ligand investigated and that the corresponding complex of this latter ligand has been modeled successfully, it is usually possible to transfer the force field directly and hence to predict the likely complexation behavior of the proposed modified system with reasonable certainty. This is, of course, of very considerable benefit in choosing (or rejecting) potential ligand modifications during the "tuning up" stage of such discrimination studies. Nevertheless, the procedure must be used with care, because the molecular mechanics method does not necessarily generate the global minimum energy structure; it is important that the minimization be carried out using a number of different starting geometries in order to help discriminate any local minima from the true global minimum. As well as using X-ray structural coordinates when available, other initial geometries have normally been arrived at after inspection of CPK and Drieding models of the complex of interest.

The molecular mechanics method also has other limitations and, as with most methods, it is almost as important to know its limitations as to know its capabilities. It is normally not possible to use molecular mechanics to decide among alternate structures that incorporate different numbers or types of bonds. Furthermore, the vast majority of calculations to date have been restricted to treating single molecules in the gas phase. Solvation or crystal-packing effects tend to be difficult to model and are not usually included in these calculations, although there has been some attempt to model each of these in isolated studies.

In a representative investigation performed in the authors' laboratory, molecular mechanics calculations indicated that the three nitrogen donors in the 17-membered ring of type *1* (Figure 5.1) will prefer to coordinate facially around three sites of an octahedral Ni(II) ion, whereas the ether oxygens occupy a further two sites with the remaining site being occupied by a halide ion or a water molecule as in *2* (Figure 5.1).[12] Such a structure was also observed to occur in the solid state using X-ray diffraction. Further molecular mechanics calculations predicted that variation of the macrocyclic ring size to yield the corresponding 19-membered ring of type *1* would result in meridional coordination of the nitrogen string.[16] A structural change of the latter type appears to be the reason for the "dislocation" in the observed stability of this complex (log $K = 6.4$) relative to the complexes of the corresponding 17- and 18-membered rings (log K values of 10.0 and 9.8, respectively).[17]

Apart from molecular mechanics, the main computational techniques currently used to treat isolated molecules are based on (in order of increasing demand upon computational resources) semi-empirical molecular orbital theory and *ab initio* orbital theory. *Ab initio* MO methods range from the simplest Hartree–Fock self-consistent field method to methods that include some treatment of the effects of electron correlation, such as those based on Moller–Plesset perturbation theory.[18]

The feasibility of using the previous methods for the modeling of metal complexes of this type is tied, in part, to the increasing power of modern computers. Not only have there been developments in the area of large,

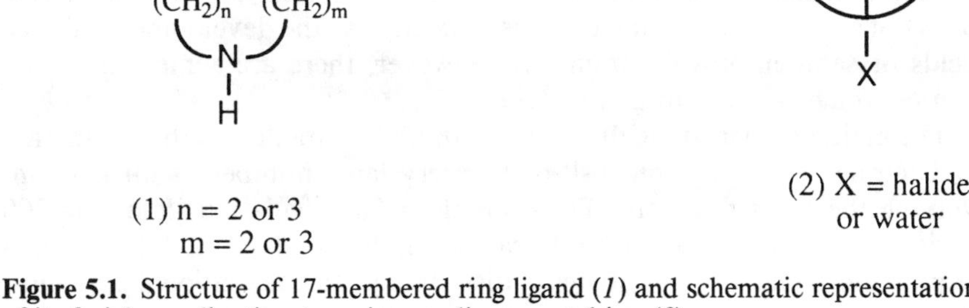

(1) n = 2 or 3
    m = 2 or 3

(2) X = halide
    or water

Figure 5.1. Structure of 17-membered ring ligand (*1*) and schematic representation of its facial coordination to a six-coordinate metal ion (*2*).

expensive supercomputers, but also many inexpensive, high-performance work stations are now available that can provide, at moderate cost, computational power that until recently had only been available on relatively expensive central mainframe computers. There is no reason to believe that this growth in computational power will not continue over the next few years and, as just mentioned, this will play an important role in aiding the widespread use of the next generation of molecular modeling programs. Coupled with these hardware developments is the continuing introduction of new and more efficient algorithms that can take advantage of the modern hardware. Some aspects of these algorithms are discussed later in this chapter.

An alternative to using actual experimental data to calibrate a molecular mechanics force field is to use properties calculated from semi-empirical or *ab initio* MO theory.[19] One such attempt in this area is that of Hagler et al.[20] for the formate anion; however, it is still not certain whether parameters obtained in this manner are transferable to other molecules. Nevertheless, even single-molecule force fields obtained in this way should prove useful in molecular dynamics simulations involving solvent molecules in order to study macroscopic properties.

Semi-empirical or approximate molecular orbital models are simplifications of *ab initio* molecular orbital theory. To compensate for the errors introduced by the approximations, adjustable parameters are incorporated into the model. These parameters are chosen so that the model will reproduce values of properties, which are observed experimentally, for a range of calibration compounds. Alternatively, the parameters may be determined by comparison with the result of *ab initio* calculation for the calibration compounds. Examples of such models are the CNDO and INDO methods of Pople[21] and the MINDO, MNDO and AM1 methods of Dewar.[22] Again these methods have proved very successful for organic molecules, but attempts to extend the parameterization to deal with compounds containing transition metals have been less successful. The difficulties faced in the parameterization process for the transition metal compounds are much the same as for molecular mechanics: a lack of experimental data and the inability to deal with a range of different bonding situations.

The technique that probably offers the greatest potential for solving problems in transition metal chemistry is *ab initio* molecular orbital theory. This does not suffer from the difficulties associated with the development of force fields or semi-empirical parameters; however, there are some other difficulties which occur with this method.

The main practical difficulty of the *ab initio* MO method is the calculation and temporary storage on disk of the very large number of integrals involving the basis functions. For example, in a calculation involving 300 basis functions more than 10^9 integrals would be involved, and these would require more than eight gigabytes of disk storage. In conventional programs, these integrals are usually calculated once and then stored because they are used many times in the course of the calculation. In the simplest SCF model, the processor time and amount of disk storage required increases as the fourth power of the number of atomic basis functions, whereas models that include election correlation scale at an even greater rate. This effectively places an upper limit on the size of molecules that can be studied.

One method that has been developed to avoid the integral storage problem is the direct SCF method.[23] In this procedure, most integrals are not stored but rather recalculated whenever needed. In this way, calculations on large molecules are possible even on computers of modest size.

Another technique that can be used to reduce the number of basis functions (and hence the number of integrals) is to replace the core electrons of the atoms with a pseudopotential.[24] This is based on the fact that the core electrons take little part in the chemical bonding, and their effect may be simulated by replacing them with an appropriate potential. Only the valence electrons are then used explicitly in the calculation; it has been shown that this approximation introduces very little error. An additional benefit is that the pseudopotential can be chosen to include the relativistic effects caused by the core electrons. This is a relatively simple way to include these effects, which become increasingly important for elements of high atomic number.

Methods based upon density functional theory[25] provide another promising approach to the calculation of the structure and properties of transition metal complexes.[26] This approach focuses on the electron density and how it varies through the molecule (and not upon the wave function as in other approaches). This method also has the advantage that it scales as the cube of the number of basis functions (in contrast to the fourth power for conventional SCF methods) and also includes some of the correlation effects. In addition, the techniques of direct SCF and relativistic pseudopotentials can also be incorporated into the method. A program package incorporating these algorithms has recently been developed in the authors' laboratory.

5.5 The Calculation of Macroscopic Properties

The calculation of macroscopic properties, such as free energy changes in solution, is a much more demanding computational problem than the calculations concerned with isolated molecules. Nevertheless, with the continuing developments in computer technology, including the use of parallel processors, much more attention will be paid to this aspect in future.

Consider the following gas phase equilibrium between a metal and a ligand:

$$M_{gas} + L_{gas} \rightleftharpoons ML_{gas}$$

The equilibrium constant K_{gas} can be obtained from the free energy change

$$\Delta G_{gas} = -RT \ln K_{gas}$$

where

$$\Delta G_{gas} = G_{ML_{gas}} - G_{M_{gas}} - G_{L_{gas}}$$

The G values for each of the species involved can, in principle, be computed using *ab initio* MO theory (see previous discussion). G can be obtained from the molecular partition function using standard expressions,[27] and the partition function in turn may be calculated from the molecular geometry, vibrational frequencies, and knowledge of the electronic energy levels. All of this information is available for an *ab initio* MO calculation.

However, gas phase equilibrium constants are generally of little interest in macrocyclic chemistry. Of greater interest is the corresponding equilibrium constant in solution, K_{solv}, associated with the equilibrium:

$$M_{solv} + L_{solv} \rightleftarrows ML_{solv}$$

This may be derived from the following thermodynamic cycle:

$$
\begin{array}{ccccc}
M_{gas} & + & L_{gas} & \overset{\Delta G_{gas}}{\rightleftarrows} & ML_{gas} \\
\big\Vert\Delta G_{M_{solv}} & & \big\Vert\Delta G_{L_{solv}} & & \big\Vert\Delta G_{ML_{solv}} \\
M_{solv} & + & L_{solv} & \overset{\Delta G_{solv}}{\rightleftarrows} & ML_{solv}
\end{array}
$$

K_{solv} can be obtained from ΔG_{solv}, which is given in turn by

$$\Delta G_{solv} = \Delta G_{gas} - \Delta G_{M_{solv}} - \Delta G_{L_{solv}} + \Delta G_{ML_{solv}}$$

This requires the additional computation of the three free energy changes $\Delta G_{M_{solv}}$, $\Delta G_{L_{solv}}$, and $\Delta G_{ML_{solv}}$ associated with the solvation of the species involved in the equilibria.

Free energy changes due to solvation can be computed by thermodynamic perturbation theory.[28] This technique can be used to calculate free energy differences ΔG between two states a and b:

$$\Delta G = G_b - G_a$$

Then

$$\Delta G = -RT \ln \left\langle e^{\frac{-\Delta H}{RT}} \right\rangle_A$$

where $\Delta H = H_b - H_a$ is the difference in the energy of the two states involved. In practice, H is replaced by a potential energy $V(r)$ given by a force field as used in molecular mechanics. The notation $\langle\ \rangle_A$ means to take an assembly average (i.e., to take an average over the configurations that the system passes through as it changes from a to b). This assembly average may be computed either by Monte Carlo[28] or molecular dynamics[29] techniques, although the latter tend to be more efficient. In molecular dynamics, the molecules are treated as systems of point masses moving under the influence of a classical force field.

If the states a and b are very different, then excessively long simulation times are required in order to reduce the statistical error in the calculations.

This can be avoided by breaking the change from a to b into a large number of steps and running the molecular dynamics simulation on each of the steps.[30] Another way to reduce excessively long simulation times is to compute $\Delta\Delta G$ values. This can be done by setting up another thermodynamic cycle with, for example, a second metal with the same ligand being present.

$$M'_{gas} \quad + \ L_{gas} \quad \overset{\Delta G'_{gas}}{\rightleftharpoons} \ M'L_{gas}$$
$$\Updownarrow\Delta G_{M'_{solv}} \quad \Updownarrow\Delta G_{L_{solv}} \quad \Updownarrow\Delta G_{M'L_{solv}}$$
$$M'_{solv} \quad + \ L_{solv} \quad \overset{\Delta G'_{solv}}{\rightleftharpoons} \ M'L_{solv}$$

Then $\Delta G'_{solv} = \Delta G'_{gas} - \Delta G_{M'_{solv}} - \Delta G_{L_{solv}} + \Delta G_{M'L_{solv}}$

and $\Delta\Delta G_{solv} = \Delta G'_{solv} - \Delta G_{solv}$

$$= \Delta G'_{gas} - \Delta G_{gas} + (\Delta G_{M_{solv}} - \Delta G_{M'_{solv}})$$
$$+ (\Delta G_{ML_{solv}} - \Delta G_{M L_{solv}})$$
$$= \Delta\Delta G_{gas} + \Delta\Delta G_{(M-M')_{solv}} + \Delta\Delta G_{(M'L-ML)_{solv}}$$

$\Delta\Delta G_{solv}$ is then the difference in free energy changes of the two equilibria in solution, which in turn gives the ratio of the two equilibrium constants; that is, this is a method for predicting discrimination behavior directly. As before, $\Delta\Delta G_{gas}$ may be obtained from *ab initio* MO theory and the $\Delta\Delta G$s of solvation by molecular dynamics simulation; however, now only two molecular dynamics simulation are required. Also, the differences between the initial and final states of the simulation are much smaller, allowing much shorter simulation times.

As yet these techniques have only just started to be used. One recent example is given by the work of Jorgensen[31] in which differences in pKa values for some simple organic anions are computed. In this case the calculated values are probably more accurate than the measured values! Another example is the investigation by Kollman et al. of host–guest complexes of 18-crown-6 with malononitrile, nitromethane, and acetonitrile.[32] In this study, the interactions of the crown and the guests with benzene were investigated in order to qualitatively evaluate the potential solvent–solute interactions that might be present.

Finally, it is noted that surveys of the potential of *ab initio* techniques for the molecular modeling of a wide range of chemical and biochemical systems have also recently appeared.[33,34]

5.6 Concluding Remarks

As the power of modern computers continues to grow, it is only a matter of time before the computational techniques described in the latter part of this chapter are used routinely for problems of interest to macrocyclic chemists. Calculations which are both more accurate and able to handle a wider range of compounds will be possible. Furthermore, the ability to closely model the solvation of metal complexes and to undertake the direct calculation of macroscopic properties, such as changes in free energy in solution, will also have a major impact. Clearly, the computational methods discussed here will aid the design of new reagents for metal ion recognition very considerably within the foreseeable future.

References

1. Lindoy, L. F. *The Chemistry of Macrocyclic Ligand Complexes;* Cambridge University; Cambridge, U.K., 1989.

2. Izatt, R. M.; Bradshaw, J. S.; Nielsen, S. A.; Lamb, J. D., Christensen, J. J.; Sen, D. *Chem. Rev.* **1985,** *85,* 271.

3. Lindoy, L. F. In *Synthesis of Macrocycles: The Design of Selective Complexing Agents;* Izatt, R. M., and J. J. Christensen, Eds.; Wiley; New York, 1987; pp 53–92.

4. Adam, K. R.; Antolovich, M.; Brigden, L. G.; Lindoy, L. F. *J. Am. Chem. Soc.* **1991,** *113,* 3346.

5. Lindoy, L. F. *Pure and Appl. Chem.* **1989,** *61,* 1575.

6. Allinger, N. L. *J. Am. Chem. Soc.* **1977,** *99,* 8127.

7. (a) Weiner, P. K.; Kollman, P. A. *J. Comput. Chem.* **1981,** *2,* 287.
   (b) Weiner, S. J.; Kollman, P. A.; Case, D. A.; Singh, U. C.; Ghio, C.; Alagona, G.; Profeta, S.; Weiner, P. J. *J. Am. Chem. Soc.* **1984,** *106,* 765.
   (c) Weiner, S. J.; Kollman, P. A.; Nguyen, D. T.; Case, D. A. *J. Comput. Chem.* **1986,** *7,* 230.

8. Brubaker, G. R.; Johnson, D. W. *Coord. Chem. Rev.* **1984,** *53,* 1.

9. Hancock, R. D. *Prog. Inorg. Chem.* **1989,** *37,* 187.

10. Wipff, G.; Weiner, P.; Kollman, P. *J. Am. Chem. Soc.* **1982,** *104,* 3249.

11. Lifson, S.; Felder, C. E.; Shanzer, A.; Libman, J. Chapter 5 in ref. 3.

12. Adam, K. R.; Brigden, L. G.; Henrick, K.; Lindoy, L. F.; McPartlin, M.; Mimnagh, B.; Tasker, P. A. *J. Chem. Soc., Chem. Commun.* **1985,** 710.

13. (a) Adam, K. R.; Donnelly, S.; Leong, A. J.; Lindoy, L. F.; McCool, B. J.; Bashall, A.; Dent, M. R.; Murphy, B. P.; McPartlin, M.; Fenton, D. E.; Tasker, P. A. *J. Chem. Soc., Dalton Trans.* **1990,** 1635.
   (b) Adam, K. R.; McCool, B. J.; Leong, A. J.; Lindoy, L. F.; Ansell, C. W. G.; Baillie, P. J.; Dancey, K. P.; Drummond, L. A.; Henrick, K.; McPartlin, M.; Uppal, D. K.; Tasker, P. A. *J. Chem. Soc., Dalton Trans.* **1991,** 2493.
   (c) Adam, K. R.; Antolovich, M.; Brigden, L. G.; Leong, A. J.; Lindoy, L. F.; Baillie, P. J.; Uppal, D. K.; McPartlin, M.; Shah, B.; Prosperio, D.; Fabbrizzi, L.; Tasker, P.A. *J. Chem. Soc., Dalton Trans.* **1991,** 2493.

14. Hancock, R. D. *Acc. Chem. Res.* **1990,** *23,* 53 and references therein.

15. (a) Martin, L. Y.; Dehayes, L. J.; Zompa, L. J.; Busch, D. H. *J. Am. Chem. Soc.* **1974,** *96,* 4046.
   (b) Searle, G. H.; Dwyer, M. *Inorg. Chim. Acta A* **1981,** *52,* 251.
   (c) Drew, M. G. B.; Hollis, S.; Yates, P. C. *J. Chem. Soc., Dalton Trans.* **1985,** 1829.
   (d) Drew, M. G. B.; Rice, D. A.; Silong, S. B.; Yates, P. C. *J. Chem. Soc., Dalton Trans* **1986,** 1081.
   (e) Drew, M. G. B.; Yates, P. C. *J. Chem. Soc., Dalton Trans* **1987,** 2563.
   (f) Schwarz, C. L.; Endicott, J. F. *Inorg. Chem.* **1989,** *28,* 4011.
   (g) Endicott, J. F.; Kumar, K.; Schwarz, C. L.; Perkovic, M. W.; Lin, W. K. *J. Am. Chem. Soc.* **1989,** *111,* 7411.

16. Adam, K. R.; Antolovich, M.; Bridgen, L. G.; Lindoy, L. F.; unpublished work.

17. Adam, K. R.; Leong, A. J.; Lindoy, L. F.; Lip, H. C.; Skelton, B. W.; White, A. H. *J. Am. Chem. Soc.* **1983**, *105*, 4645.

18. Hehre, W. J.; Radom, L.; Schleyer, P. R.; Pople, J. A. *Ab initio Molecular Theory;* Wiley; New York; 1986.

19. Badertscher, M.; Musso, S.; Welti, M.; Pretsch, E.; Maruizumi, T.; Ha, T.-K. *J. Comput. Chem.* **1990**, *11*, 819.

20. Maple, J. R.; Dinur, U.; Hagler, A. T. *Proc. Natl. Acad. Sci. USA* **1988**, *85*, 5350.

21. Pople, J. A.; Beveridge, D. L. *Approximate Molecular Orbital Theory;* McGraw-Hill; New York, 1970.

22. (a) Bingham, R. C.; Dewar, M. J. S.; Lo, D. H. *J. Am. Chem. Soc.* **1975**, *97*, 1294, 1302, 1307.
    (b) Dewar, M. J. S.; Thiel, W. *J. Am. Chem. Soc.* **1977**, *99*, 4899, 4907.
    (c) Dewar, M. J. S.; Zoebisch, E. G.; Healy, E. F.; Stewart, J. J. P. *J. Am. Chem. Soc.* **1985**, *107*, 3902.

23. (a) Almlof, J.; Faegri, K.; Korsell, K. *J. Comput. Chem.* **1982**, *3*, 385.
    (b) Haeser, M.; Ahlrichs, R. *J. Comput. Chem.* **1989**, *10*, 104.

24. Krauss, M.; Stevens, W. J. *Ann. Rev. Phys. Chem.* **1984**, *35*, 357.

25. Parr, R. G. *Ann. Rev. Phys. Chem.* **1983**, *34*, 631.

26. (a) Andzelm, J.; Radzio, E.; Salahub, R. *J. Chem. Phys.* **1985**, *83*, 4573.
    (b) Rosch, N.; Jorg, H.; Dunlap, B. I. In *Quantum Chemistry: The Challenge of Transition Metals and Coordination Chemistry;* Veillard, A., Ed.; D. Reidel; Dordrecht, 1986,; p 179.
    (c) Ziegler, T.; Nayle, J. R.; Snijders, J. G.; Baerends, E. J. *J. Am. Chem. Soc.* **1989**, *111*, 5631.
    (d) Ziegler, T.; Tschinke, V.; Fan, L.; Becke, A. D. *J. Am. Chem. Soc.* **1989**, *111*, 9177.
    (e) Ziegler, T.; Tschinke, V.; Baerends, E. J.; Snijders, J. G.; Ravenk, W. *J. Phys. Chem.* **1989**, *93*, 3050.

27. Atkins, P. W. *Physical Chemistry;* Oxford University; Oxford, 1978.

28. (a) Bash, P. A.; Field, M. J.; Karplus, M. *J. Am. Chem. Soc.* **1987**, *109*, 8092.
    (b) Singh, U. C.; Brown, F. K.; Back, P. A.; Kollman, P. A. *J. Am. Chem. Soc.* **1987**, *109*, 1607.
    (c) Jorgensen, W. L. *Acc. Chem. Res.* **1989**, *22*, 184.

29. Jorgensen, W. L. *J. Phys. Chem.* **1983**, *87*, 5304.

30. Van Gunsteren, W. F.; Berendsen, H. J. C. *Angew. Chem. Int. Ed. Engl.* **1990**, *29*, 992.

31. Jorgensen, W. L. *J. Am. Chem. Soc.* **1989**, *111*, 4190.

32. Grootenhuis, P. D. J.; Kollman, P. A. *J. Am. Chem. Soc.* **1989**, *111*, 4046.

33. Kollman, P. A.; Merz, K. M. *Acc. Chem. Res.* **1990**, *23*, 246.

34. Hillier, I. H.; Palmer, M. H.; Price, S. L.; Reynolds, C. A. *Chem. Brit.* **1990**, 1075.

CHAPTER 6

Design of New Macrocyclic Polyamine Ligands Beyond Cyclam to Control Redox Properties of Metal Ions

Eiichi Kimura
School of Medicine, Hiroshima University
1-2-3 Kasumi
Minami-ku Hiroshima, 734
Japan

6.1 Introduction

A great number of ligands have been devised as agents to control the redox properties of metal ions. However, there have been few ligand systems that were successfully designed to disclose the variation of metal behaviors with different oxidation states and to establish basic principles in reaching the optimum ligand structures. Macrocyclic tetraamines such as those shown in Figure 6.1 were developed by Busch and his co-workers and are one of the few examples that met with some success in maneuvering Ni(I) $\rightleftarrows$ Ni(II)

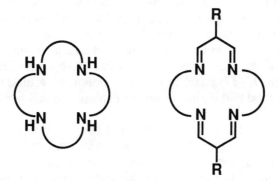

Figure 6.1. Typical macrocyclic tetraamines conventionally studied.

$\rightleftharpoons$ Ni(III) in the metal complexes.[1] Macrocyclic structures have two major advantages: (1) their metal complexes are thermodynamically as well as kinetically very stable, which is extremely useful for redox conversion of the enclosed metal ions without decomplexation; (2) modification and attachment of various functional groups is relatively easy while retaining the basic macrocyclic geometries, which then allows evaluation of a step-by-step approach.

We have been constructing new cyclams (or its homologues) by adding various devices. Typical derivatives from cyclam are presented, along with their abbreviated names in Scheme I (see Figure 6.2). We have found that with slight structural modifications, cyclam *1* can be more intelligent. For instance, the monooxocyclam *2* can incorporate Zn(II) (to yield [Zn(II)H$_{-1}$L]$^+$ complex) but not Cd(II).[2] Such a clear-cut distinction of Zn(II) from Cd(II) was impossible to achieve with cyclam and its congeners; Dioxocyclam *3* is a ligand to yield square planar (N$^-$)$_2$ N$_2$ complexes [M(II)H$_{-2}$L]^0 with concomitant deprotonation from the two amides.[3] It has an exclusive affinity for Cu(II) with kinetics also taken into consideration. Accordingly, a lipophilic dioxocyclam attached to a long alkyl chain (C$_{16}$H$_{33}$-) is useful for organic solvent extraction of Cu(II).[4] And dithiadioxocyclam *10* is a new, selective chelating agent that incorporates only noble metal ions, Pt(II) and Pd(II), to yield square planar (N$^-$)$_2$S$_2$ complexes [M(II)H$_{-2}$L]^0, thereby excluding other metal ions Cu(II), Ni(II) or Co(II).[5]

In this article, another remarkable intelligent function, endowed by the macrocyclic modifications to control metal redox properties, will be introduced.

6.2 Axial Coordination by Pendant Donors

Cyclam *1*, with an appropriate 14-membered ring size like biological tetraamines, porphyrins (unsaturated 16-membered ring), or corrins (15-membered ring), tends to incorporate metal ions into its cavities and forms stable, square-planar N$_4$ complexes. The biological macrocyclic tetraamines are well functionalized for their specific activities by the specific proximate donor ligands (from neighboring amino acid residues) such as imidazole, phenolate, or cystein at an axial position. By the same token, placement of intramolecular axial donor groups to cyclam could attach various enzyme functions to these metal complexes. We, therefore, have synthesized a series of new cyclams (*8, 12–17*) attached to potent donor molecules, using a novel annulation method that we had originally explored. (See Scheme II in Figure 6.3.)

It was anticipated that the possible axial coordination of a phenolate,[15] pyridyl,[12] or imidazole pendant[18] not only enhances metal-enclosures kinetically as well as thermodynamically but also tends to stabilize the enclosed metal ions at higher or lower oxidation states. A strong interaction of these axial pendants was well demonstrated by the almost 100% high-

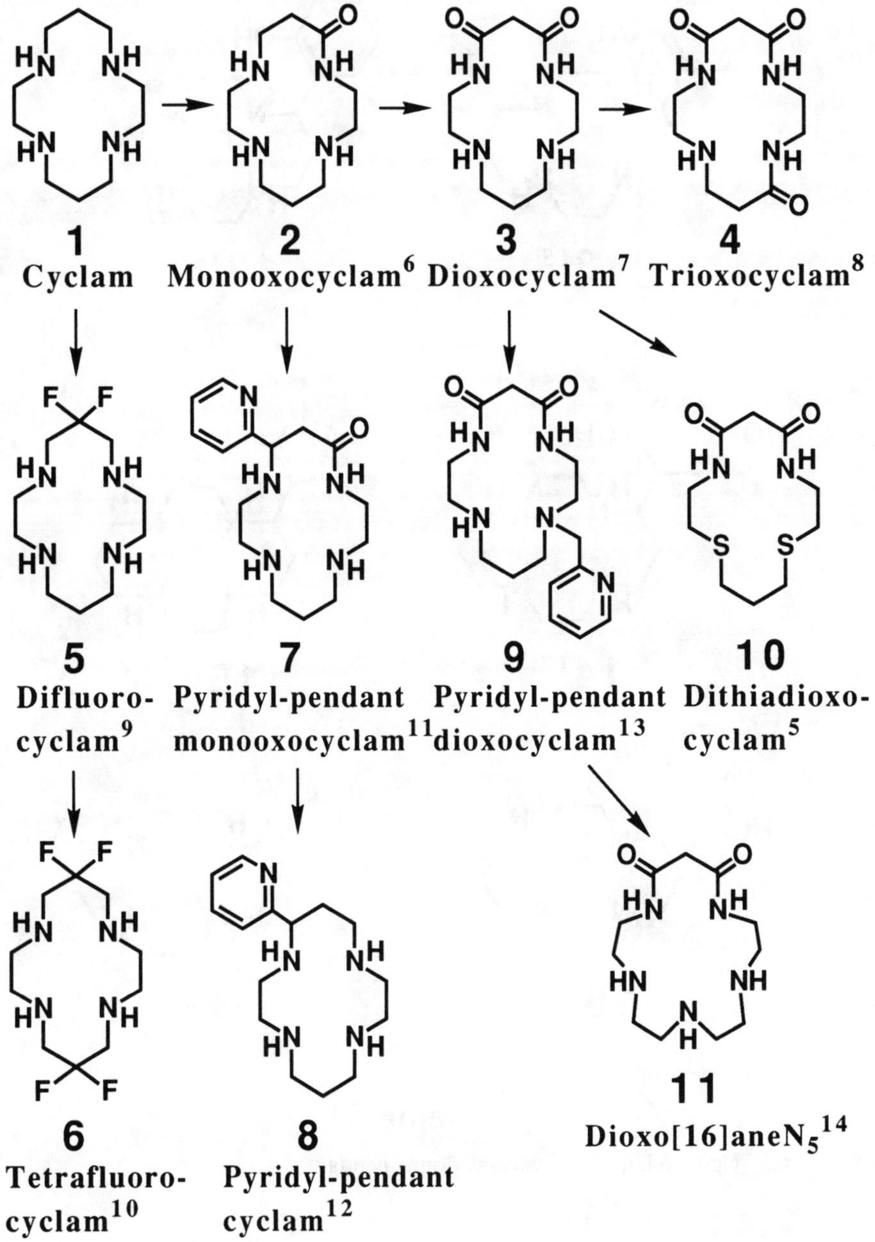

Scheme I

Figure 6.2. Newly developed cyclam derivatives.

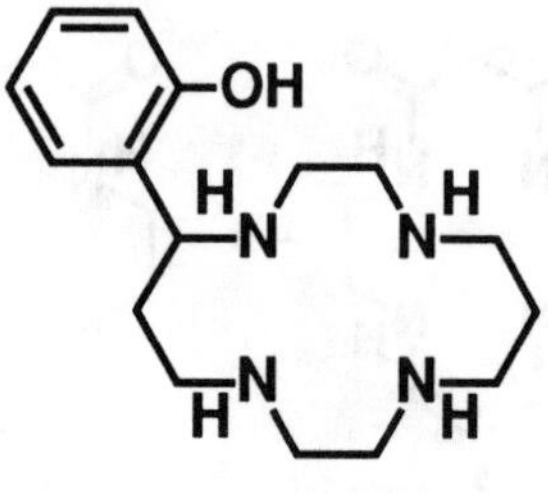

12[15]

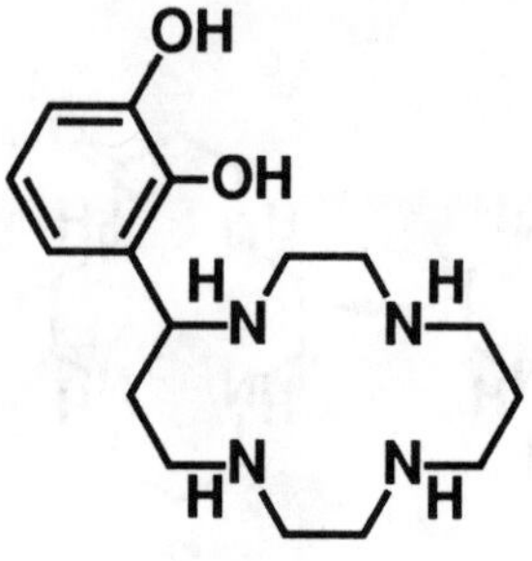

13[16]

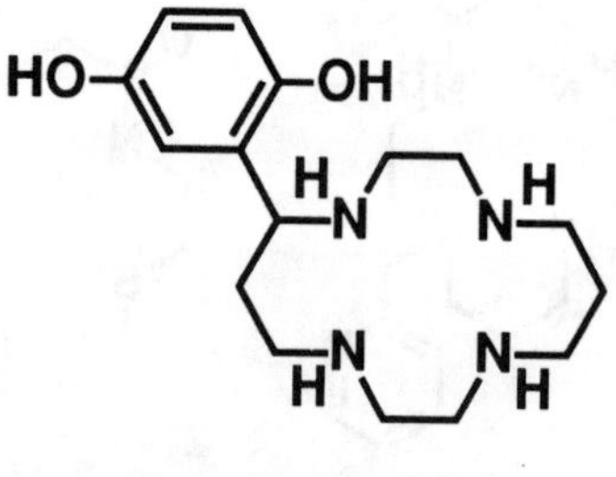

14[17]

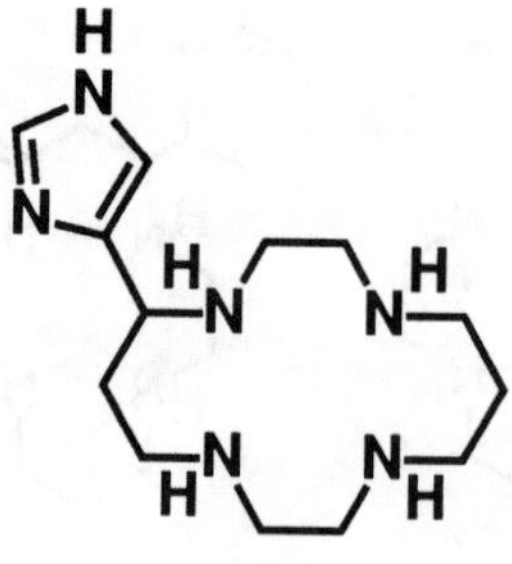

15[18]

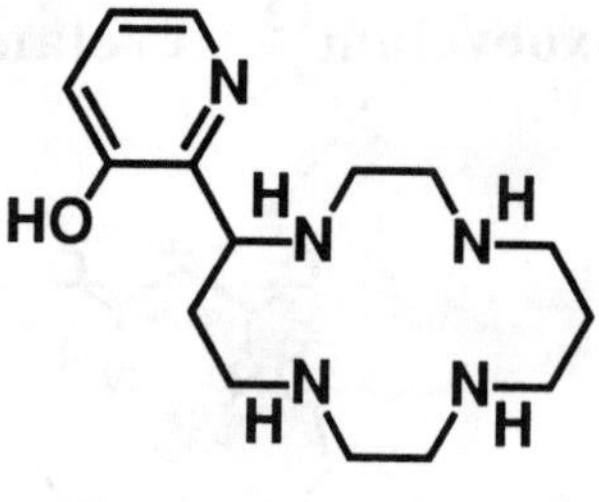

16[19]

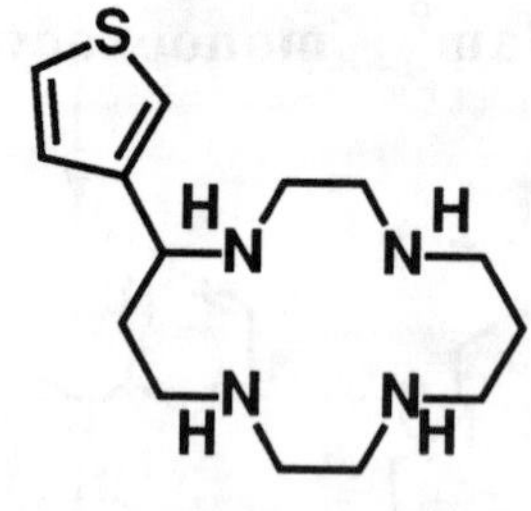

17[23]

Scheme II

Figure 6.3. New cyclams with various donor pendant.

spin state of the Ni(II) complexes *19–21,* while in their absence the Ni(II) ion in *18* is in equilibrium between high-spin and low-spin states.[20] Furthermore, in the case of the phenol pendant complexes,[15,21,22] the phenol pK_a values of ~6 in the metal complexes *19* are much lower than the normal value of ~10. While cyclam *1* without the donor side arms fails to encapsulate Fe^{3+} in aqueous solution, the intramolecular, axial coordination of these pendants makes a new type of Fe^{3+} sequestering agents.[12,15,18] The Fe complexes of these cyclams have shown all quasi-reversible ($1e^-$ redox) cyclic voltammograms with the redox potential (for Fe[III/II]) at -0.16 V, $+0.12$ V, and 0.00 V versus SCE for *19*, *20*, and *21*, respectively. Evidently, the better the σ-donor ability is, the more stabilized the $+3$ oxidation state of Fe. In addition, the pyridyl pendant may have a π-acceptor character so as to contribute to the stability of Fe^{2+}.

The apical phenolate (*19*),[21] pyridyl (*20*),[12] and imidazole coordinations (*21*)[18] were unequivocally proven by X-ray crystal analyses of Ni(II) complexes. The four N atoms in the cyclams are coplanar with Ni(II) staying in this plane. The Ni–N bond distances are in the normal range for the high-spin Ni(II)-N (~2.08 Å). In *19*, the phenolate oxygen donor is almost at the apex of the pyramid with a very short Ni–O bond of 2.015 Å. The cyclam moiety takes the most stable trans III configuration. The analogous coordinate geometries are more or less similar to the pyridyl (*20*) and imidazole cyclam complexes (*21*), although the axial coordinations are a little more tilted due to the shorter linkages between the cyclam and the donor atoms. (See Figure 6.4.)

Hence, redox potentials of these metal complexes are regarded as direct indicators of the axial donor effects (see Table 6.1).

It is noteworthy that with the phenol-pendant cyclam complexes, the Ni(III/II) redox potentials are pH-controllable; below pH 6 where the phenol is undissociated (*22*), $E_{1/2}$ is $+0.50$ V, and above pH 8 where the dissociated phenol interacts with Ni (*19*), it is $+0.35$ V.[21] (See Figure 6.5.)

It is now readily conceivable that the attachment of other donor pendants would further expand the scope of this strategy to control the metal redox properties.

If the pendant donor ligand is also redox-active, the redox-coupling between the axial donor and the metal ions is expected to occur. This is indeed

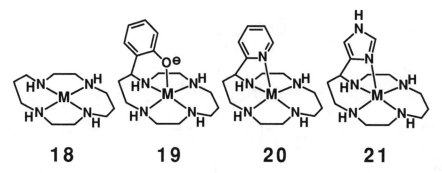

18   **19**   **20**   **21**

Figure 6.4. Cyclam complexes without (18) and with apical phenolate (19), pyridyl (20) and imidazol (21) coordinations.

the case for the Fe(II)-catechol-pendant cyclam complex 23,[16] which offered the first prototype for synergistic intramolecular redox-coupling between monodentate catecholate and metal ions to render the catechol unusually vulnerable to oxidation. Further modification of metal ions or the macrocyclic structure with a catechol pendant would yield a novel redox system. Moreover, the reactivity of the remaining sixth axial position in 23 would be interesting as a catalytic site. (See Figure 6.6.)

Table 6.1. Comparison of redox potentials (V vs. SCE) for M(III/II) in aqueous solution at 25°C.

	18	19	20	21
Ni(III/II) $I = 1.5$ (Na$_2$SO$_4$)	+0.50	+0.35	+0.61	+0.54
Fe(III/II) $I = 0.1$ (NaClO$_4$)	unstable	−0.16	+0.12	0.00

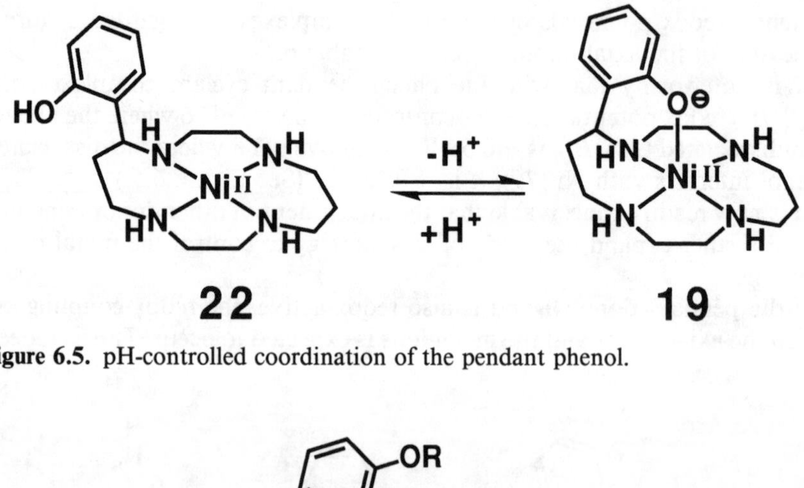

22   **19**

Figure 6.5. pH-controlled coordination of the pendant phenol.

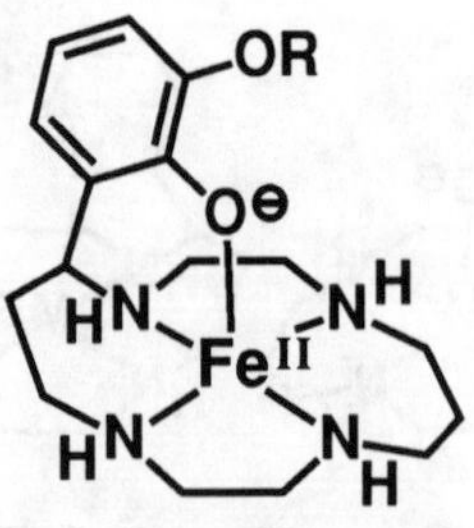

23

Figure 6.6. Fe(II)-catechol pendant cyclam complex.

6.3 Modification of Equatorial Donors

6.3.1 Direct Interaction Sites

Dioxocyclam *3* was originally viewed as a hybrid ligand of cyclam and oligopeptides such as triglycine *25*.[24,25] *3* and *25*[24,25] interact with M(II) (e.g., Cu(II), Ni(II), Co(II), Pd(II), Pt(II)) ions with concomitant dissociation of the amide protons to accommodate them, normally yielding square-planar complexes, *24* and *26*, respectively. Just as were reported for the Cu(II)[24] and Ni(II) complexes of *26*,[25] the imide anion donors of *24* stabilize Cu(III), which was diagnosed by the lowered oxidation potentials with introduction of the oxo functions into cyclam.[3] The Cu(III/II) redox potentials (V versus SCE) are summarized in Figure 6.7. We see here that as the number of imide

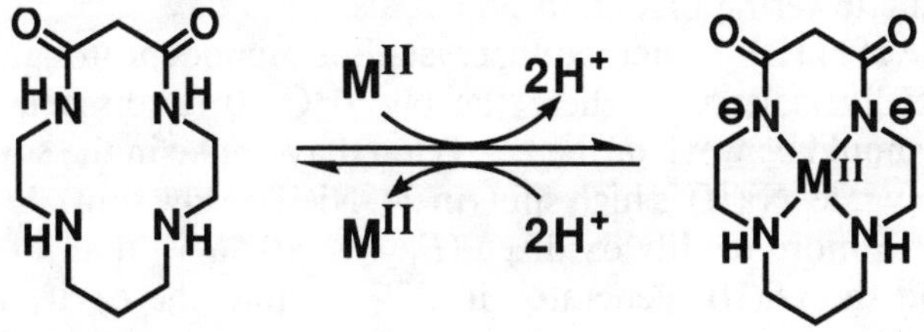

3
Dioxocyclam

24

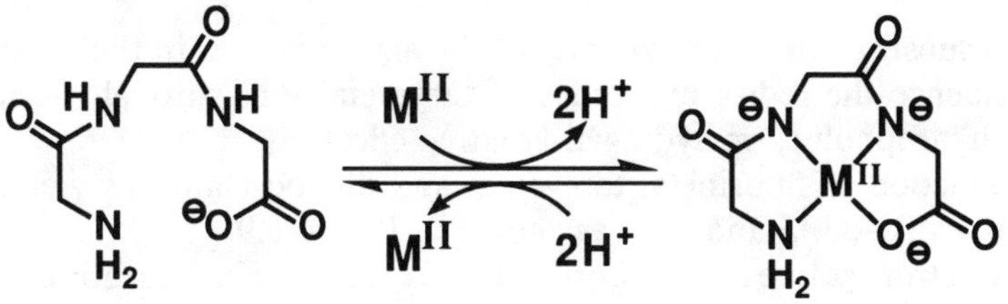

25
Triglycine

26

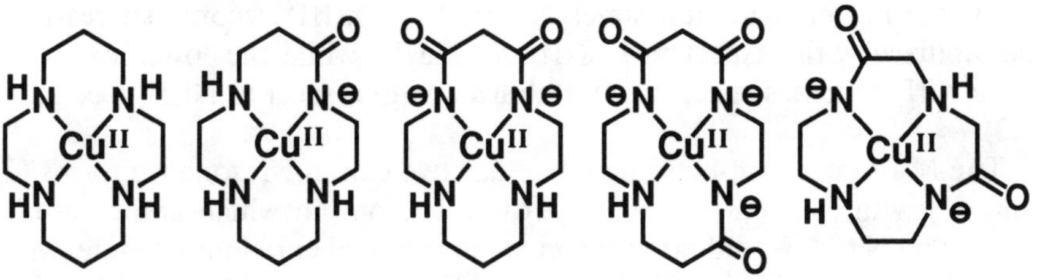

>+1.0 V **+0.86 V** **+0.64 V** **+0.42 V** **+0.42 V**

Figure 6.7. Redox potentials (*vs.* SCE in aqueous solution at 25°C, I=0.2) for Cu(III)/Cu(II) in cyclam complexes.

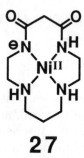

27

Figure 6.8. Nickel(II)-monodeprotonated dioxocyclam complex.

anions increases in cyclam skeleton, Cu(III) is easier to attain, as demonstrated by the lowering oxidation potentials.[8]

In the case of Ni(II)–*3* interaction, crystalline monodeprotonated complex *27* was initially isolated,[26] where the Ni-NHC=O (undissociated amide) interaction should be weak or the Ni(II) ion may not be in the square-planar cavity. At any rate, Ni(II) is high spin (in *24*, Ni(II) is low spin). Interestingly, Ni(II) in *27* is more readily oxidized ($E_{1/2}$ = +0.56 V) than in *24* ($E_{1/2}$ = 0.81 V), and the Ni(III) generated in *27* goes into the cavity of *24*. (See Figure 6.8 above.)

6.3.2 Noninteraction Sites

Various substituents in the vicinity of the amide donors in the oxocyclams can influence the redox properties of the metal ions through steric, electronic, hydrophobic, or hydrogen bonding effects.

Introduction of fluorine(s) to cyclam and dioxocyclam has yielded new derivatives *28–31, 5* and *6*,[9,10] as shown in Figure 6.9.

The electronegative and hydrophobic effects of the F substitution are most evident in electrochemical properties of Cu and Ni complexes. The fluorinated [NiL]$^{2+}$ and [CuH$_{-2}$L]^0 gave quasi-reversible cyclic voltammograms for Ni(III/II), Ni(II/I), and Cu(III/II), respectively. In either Cu or Ni systems, the higher oxidation states Ni(III) and Cu(III) become successively destabilized with respect to Cu(II) and Ni(II), while the lower oxidation state Ni(I) becomes successively stabilized with respect to Ni(II) (see Table 6.2).

The M-imide anion-bonding mode has two extreme resonance forms (*24* and *32*) with different donor properties. The one in which anions are localized on N (*24*) would favor the metal ions of higher oxidation state, while the other one in which anions are on O (*32*) would favor the lesser oxidized metal ions. (See Figure 6.10.) Any factors shifting this resonance would affect the redox potentials of the metal ions. The amine group on the malonate residue of dioxocyclam can change this resonance (in favor of *34* over *33*),

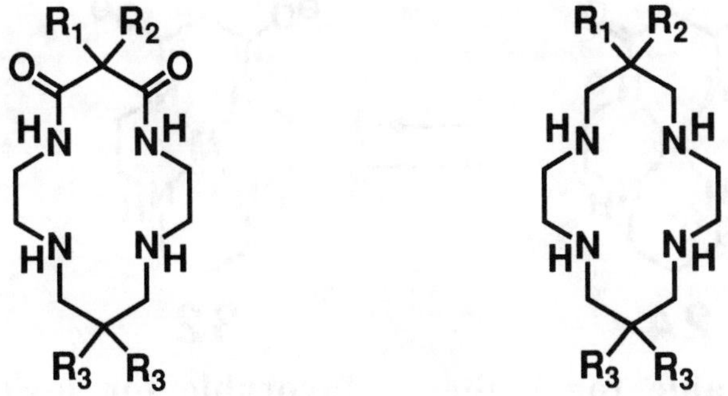

28; R_1=F, R_2=R_3=H   31; R_1=F, R_2=R_3=H
29; R_1=R_2=F, R_3=H   5; R_1=R_2=F, R_3=H
30; R_1=R_2=R_3=F   6; R_1=R_2=R_3=F

Figure 6.9. Various fluorinated cyclams.

Table 6.2. Redox potentials of fluorinated macrocyclic complexes for Cu(III/II),
Ni(III/II), and Ni(II/I).

	$E_{1/2}$ (V vs. SCE)[a]		
Ligand	Cu(III/I)[b]	Ni(III/II)[c]	Ni(II/I)[d]
3	+0.64	—	—
28	+0.69	—	—
29	+0.83	—	—
30	e	—	—
1	—	+0.50	−1.56
31	—	+0.52	−1.52
5	—	+0.63	−1.46
6	—	+0.81	−1.42

[a] All solutions were deaerated by purified Ar.
[b] 0.5 M (NaSO$_4$), 25°C, pH 7.0. Working electrode (WE): glassy carbon.
[c] 0.5 M (Na$_2$SO$_4$), 25°C, pH = 6~7. WE: glassy carbon.
[d] 0.1 M (NaClO$_4$), 25°C, pH = 7.0. WE: hanging mercury drop electrode.
[e] The oxidation process did not occur in water. In DMF, $E_{1/2}$ = +0.85 V vs. SCE (0.1 M [NaClO$_4$], 25°C.
WE: glassy carbon).

but when the amine is protonated, the form with the hydrogen bondings
36 is favored over *35*.[27] (See Figure 6.11.) The same argument is applied to
the dioxocyclen system (*38* versus *37*).[28] (See Figure 6.12.)

Modification of the environmental polarity around metal ions would se-
riously affect the redox potentials as well. In general, substituents enhancing
hydrophobicity (e.g., *39, 30,* and *40*[13]) are likely to stabilize lower oxidation
states of metal.

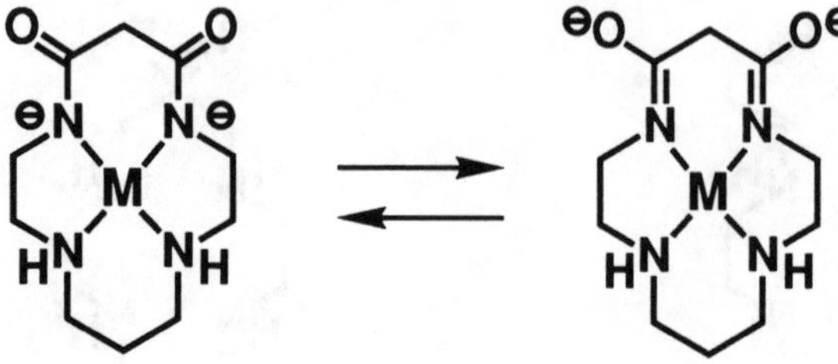

24

favorable for higher
oxidation state

32

favorable for lower
oxidation state

Figure 6.10. Doubly deprotonated dioxocyclam complex in two extreme resonance forms.

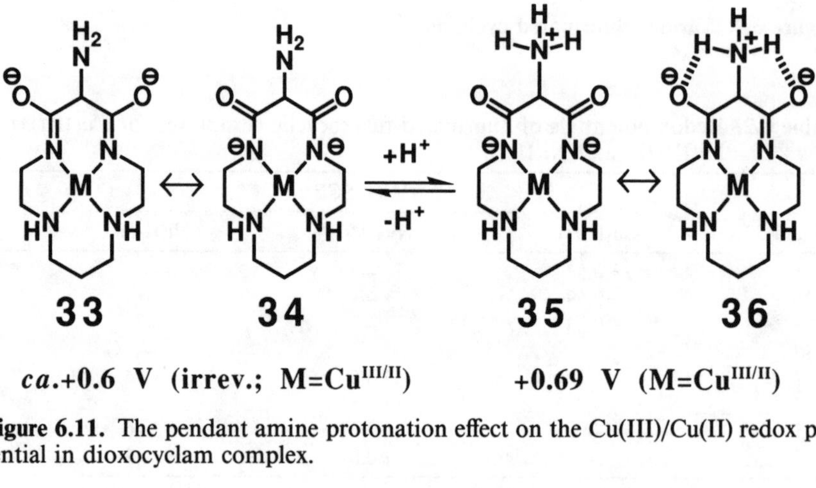

33   **34**   **35**   **36**

$ca.$+0.6 V (irrev.; M=Cu$^{III/II}$)   +0.69 V (M=Cu$^{III/II}$)

Figure 6.11. The pendant amine protonation effect on the Cu(III)/Cu(II) redox potential in dioxocyclam complex.

6.4 Simultaneous Modification of Equatorial and Axial Donors

Attachment of donor pendants to one of the secondary amines in dioxocyclam, *40–43,* dramatically changes the Ni(III/II) redox potentials, as shown in Figure 6.13.[26]

6.5 Synergistic Effects of Equatorial and Axial Modification: Dioxo[16]aneN$_5$

Although Ni(II)–dioxocyclam complex *24* has $E_{1/2}$ of + 0.81 V and Ni(II)-[16]aneN$_5$ complex *44* has $E_{1/2}$ of +0.66 V for the Ni(III/II) couple, the

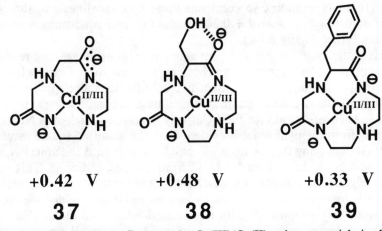

Figure 6.12. The pendant effects on the Cu(III)/Cu(II) redox potentials in dioxocyclam.

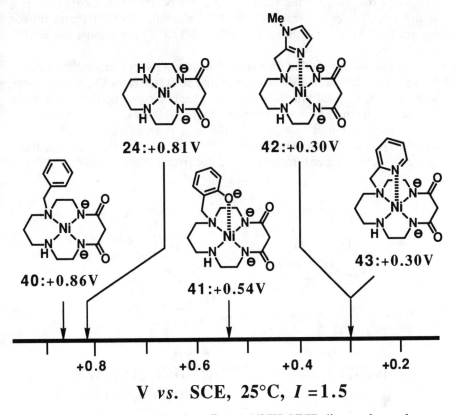

Figure 6.13. The axial coordination effect on Ni(III)/Ni(II)-dioxocyclam redox potentials.

dioxo[16]aneN$_5$ complex *45* combines those functionalities and shows an abnormally low $E_{1/2}$ value of + 0.24 V under the same conditions in aqueous solution.[14] (See Figure 6.14.)

This is the lowest-reported $E_{1/2}$ in aqueous solution, suggesting ready attainment of the Ni(II) species. An X-ray crystal structure of the Ni(II) complex *45* shows that it has a distinct five-coordinate, square-pyramidal geometry with the two deprotonated amide nitrogens coordinating at the basal plane.[29] The steric strain for the (relatively large-sized) high-spin Ni(II) ion suffering from the tight space is evident from the distorted square pyramid, the Ni(II) ion lying 0.22 Å from the basal plane toward the apical N, and the apical Ni–N bond bent by 18.4° from the perpendicular. This steric factor originating from the macrocyclic structure contributes to the observed extremely low oxidation potential. Indeed, air oxidation yielded the brown-colored Ni(III) complex *46* (with R = benzyl-substituted dioxo[16]aneN$_5$), an unprecedented case for Ni(II) complexes.[30] The crystal structure of *46* has proven to have less steric strain, with a more ideal square-pyramidal structure.[30] Hence, we conclude that reduction of the steric strain plays an important role in lowering the $E_{1/2}$ value for Ni(III/II) in dioxo[16]aneN$_5$ complexes. It is of interest that the $E_{1/2}$ values for Cu(III/II) with dioxo-cyclam (+0.64 V) and dioxo[16]aneN$_5$ (+0.68 V) are almost the same.[3] (See Figure 6.15.).

Most interesting of all in regard to *45* is that at room temperature it binds with O$_2$ in a 1:1 adduct *47*, and by doing so it activates O$_2$, which reacts with benzene to yield phenol, where the phenol oxygen derives 100% from O$_2$ and not from H$_2$O,[31] as shown in Figure 6.16. Two new interesting research themes should emerge, one of which is mechanistic study of this novel Ni–O$_2$ activation, which may or may not be relevant to the reactions of the O$_2$-activating enzymes such as cytochrome P450. Another will focus

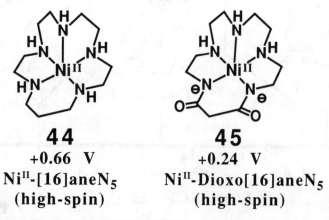

44

+0.66 V

Ni^{II}-[16]aneN$_5$

(high-spin)

45

+0.24 V

Ni^{II}-Dioxo[16]aneN$_5$

(high-spin)

Figure 6.14. Comparison of the Ni(III)/Ni(II) potentials with or without oxo functions in [16]aneN$_5$ complexes.

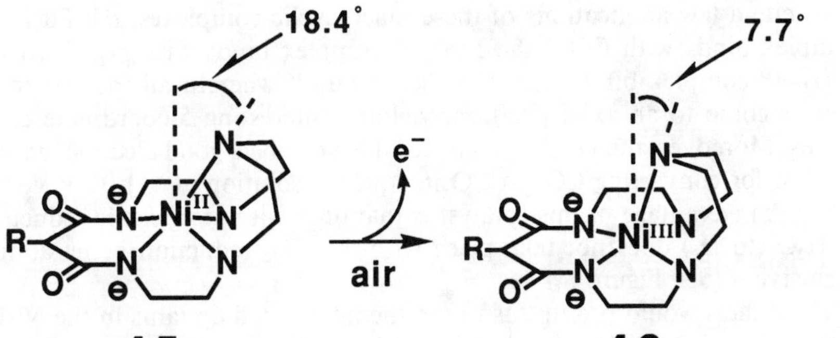

45
More steric strain

46
Steric strain relieved

Figure 6.15. Schematic representation of Ni(II)- and Ni(III)-dioxo[16]aneN$_5$ structures.

45

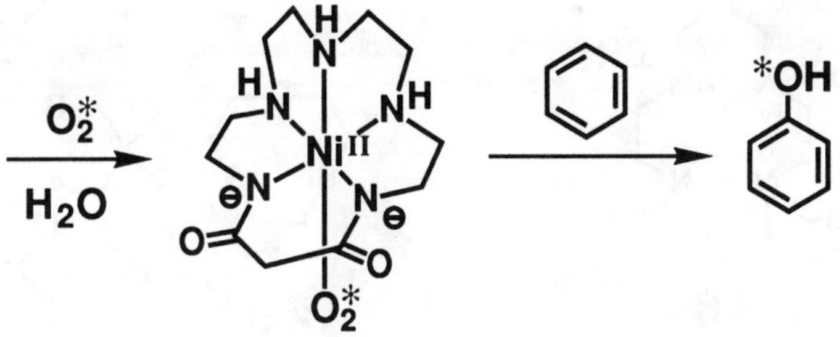

47

Figure 6.16. Reaction of *45* with O_2 for monooxygenation of benzene.

on its applications as a catalyst for direct synthesis of phenols from benzene (or substituted benzenes) and O_2 at room temperature and atmospheric pressure, or as a reagent for DNA cleavage with the activated O_2 like Fe–bleomycin complexes.[32]

6.6 Perspectives and Applications of Redox-Controlling Macrocyclic Polyamine Ligands

6.6.1 For Lower Oxidation States of Metal Ions

Among metal ions at lower oxidation states are Ni(I), Pd(I), Pt(I), and so on that are often kinetically too labile to hold for further applications. Cyclam *1*, tetramethyl cyclam *48*, thiacrown ligands (e.g., *49, 50*) and the like are now proven to be appropriate ligands for those metal ions.

To cite a few applications of these macrocyclic complexes, the Pd(I)–*48* complex binds with CO,[33] the Mo⁰–*50* complex binds with N_2,[34] and the Ni(I)–*48* complex binds with the alkyl group,[35] wherein all the incoming donors come to an axial position, yielding interesting 5-coordinate complexes. Moreover, the Ni(I)–cyclam complex can be a good electrochemical catalyst for converting CO_2 to CO in aqueous solution at −1.05 V versus NHE. An advantage of this catalyst is that otherwise more facile reduction of H_2O (to H_2) does not take place with the CO_2 reduction being almost exclusive.[36] (See Figure 6.17.)

These facts would prompt us to use the fluorinated cyclams in the Ni(I)–cyclam (for $CO_2 \rightarrow CO$) or Co(III)–cyclam catalysts (for $NO_3^- \rightarrow NH_3$).[37] Moreover, N-methylated fluorocyclams (such as *51*) would stabilize Ni(I), Pd(I) or Rh(II) more dramatically than fluorinated cyclams due to the inductive effect as well as the hydrophobic effects of F, and accordingly new chemistry as well as applications would be evolved. (See Figure 6.18.)

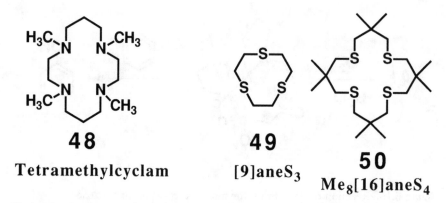

48

Tetramethylcyclam

49

[9]aneS₃

50

Me₈[16]aneS₄

Figure 6.17. Macrocyclic ligands to stabilize lower oxidation states of metal ions.

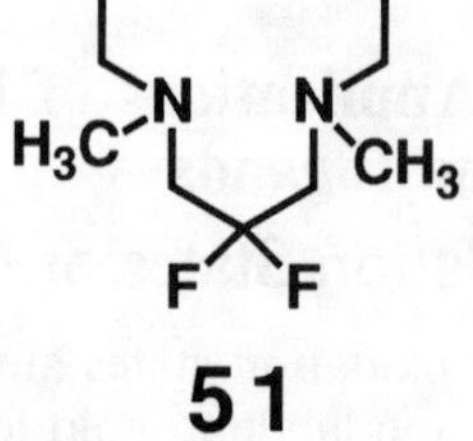

51

Figure 6.18. A hypothetical cyclam appropriate for lower oxidation states of metal ions.

6.6.2 For Higher Oxidation States of Metal Ions

The N-alkylated cyclams stabilize not only reduced forms of metal ions but also higher-valent Ru[38] and Os.[39] Electrochemical oxidation of *trans*-[M(III)(*48*)X$_2$]$^+$ (where M = Ru, Os) yields *trans*-[M(IV)(*48*)X$_2$]$^{2+}$. Chemical oxidation of the M(III) complex produced various M^{n+}–oxo complexes, including well-characterized crystalline *trans*-[Ru(IV)(*48*)O(MeCN)]$^{2+}$.[40] The Ru(IV)–dioxo complex *52* is shown to be a selective reagent for oxidation of alcohol (to aldehyde) and to activate the C–H bond.[38] Hence, in these complexes *48* may well be replaced by the tetramethyl fluorocyclam *51*.

As described in the previous section, Ni(II) in *45* is oxygenated. The resulting O$_2$ adduct attacks benzene to yield phenol. The yield of phenol, however, is neither quantitative or catalytic, due to the degradation of the ligand *11* during the oxygen transfer. This decomposition product is independently shown to be *53* both by Martell[41] and by us.[42] (See Figure 6.19.) The Fe(II)–bleomycin complex *54*, shown in Figure 6.20, that also activates O$_2$ similarly suffers from self-degradation in the course of DNA cleavage.[32] The O$_2$ bound to Fe(III)–bleomycin attacks cancer DNA (especially G–C base pairs) to abstract the H radical at C-4 ribose to pave the way for the succeeding O$_2$ attack, leading to eventual ribose degradation.[32] With our dioxo[16]aneN$_5$, we will be able to attach various functionalities so as to increase the substrate selectivity, affect the Ni(III/II) redox potentials, modulate the coordination environments, or protect the ligand from self-decomposition. We currently believe that the mode of O$_2$ activation by the Ni(II)-complex *45* is different from the one by Fe(II)–bleomycin, because the latter cannot yield phenol from benzene and air. Our system may be closer to cytochrome P450, which can monooxygenate benzene. In the previous P450 models using porphyrin derivatives, the monooxygenation of benzene has been unsuccessful, although examples of epoxidation of olefins, another typical P450 reaction, are abundantly documented.[43] Thus, future exploi-

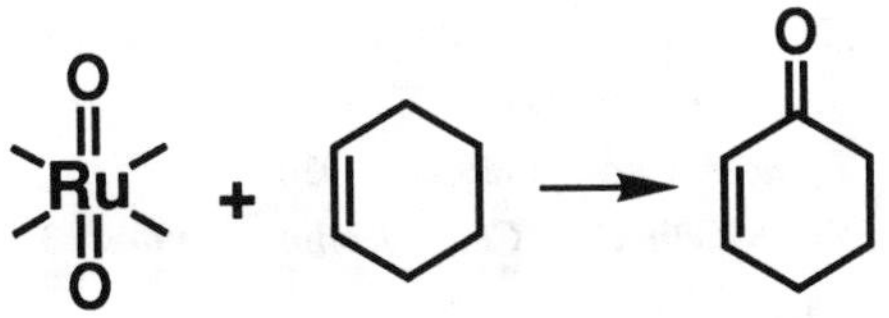

52

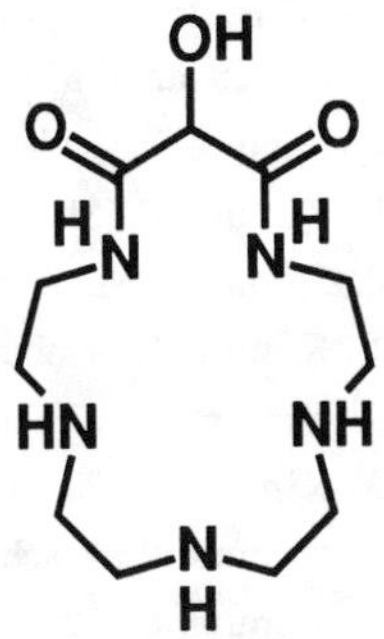

53

Figure 6.19. Oxygen transfer to an allyl position and an active methylene.

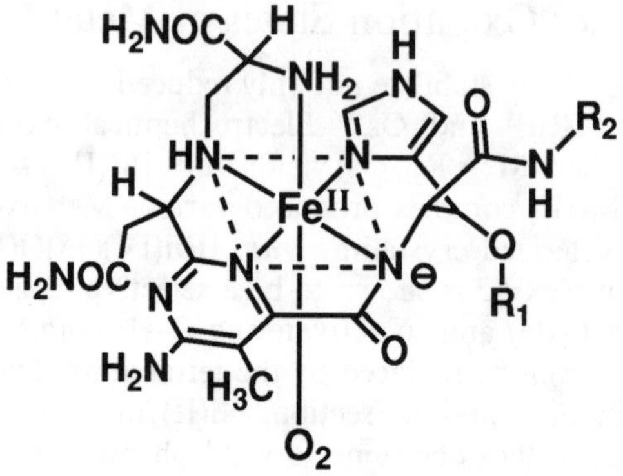

54

Figure 6.20. A proposed O_2 adduct of bleomycine-Fe(II) complex.

tation of our Ni(II)–dioxo[16]aneN$_5$ *45* will offer quite a new and exciting subject for research and application.

References

1. Lovecchio, F. V.; Gore, E. S.; Busch, D. H. *J. Am. Chem. Soc.* **1974,** *96,* 3109. For review: Busch, D. H. *Acc. Chem. Res.* **1978,** *11,* 392.

2. Kimura, E.; Koike, T.; Shiota, T.; Iitaka, Y. *Inorg. Chem.* **1990,** *29,* 4621.

3. Kimura, E. *J. Coord. Chem.* **1986,** *15,* 1.

4. Kimura, E.; Dalimunte, C. A.; Yamashita, A.; Machida, R. *J. Chem. Soc., Chem. Commun.* **1985,** 1041.

5. Kimura, E.; Kurogi, Y.; Wada, S.; Shionoya, M. *J. Chem. Soc., Chem. Commun.* **1989,** 781.

6. Machida, R.; Kimura, E.; Kodama, M. *Inorg. Chem.* **1983,** *22,* 2055.

7. (a) Kodama, M.; Kimura, E. *J. Chem. Soc., Dalton Trans.* **1979,** 327.
   (b) Kodama, M.; Kimura, E. *J. Chem. Soc., Dalton Trans.* **1979,** 694.

8. Kimura, E.; Koike, T.; Machida, R.; Nagai, R.; Kodama, M. *Inorg. Chem.* **1984,** *23,* 4181.

9. Kimura, E.; Shionoya, M.; Okamoto, M.; Nada, H. *J. Am. Chem. Soc.* **1988,** *110,* 3679.

10. Shionoya, M.; Kimura, E.; Iitaka, Y. *J. Am. Chem. Soc.* **1990,** *112,* 9237.

11. Kimura, E.; Koike, T.; Nada, H.; Iitaka, Y. *Inorg. Chem.* **1988,** *27,* 1036.

12. Kimura, E.; Koike, T.; Nada, H.; Iitaka, Y. *J. Chem. Soc., Chem. Commun.* **1986,** 1322.

13. Kimura, E. *Pure & Appl. Chem.* **1989,** *61,* 823.

14. Kimura, E.; Machida, R.; Kodama, M. *J. Am. Chem. Soc.* **1984,** *106,* 5497.

15. Kimura, E.; Koike, T.; Takahashi, M. *J. Chem. Soc., Chem. Commun.* **1985,** 385.

16. Kimura, E.; Joko, S.; Koike, T.; Kodama, M. *J. Am. Chem. Soc.* **1987,** *109,* 5528.

17. Kimura, E. *Pure & Appl. Chem.* **1986,** *58,* 1461.

18. Kimura, E.; Shionoya, M.; Mita, T.; Iitaka, Y. *J. Chem. Soc., Chem. Commun.* **1987,** 1712.

19. Kimura, E.; Kotake, Y.; Koike, T.; Shionoya, M.; Shiro, M. *Inorg. Chem.* **1990**, *29*, 4991.

20. Fabbrizzi, L.; Micheloni, M.; Paoletti, P. *Inorg. Chem.* **1980**, *19*, 535.

21. Iitaka, Y.; Koike, T.; Kimura, E. *Inorg. Chem.* **1986**, *25*, 402.

22. Kimura, E.; Koike, T.; Uenishi, K.; Hediger, M.; Kuramoto, M.; Joko, S.; Arai, Y.; Kodama, M.; Iitaka, Y. *Inorg. Chem.* **1987**, *26*, 2975-2983.

23. Kimura, E., unpublished results.

24. Boss, F. P.; Chellappa, K. L.; Margerum, D. W. *J. Am. Chem. Soc.* **1977**, *99*, 2195.

25. Boss, F. P.; Margerum, D. W. *Inorg. Chem.* **1977**, *16*, 1210.

26. Kimura, E.; Kurosaki, H., unpublished results.

27. Kimura, E.; Koike, T., unpublished results.

28. Kimura, E.; Umeyama, H., unpublished results.

29. Kushi, Y.; Machida, R.; Kimura, E. *J. Chem. Soc., Chem. Commun.* **1985**, 216.

30. Machida, R.; Kimura, E.; Kushi, Y. *Inorg. Chem.* **1986**, *25*, 3461.

31. Kimura, E.; Machida, R. *J. Chem. Soc., Chem. Commun.* **1984**, 499.

32. Stubbe, J.; Kozarich, J. W. *Chem. Rev.* **1987**, *87*, 1107.

33. Blake, A. J.; Gould, R. O.; Hyde, T. I.; Schröder, M. *J. Chem. Soc., Chem. Commun,* **1987**, 431.

34. Yoshida, T.; Adachi, T.; Kaminaka, M.; Ueda, T. *J. Am. Chem. Soc.* **1988**, *110*, 4872.

35. D'Aniello, Jr., M. J.; Barefield, E. K. *J. Am. Chem. Soc.* **1976**, *98*, 1610.

36. Collin, J. P.; Sauvage, J. P. *Coord. Chem. Rev.* **1989**, *93*, 245.

37. Taniguchi, I.; Nakashima, N.; Yasukouchi, K. *J. Chem. Soc., Chem. Commun.* **1986**, 1814.

38. Che, C.-M.; Lai, T.; Wong, K. *Inorg. Chem.* **1987**, *26*, 2289.

39. Che, C.-M.; Cheng, W. *J. Am. Chem. Soc.* **1986**, *108*, 4644.

40. Che, C.-M.; Wong, K.; Mak, T. C. W. *J. Chem. Soc., Chem. Commun.* **1985**, 546.

41. Chen, D.; Martell, A. E. *J. Am. Chem. Soc.* **1990**, *112*, 9412.

42. Kimura, E.; Shionoya, M. unpublished results.

43. (a) White, R. E.; Coon, M. J. *Annu. Rev. Biochem.* **1980**, *49*, 315.
    (b) Groves, J. T. *Adv. Inorg. Biochem.* **1979**, 119.

CHAPTER 7

Preorganization and Molecular Recognition in Binuclear Macrocyclic and Macrobicyclic Complexes

Arthur E. Martell
Department of Chemistry
Texas A&M University
College Station, Texas 77843-3255

Macrocyclic ligands may be employed to develop preorganization of binuclear complexes to a higher degree than has been achieved by the use of open-chain binucleating ligands. This property is developed to an even greater extent in the macrobicyclic (cryptand) binucleating ligands.

7.1 Binuclear Macrocyclic Complexes

7.1.1 BISDIEN and Its Analogs

The macrocyclic ligand BISDIEN (*1*) (see Figure 7.1) has been shown to form binuclear complexes of first-row transition metal ions, which act as hosts to bind secondary bifunctional donors (guests) as bridging groups.[1,2,3] The bridging groups have been labeled *cascade* complexes by Lehn.[3] Ex-

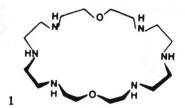

Figure 7.1. Structure of [24]aneN$_6$O$_2$, BISDIEN (1).

amples of secondary anionic guests in binuclear BISDIEN complexes are hydroxide ion, imidazolate anion, and dioxygen, for which quantitative binding constants have been reported.[1] In addition, the crystal structure of an imidazolate-bridged binuclear copper(II)–BISDIEN complex has been reported by Lippard and co-workers.[4,5]

A hydroxide ion-bridged dicopper(II) center incorporated in BISDIEN was found to have magnetic and spectroscopic properties closely resembling those of binuclear copper centers in enzymes, and the hydroxide anion was therefore suggested as the bridging ligand in several binuclear enzymic systems.[6,7]

The quantitative stability studies of Martell et al.[1] have shown that the dinucleating tendencies of BISDIEN are not fully developed because of the flexibility of the 24-membered macrocyclic ring. Thus, metal ions such as Ni(II) or Co(II) having a coordination sphere of octahedral geometry form highly stable 1:1 (mononuclear) complexes, in which amino nitrogen donors from both sides of the macrocycle are involved, requiring considerable folding of the ring; the tendency to add a second metal ion to form a binuclear complex is actually quite weak. However, the binuclear complexes of BISDIEN are greatly stabilized in solution by a bridging hydroxide ion (2) as well as an unusual bridging bifunctional imidazolate anion (3), shown in Figure 7.2. The dissociation of the hydrogen ion from imidazole to form an anion is an indication of the high degree of stabilization achieved in the binuclear copper(II)–imidazolate structure.

Comparison of the dioxygen affinities of the dicobalt(II) complexes of BISDIEN (1) with those of the analogous ligands illustrated in Figure 7.3, BISBAMP (4) and O2-BISBAMP (5) demonstrate how the nature and geometry of the macrocycle influence the degree of recognition of its binuclear cobalt complex for molecular oxygen. The dioxygen affinity of Co_2L to form hydroxo- and peroxo-bridged binuclear cobalt complexes is sufficiently high for L = BISDIEN that they constitute the major constituents of the solution (~100%) above pH 6. The dioxygen complex of binuclear dicobalt BISBAMP, however, is much less stable and forms in appreciable concentrations only at pH ~9 and above.[8] The lower stability is considered to be due in part to the lower basicity of the pyridine nitrogens relative to the aliphatic

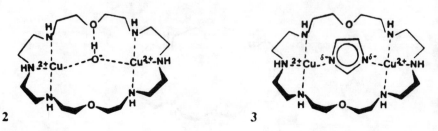

Figure 7.2. Structures of $Cu_2(OH)BISDIEN^{3+}$ (2) and $Cu_2(imid)BISDIEN^{3+}$ (3).

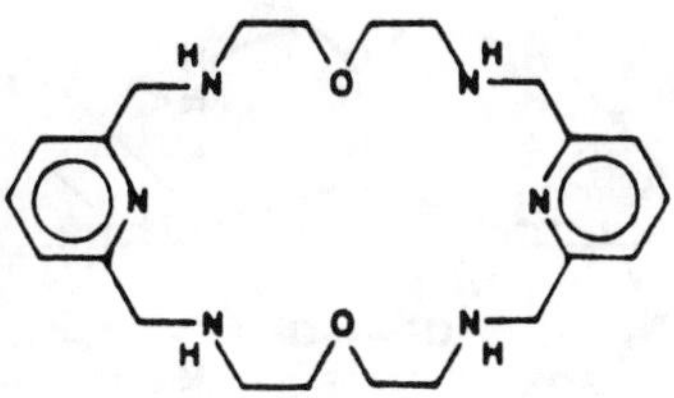

Figure 7.3. Structures of BISBAMP (4) and O2-BISBAMP (5).

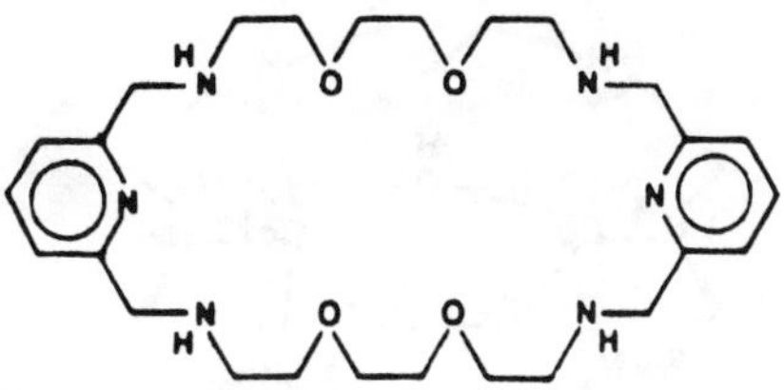

amino groups in BISDIEN and especially to the greater distance of separation of the cobalt centers due to the rigidity of the pyridine rings. Thus the folding of the marocyclic ring necessary to accommodate the donor groups of the host to fit the size (and shape) of the guest occurs much more readily with the flexible macrocycle BISDIEN than with the more rigid macrocycle BISBAMP. This interpretation is further supported by the very low dioxygen affinity of O2-BISBAMP (5), which is so weak that the binuclear peroxo-bridged complex is never more than a minor constituent in solution.[9] This result is interpreted as being due to the greater distance between the Co(II) centers in the binuclear complex of 5 and to the greater folding of the macrocyclic ring, which is necessary to bring the metal ions close enough to each other to simultaneously coordinate dioxygen to form a bridging peroxo group.

7.1.2 Molecular Recognition and Catalysis in Binuclear BISDIEN Complexes

Inspection of the probable structure and coordinate bonding modes of the dioxygen complex of the binuclear dicobalt BISDIEN complex indicates that the most stable species formed in solution, the hydroxo and peroxo-bridged complex (6) [$Co_2LO_2(OH)^{3+}$ (L = BISDIEN)], has an additional site on each metal center occupied by a coordinated water molecule (7). This complex is converted at high pH to the dihydroxo $Co_2LO_2(OH)_2^{2+}$ and eventually the trihydroxo $Co_2LO_2(OH)_3^+$ forms. The existence of these two additional coordination sites on the formally octahedral Co(III) centers brings up the possibility of adding a third bridging bifunctional donor group to 6. Accordingly, quantitative potentiometric studies of 6 in the presence of oxalic acid and catechol by Motekaitis and Martell[10] and by Szpoganicz et al.[11] have demonstrated the formation of the tribridged binuclear cobalt(III) macrocyclic complexes 8 [$Co_2LL'O_2(OH)^+$] and 9 [$Co_2LL''O_2(OH)^+$], respectively (where L = BISDIEN, H_2L' = oxalic acid, and H_2L'' = catechol). (See Figure 7.4.) Determination of the stabilities of all complex species in solution of the system Co(II), O_2, BISDIEN, and oxalate showed that the tribridged complex species (8) is present in significant concentrations from

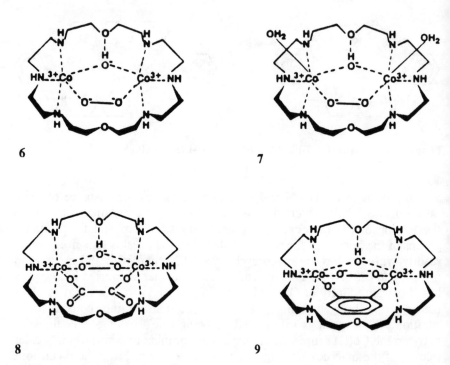

Figure 7.4. Structure of $Co_2(OH)(O-O)BISDIEN^{3+}$ (6), additional coordination sites on $Co_2(OH)(O-O)BISDIEN^{3+}$ (7), structures of $Co_2(OH)(O-O)(ox)BISDIEN^+$ (8) and $Co_2(OH)(O-O)(cat)BISDIEN^+$ (9).

pH 7–10, as is indicated by the species distribution curves in Figure 7.5.[10] Similarly Szpoganicz et al. have demonstrated[11] that an analogous triply bridged complex is formed, as indicated in Formula 9, when the third bridging bifunctional donor is the catecholate ion. The "recognition" of the third bifunctional donor by the hydroxo- and peroxo-bridged dicobalt(III) complex (6,7) is very sensitive to the size and shape of the third bridging group because it is highly preorganized by the two bridging groups already present as well as by the macrocyclic ring. While this complex recognizes and binds both oxalate[10] and catecholate[11] as bridging groups, it does not recognize the dianion of hydroquinone, which except for the geometrical positioning of its negative oxygen donors is similar in properties to the dianion of catechol.

Encapsulation of both dioxygen and oxalate within the same macrocyclic cavity, as in 8, has been shown[10] to result in the facile oxidation of oxalate to carbon dioxide by the coordinated dioxygen. There is a mismatch in the stoichiometry of the oxidation of oxalate versus the reduction of oxygen; it takes two electrons to oxidize the oxalate, whereas four electrons are required to reduce the oxygen to water. The result is the conversion of the oxalate–oxygen complex to the binuclear cobalt(III) species, and the reaction

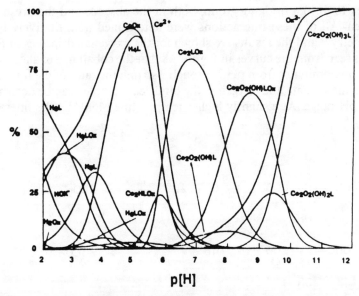

Figure 7.5. Distribution of species as a function of pH for a solution containing a 1:1:2 molar ratio of BISDIEN–oxalate–Co(II) at 25.0°C, $\mu = 0.100$ M under 1.00 atm oxygen; [BISDIEN] = 0.00200 M.

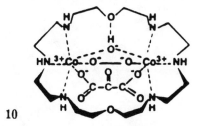

10

Figure 7.6. Structure of $Co_2(OH)(O\text{-}O)(mes)BISDIEN^+$ (10).

ends at that point because cobalt(III) is inert and cannot undergo substitution in a reasonable length of time. Therefore, a four-electron reductant is needed, and mesoxalate seems to fit the requirements as indicated by the following equations (Fig. 7.7). Either mesoxalate is reduced directly and oxygen is inserted in the central carbonyl carbon atom, or the hydrate is oxidized to CO_2 and H^+ ion is released. In any case, four electrons are required to oxidize the mesoxalate, and the end result is that the binuclear cobalt complex retains its original oxidation state of $+2$. The dicobalt–BISDIEN–oxygen complex, in which mesoxalate forms a third bridging group (Figure 7.6), required the determination of the interaction of mesoxalate with the protonated BISDIEN itself and with the binuclear cobalt complex to form a bridging group, both in the absence and the presence of dioxygen. All of these interactions were determined quantitatively by potentiometry,[12] and the results are shown for the oxygenated species in Figure 7.8. It is seen from the curves in Figure 7.8 that the cobalt mesoxalate oxygen species predominates from pH 7 through 9 so that the rates of the interaction between mesoxalate and oxygen in the binuclear complex can be determined in that pH range. At a slightly higher temperature (45.0 °C) the interaction

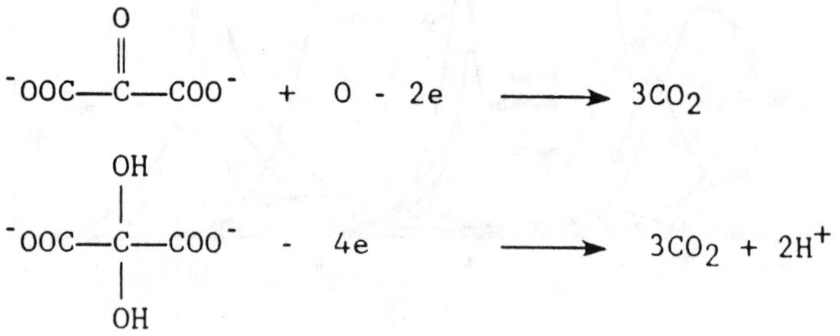

$$^-OOC\text{---}\overset{\overset{\textstyle O}{\|}}{C}\text{---}COO^- \ + \ 0 - 2e \ \longrightarrow \ 3CO_2$$

$$^-OOC\text{---}\overset{\overset{\textstyle OH}{|}}{\underset{\underset{\textstyle OH}{|}}{C}}\text{---}COO^- \ - \ 4e \ \longrightarrow \ 3CO_2 \ + \ 2H^+$$

Figure 7.7. Possible oxidation pathways of mesoxalate.

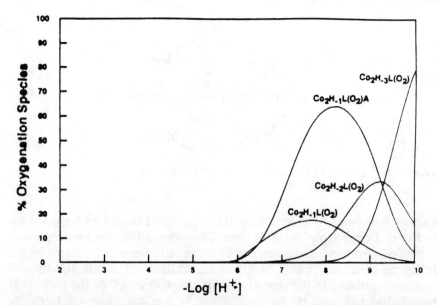

Figure 7.8. Distribution of dioxygen complex species present as a function of pH in a solution containing a 1:1:2 molar ratio of BISDIEN (L), mesoxalic acid (H_2L), and cobalt(II), respectively, under 1.00 atm of dioxygen at 25.0°C and $\mu = 0.100$ (KCl). $T_L = T_A = 1/2T_{Co(II)} = 2.0 \times 10^{-3}$ mol. Species not containing dioxygen are omitted for clarity. Negative subscripts indicate the number of hydroxo donor groups present, i.e., the number of protons abstracted from coordinated water molecules.

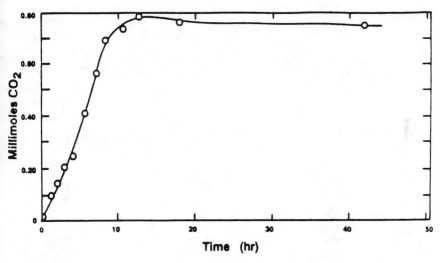

Figure 7.9. Amounts of carbonate (as CO_2) formed in the reaction between coordinated O_2 and mesoxalate in the dinuclear BISDIEN–Co(II) complex at 45.0°C and $\mu = 0.100$ mol (KCl). Appropriate amounts of 1.00 molar KOH were supplied as necessary to maintain a constant pH of 8.5.

is seen to occur not only stoichiometrically but catalytically as in Figure 7.9, which shows the results obtained by treating a tenfold excess of mesoxalate with the cobalt–BISDIEN–dioxygen complex, in which the mesoxalate is oxidized to CO_2 and the dioxygen is reduced to water. The redox reaction is catalytic, with the number of turnovers seemingly limited only by the amount of mesoxalate supplied to the reaction mixture.

7.1.3 A New Tyrosinase Model

The macrocyclic tetra-Schiff base *11*, formed by the dipodal condensation of *m*-benzenedicarboxaldehyde with diethylenetriamine, is a dinucleating ligand that coordinates two Cu(I) ions, as indicated by *12*. This complex reacts with dioxygen[13] to form a binuclear Cu(II) complex of the Schiff base with hydroxide- and phenoxide-bridging groups, *14*. (See Scheme in Figure 7.10.) The latter is formed by insertion of one of the oxygen atoms of the dioxygen into the aromatic ring. The fact that all of the inserted oxygen comes from molecular oxygen was demonstrated by the use of $^{18}O_2$ tracer[13] and lends support for the suggestion that the tyrosinase-like oxygenase insertion reaction occurs through the formation of an intermediate binuclear (peroxo-bridged) dioxygen complex (*13*), which has a very short lifetime under the reaction conditions employed.

It is of interest to point out that reaction conditions analogous to *11* → *14*, with Co^{2+} in place of Cu$^+$, gives a dioxygen complex stable enough to

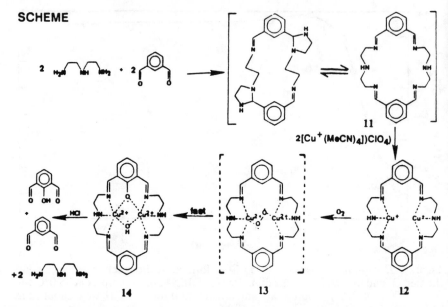

Figure 7.10. Cu(II)-mediated hydroxylation of m-benzenedicarboxaldehyde.

be well characterized at ambient temperature that undergoes metal-centered degradation to the corresponding inert binuclear Co(III) complex. No dioxygen insertion occurs.[13]

7.1.4 Copper(I) Dioxygen Complexes

Ngwenya et al.[14] prepared a macrocycle in which an electrophilic benzene ring was not available for oxygen insertion from the dicopper(I) dioxygen complex by basing the bridging groups on furan-2,5-dialdehyde instead of benzene-1,3-dialdehyde. The resulting macrocycle contains an oxygen in place of an aromatic CH group and did not undergo oxygen insertion but rather formed an unstable copper(I)–dioxygen complex designated as a Cu(II)–peroxo-bridged species (*15*). (See Figure 7.11.) The spectrum of this complex, illustrated in Figure 7.12, was taken at 5.0 °C and showed that it would exist for several hours before degradation eventually took place, giving the corresponding copper(II) dibridged complex illustrated by formula *16*. (See Figure 7.11.) The spectrum in Figure 7.12 indicates that the complex is similar to that prepared from the binuclear copper(I) complex of the binucleating ligand derived from *m*-xylyl-bis-benzimidazole.[15] This and similar dioxygen complexes of binuclear copper(I) might be of importance in the oxidation or hydroxylation of various substrates.

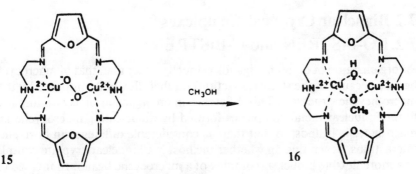

Figure 7.11. Decomposition of a Cu(I) dioxygen complex (15) to form the di-μ-hydroxo-bridged Cu(II) complex (16).

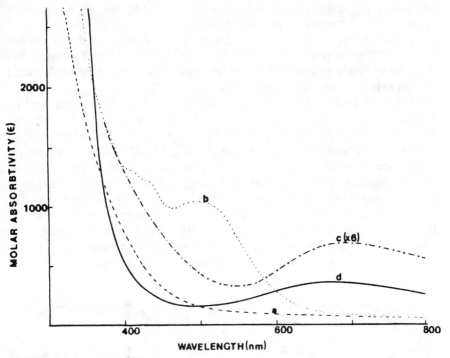

Figure 7.12. Absorption spectra of the copper–$FD_2(DIEN)_2$ complexes in 1:3 acetonitrile:methanol solution: a (-----) 1.84×10^{-4} molar Cu(I) complex of $FD_2(DIEN)_2$ prior to oxygenation; b (·····) system a during oxygenation; c (-·-·-) the irreversible degradation product of b, 1.05×10^{-3} molar; and d (———) the Cu(II) complex of $FD_2(DIEN)_2$, 1.84×10^{-4} molar.

7.2 Binuclear Cryptand Complexes

7.2.1 O-BISTREN and C-BISTREN

Preorganization of cryptand ligands is much greater than that of macrocyclic ligands because of the steric restrictions on their flexibility imposed by their macrobicyclic structures. This difference in preorganization is also imparted to the binuclear metal complexes formed by dinucleating macrocyclic and macrobicyclic ligands, so that there is considerable difference in host–guest interactions, depending on whether the host is a binuclear cryptate complex or a more flexible binuclear complex of a macrocyclic ligand. In this section the macrobicyclic (cryptand) polyamine ligands O-BISTREN (*17*) and C-BISTREN (*18*), illustrated in Figure 7.13, are compared with respect to how their varying degrees of preorganization influence the degree of molecular recognition for various bridging donor ions and molecules as guests. The macrobicyclic ligands O-BISTREN and C-BISTREN show strong dinucleating tendencies,[16-18] with or without suitable coordinating bridging groups. It has been pointed out[18] that the cavities in the C-BISTREN structure seem to be less preorganized for binuclear complex formation than those of O-BISTREN, probably because of the tendency of the hydrocarbon bridges to self-associate through hydrophobic bonding.

There is an interesting reversal in the relative magnitudes of the successive Cu(II) binding constants of O-BISTREN and C-BISTREN. The latter has more basic donor groups and would be expected to form more stable metal complexes; however, its 1:1 complex with Cu(II) is considerably weaker than that of O-BISTREN. On the other hand, its second metal-binding constant is much stronger than that of O-BISTREN, with the result that the overall binding constant ($\beta = [M_2L]/[M]^2[L]$) is somewhat larger for C-BISTREN.[18] Coordination of the first metal ion by C-BISTREN would require opening up the cryptand cavity and disrupting the hydrophobic bonding associations of the hydrocarbon bridges. Thus, the ligand would then be more prepared (preorganized) for coordination of the second metal ion, which would take place with a relatively high stability constant. Overall, formation of the binuclear copper(II)–C-BISTREN cryptate complex involves two effects operating in opposite directions—the breaking up of hydrophobic bonding, which would cost energy and the greater basicity of the

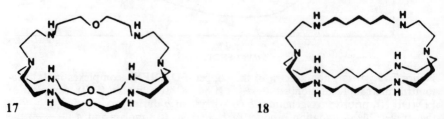

Figure 7.13. Structures of O-BISTREN (17) and C-BISTREN (18).

amino groups, which favors higher stability. The latter effect is slightly predominant.

The binding constants of (isoelectronic) hydroxide and fluoride ions to the binuclear Cu(II) complexes of O-BISTREN and C-BISTREN (Fig. 7.14) are presented in Table 7.1. Of the two binuclear Cu(II) cryptates, that of the more basic C-BISTREN would be expected to have stronger Cu–N coordinate bonds and therefore weaker coordinate bonds to secondary ligands such as F^- and OH^-. The fluoride ion is seen to behave normally, with approximately a factor of ten difference in its binding constants; Cu_2O-BISTREN^{4+} has the higher affinity for the fluoride anion as predicted (Table 7.1, Formulas *21* and *22*). This interpretation, however, is inadequate to

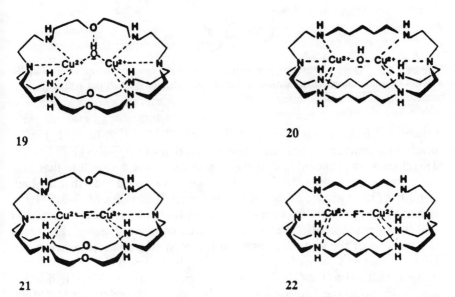

Figure 7.14. Structures of hydroxide- and fluoride-bridged Cu(II) complexes of O- and C-BISTREN (see text).

Table 7.1. Copper(II) Cryptate Stability Constants of O-BISTREN and C-BISTREN, $\mu = 0.100$ M (NaClO$_4$), $t = 25.0°$C.

Complexes	Equilibrium Quotient, Q	Log Q	
		O-BISTREN[a]	C-BISTREN[a]
	$[CuL^{2+}]/[Cu^{2+}][L]$	17.59	15.39
	$[Cu_2L^{4+}]/[CuL^{2+}][Cu^{2+}]$	10.73	13.37
19, 20	$[Cu_2(OH)L^{3+}][H^+]/[Cu_2L^{4+}]$	−3.89	−7.59
19, 20	$[Cu_2(OH)L^{3+}]/[Cu_2L^{4+}][OH^-]$	9.89	6.19
21, 22	$[Cu_2FL^{3+}]/[Cu_2L^{4+}][F^-]$	4.5	3.3[b]

[a]Ref. 18.
[b]$\mu = 0.100$ M (0.090 M NaClO$_4$ + 0.010 M NaF)

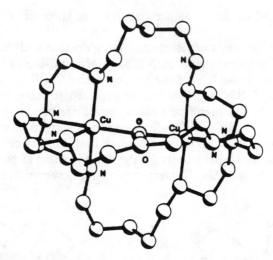

Figure 7.15. Diagram illustrating structure of hydroxo-bridged dicopper(II)–O-BIS-TREN cryptate showing proximity of OH to an ether oxygen of the ligand.

explain the fact that the affinity of Cu_2O-BISTREN^{4+} (*19*) for OH$^-$ ions is nearly four orders of magnitude greater than that of Cu_2C-BISTREN^{4+} (*20*). Therefore it was concluded that the former may have a structure that provides stabilization not available to the latter. It was suggested[16] that the high stability of the hydroxo bridge of the dicopper(II) complex of O-BISTREN is due to hydrogen bonding to one of the ether bridging groups, an interaction that is not possible with the fluoride-bridged complex. Figure 7.15, which is a drawing of the structure based on the crystallographic parameters obtained,[19] confirms the hydrogen bonding of the hydroxo bridge to an ether oxygen, with a bent Cu-OH-Cu angle. The greater basicity of the hydroxide ion compared to the fluoride ion is readily seen in the relative pKas of the two bridging ligands; the hydroxide ion has a pKa of approximately 10^{13} to 10^{14}, whereas the fluoride ion has a pKa of 3.0. Thus the hydroxide should be, and is, bound to the copper ion much more strongly than the fluoride ion.

7.2.2 Dioxygen Complexes

The differences in stabilities of the dioxygen adducts of the binuclear cobalt complexes of O-BISTREN (*17*) and BISDIEN (*1*) also deserve consideration. The higher basicity sum of the eight amino groups of O-BISTREN over those of the six amino groups of BISDIEN would lead one to expect the former to have a much higher dioxygen affinity than the latter.[20,21] It turns out, however, that the oxygenation constant associated with the formation of *23* is about three orders of magnitude lower than that of *6*. The explanation offered for this interesting reversal of the expected relative magnitudes of the oxygenation constants is based on possible steric crowding in the cavity of *23*, which interferes with metal–dioxygen bond formation. This complex turns out to be of interest for oxygen separation processes because of its rapid reversibility at moderate temperatures, the resistance of the ligand to oxidative attack by coordinated dioxygen, and the fact that the dioxygen complex is not converted to an inert cobalt(III) complex at a measurable rate at moderate temperatures.

Although Co(II)–O-BISTREN (*23*), illustrated in Figure 7.16, is a highly successful dioxygen carrier and perhaps the most effective oxygen carrier discovered up to the present time,[20,21] the difficulty of synthesizing the ligand, which is well documented in the literature,[16,17] makes its application for that purpose highly impractical. The search for similar (yet less expensive to prepare) dinucleating ligands as alternates for O-BISTREN has led to studies of the dioxygen affinities of dicobalt complexes of C-BISTREN (*18*) and PXBISTREN (formula *8* in Table 7.2). It has been found[22] that the cobalt(II) binuclear complexes of these alternative ligands do not carry oxygen in aqueous solution, perhaps for quite different reasons. Although C-BISTREN is probably as flexible as O-BISTREN, the pocket between the two metal ions is perhaps too hydrophobic to readily accommodate the highly polar dinegative peroxide anion as a bridging group, in view of the high solvation requirement of the peroxide anion. The structure of the ligand PXBISTREN must be considerably preorganized in view of the rigidity of the three six-carbon bridges imposed by the phenyl rings. Therefore it seems that the two TREN subunits of the ligand are probably held too far apart to allow the two-atom peroxo group to coordinate both cobalt(II) ions simultaneously. The negative evidence of the oxygen-carrying properties of these binuclear cryptates suggests a reasonable rationale for the success of dicobalt(II)–O-BISTREN as an oxygen carrier: considerable folding of the cryptand ligand is necessary to accommodate a bridging peroxo group. This concept may account for the low thermodynamic stability of the dioxygen complex, as well as the ease with which it returns to the oxygen-free binuclear Co(II) complex.

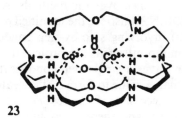

23

Figure 7.16. Structure of [Co$_2$(OH)(O-O)O-BISTREN]$^{3+}$ (23).

7.2.3 An Inert Copper(I) Cryptate

A novel macrobicyclic hexa Schiff base complex was synthesized by a template silver nitrate condensation of a 3:2 molar ratio of benzene isophthaldehyde and tris(3-aminopropyl)amine, producing a complex that was converted to the copper(I) hexa Schiff base cryptate by transmetallation.[23] Surprisingly, the binuclear copper(I) complex of MX_3TAPN_2 (see page 127) showed no reactivity toward the reagents that usually combine with copper(I) Schiff base complexes. It would not react with carbon monoxide, the usual anions such as thiocyanate, or even with dioxygen but rather remained as the binuclear copper(I) complex of the hexa Schiff base, which is apparently a very stable configuration. Crystal structure determination showed that the copper(I) is fully coordinated by three Schiff base nitrogens and a tertiary amino nitrogen in an approximately trigonal pyramidal arrangement. Apparently the arms of the tris(3-aminopropyl)amine are sufficiently long to promote a relatively strain-free hexacoordinated Cu(I) complex that prefers to remain as Cu(I) in the arrangement described above (rather than as a 5-coordinate carbonyl complex or a Cu(II)–dioxygen complex) even with phenolic and hydroxyl bridges, which are possible in the oxidized form of the ligand. It should be noted that the Cu(I) in this inert complex is tetracoordinated, whereas it is tri- or di-coordinated in all of the Cu(I) complexes of the macrocyclic ligands described above, in which hydroxylation of an aromatic ring takes place or an oxygen complex is formed.

7.3 Methods of Synthesis

7.3.1 Nontemplate Method

In recent papers, Menif and Martell[13,24,25] and Chen and Martell[26] describe a general reaction involving the $2+2$ dipodal condensation of a rigid dialdehyde and diethylenetriamine and the $2+3$ tripodal condensation of a rigid dialdehyde with tris(aminoethyl)amine (TREN) to produce macrocyclic and macrobicyclic Schiff bases, respectively, in good yield. The tetra and hexa Schiff bases were also reduced with sodium borohydride nearly quantitatively to the corresponding saturated macrocycles and cryptand ligands. The total number of ligands described in these papers is 22; they are listed in Table 7.2. In cases where this method of synthesis works, it constitutes a considerable advance over the methods used for the preparation of BISTREN and BISDIEN,[27] (also see the modified procedure for BISTREN synthesis by Murase[17]) involving step-by-step condensation of functional groups in which the amino nitrogens are protected by tosylation. The new synthetic procedure involves a single-step condensation of the dialdehyde with the bis or tris primary amine to form a binucleating tetra or hexa Schiff base. A second step is necessary if the reduced macrocyclic or cryptand ligands are to be prepared, and these are obtained by nearly quantitative hydrogenation of the corresponding Schiff bases. The procedure followed

Table 7.2. Macrocyclic and Macrobicyclic Ligands Synthesized by the Nontemplate Method.

Aldehyde	Polyamine	Product	Product (NaBH$_4$ reduce)	Overall Yield
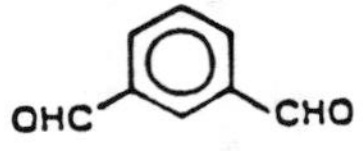	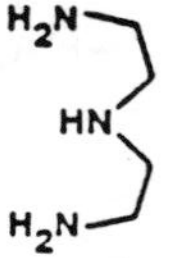	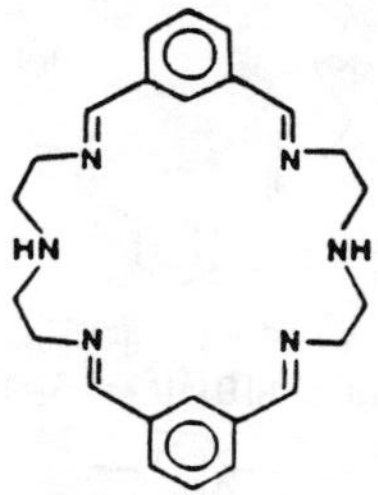 *1* Bis-isophthaldehydeBISDIEN Schiff base	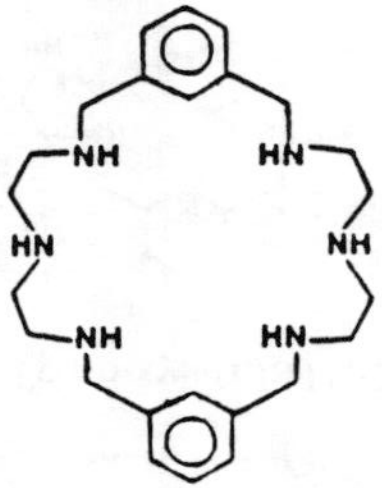 *2* Bis-isophthaldehydeBISDIEN	46%
	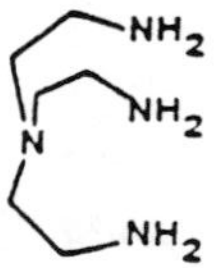	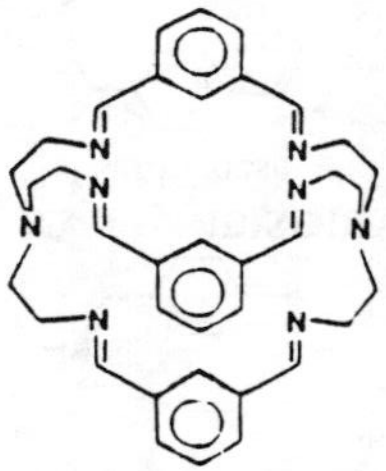 *3* Tris-isophthaldehydeBISTREN Schiff base	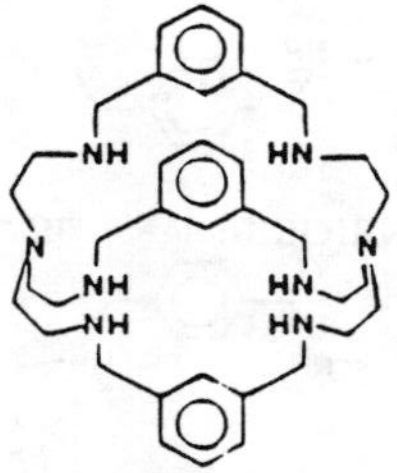 *4* Tris-isophthaldehydeBISTREN	60%

Table 7.2. *(continued)*

Aldehyde	Polyamine	Product	Product (NaBH₄ reduce)	Overall Yield

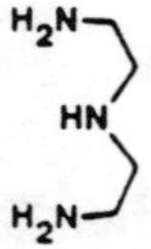

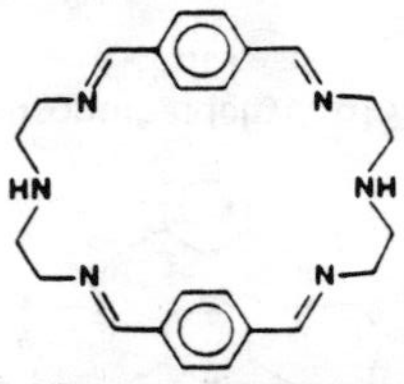

5 Bis-p-xylylBISDIEN Schiff base — *6* Bis-p-xylylBISDIEN — 56%

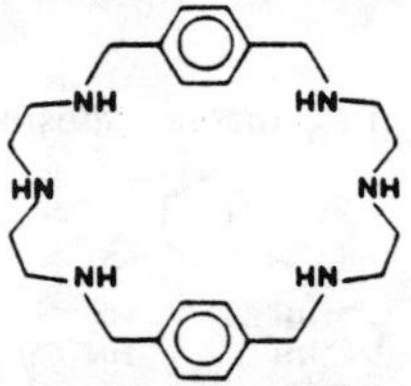

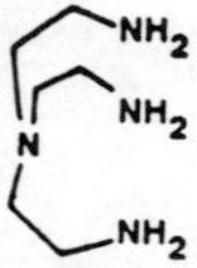

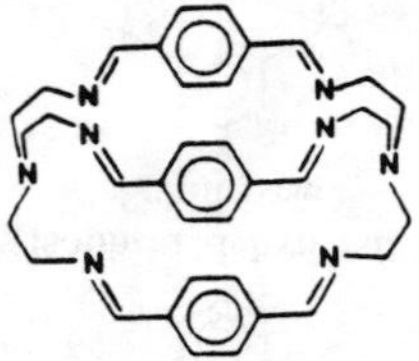

7 Tris-p-xylylBISTREN Schiff base — *8* Tris-p-xylylBISTREN — 40%

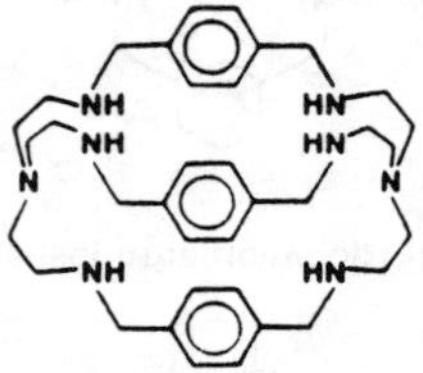

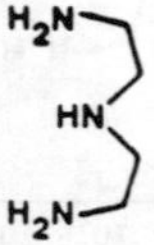

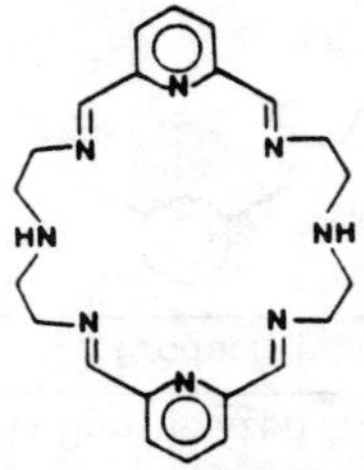

9 BispyridineBISDIEN Schiff base — *10* BispyridineBISDIEN — 70%

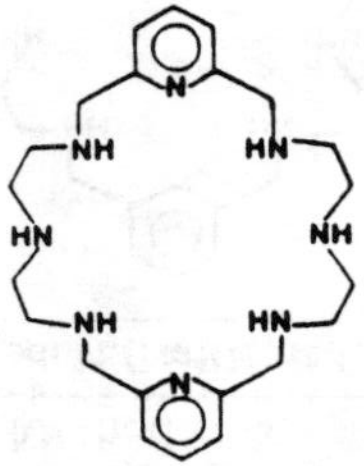

Table 7.2. *(continued)*

Aldehyde	Polyamine	Product	Product (NaBH₄, reduce)	Overall Yield
	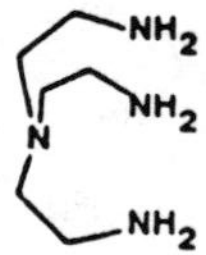	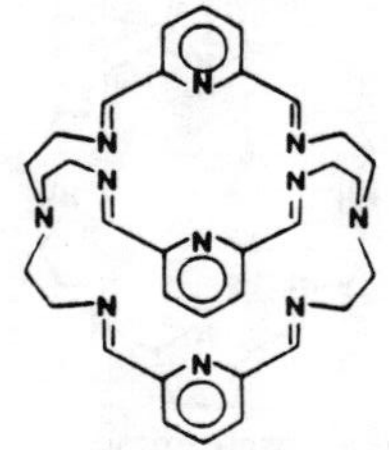 *11* TrispyridineBISTREN Schiff base	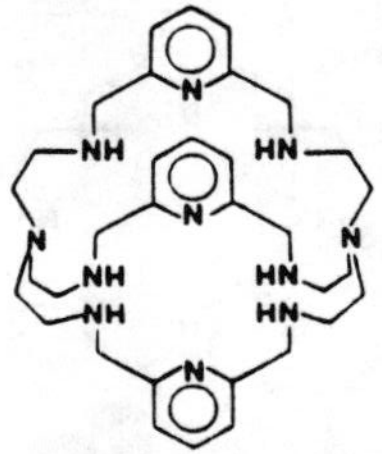 *12* TrispyridineBISTREN	40%
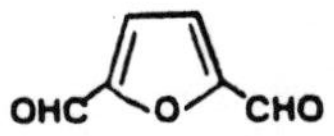		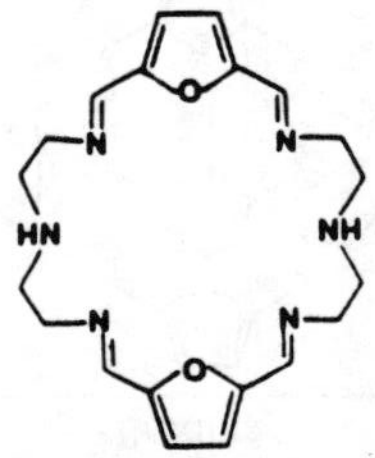 *13* BisfuranBISDIEN Schiff base	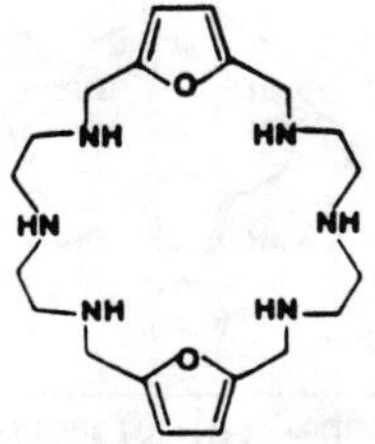 *14* BisfuranBISDIEN	45%

Table 7.2. *(continued)*

Aldehyde	Polyamine	Product	Product (NaBH₄ reduce)	Overall Yield
	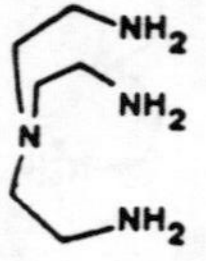	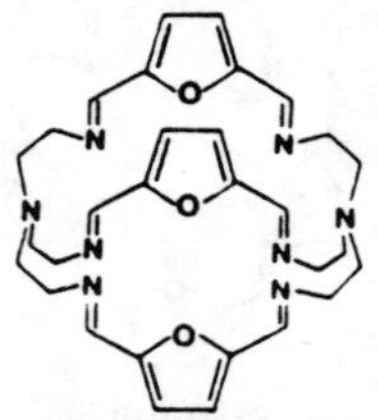 *15* TrisfuranBISTREN Schiff base	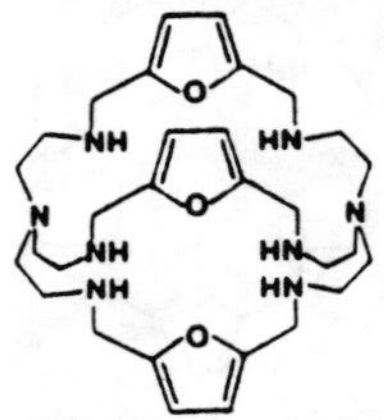 *16* TrisfuranBISTREN	38%
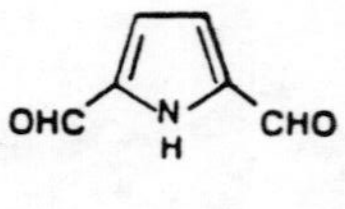	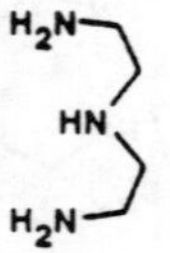	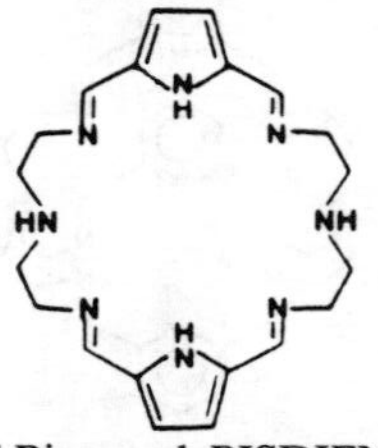 *17* BispyrroleBISDIEN Schiff base	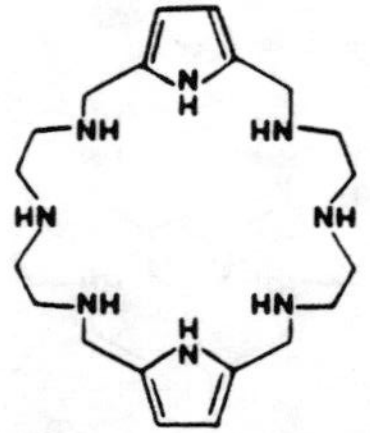 *18* BispyrrolylBISDIEN	40%

Table 7.2. (continued)

Aldehyde	Polyamine	Product	Product (NaBH₄ reduce)	Overall Yield
	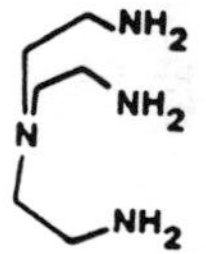	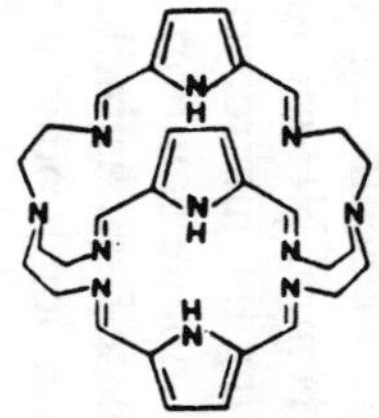 19 TrispyrroleBISTREN Schiff base	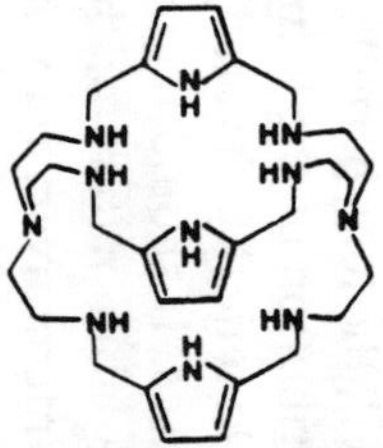 20 TrispyrroleBISTREN	34%
	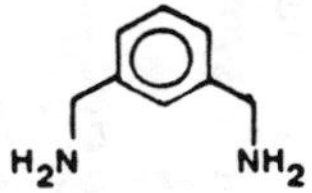	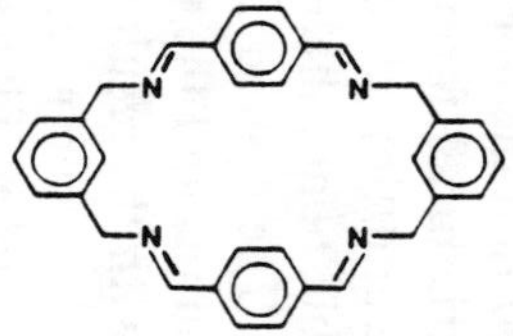 21 Bis-p-xylylbis-m-xylyldiamine Schiff base	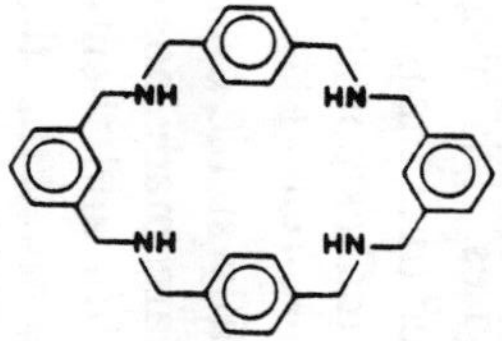 22 Bis-p-xylylbis-m-xylyldiamine	36%

was similar to that of Jazwinski et al.[28] to carry out the dipodal 2+2 condensation of an aldehyde with a polyamine.

The reaction described above is relatively new; the few examples that have been published previously are to be found in the work of Alcock et al.,[29] Jazwinski et al.,[30] McDowell and Nelson,[31] and Drew et al.[32] As indicated in Table 7.2, the yields seem to be especially high with isophthaldehyde condensations (*1–4*), and this has been supported by McKee et al.[33]

As a rationalization for the strong predomination of cyclic condensation over the normally expected linear condensation in the production of polymeric Schiff bases, it is suggested that the cyclic product is thermodynamically preferred over the polymer. As well, the long period of time allowed for the reaction makes possible the redistribution of the Schiff base species from the oligmers (which are probably kinetically favored) to the more thermodynamically stable cyclic Schiff bases. It should be noted that such rearrangements are made possible by the labile nature of the Schiff bases formed in solution and their ability to interconvert into one another. It should be further noted that the high yields in these syntheses were obtained without the use of metal ions as templates for the condensations.

7.3.1.1 Isomerization of the Macrocycles

It has been noted that the macrocycle *1* in Table 7.2 (from isophthaldehyde and diethylenetriamine) formed a mixture of Schiff bases that consisted primarily of the addition products of the central NH group to an adjacent Schiff base double bond, thereby contracting the macrocyclic ring from 24 to 18 members. The system consists of a mobile equilibrium between the addition compounds and the tetra Schiff base itself as indicated in Scheme of Figure 7.10, with the result that crystallization produced the isomer with the reduced macrocyclic ring. However, the solution was in mobile equilibrium and behaved as if it were the tetra Schiff base when Cu(I) was added to the reaction mixture. Similar behavior was noted for furandicarboxaldehyde and diethylenetriamine (Table 7.2, formula *13*.)[13,24] This type of reaction involving a contraction of the macrocyclic ring is described below for the metal template syntheses of macrocyclic compounds.

7.3.2 Template Syntheses

The template method, involving the organizing ability of a metal ion through coordination to form macrocyclic complexes, is well established and dates back from the time of Curtis,[34,35] Busch,[36] and others who published synthetic procedures that have been described in detail. This section focuses on the template synthesis of binuclear metal complexes of macrocyclic and macrobicyclic ligands. A summary of the work completed thus far is given in Table 7.3. The template synthesis of macrocyclic ligands dates back to the late 1970s with the work of Nelson, Fenton, and co-workers.[47] It was not

Table 7.3. Template Synthesis

Aldehyde or Ketone	Polyamine	Metal Ions	Product	Overall Yield	Reference
		Ba^{2+} then Cu^{2+}		60%	37
		Ba^{2+} then Cu^{2+}		65%	38
		Pb^{2+}		64%	39

Table 7.3. (continued)

Aldehyde or Ketone	Polyamine	Metal Ions	Product	Overall Yield	Reference
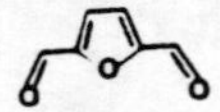	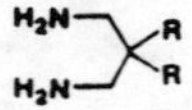	Ba^{2+} then Cu^{2+}	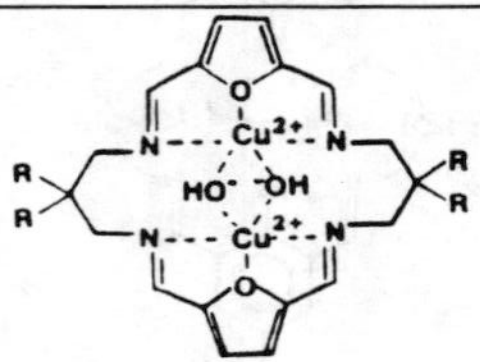	R=H ~50%	40–43
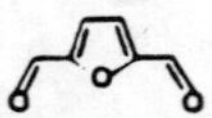	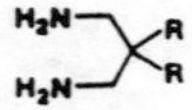	Ba^{2+} then Cu$^+$	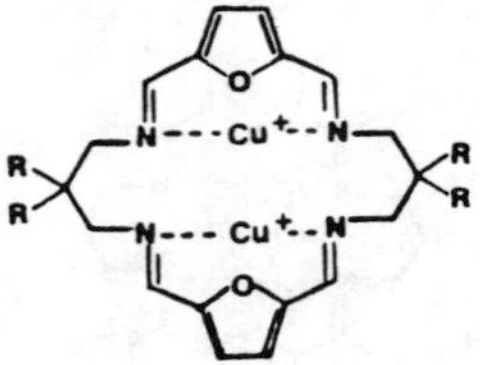	R=H 48% R=CH$_3$ 50%	43
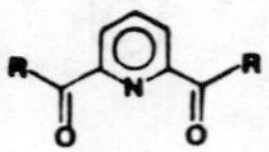		Ba^{2+} then Cu^{2+}	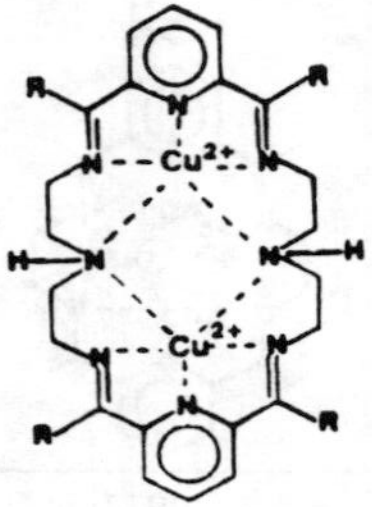	R=CH$_3$ 20% R=H 20% R=CH$_3$ 38%	44 45

Table 7.3. *(continued)*

Aldehyde or Ketone	Polyamine	Metal Ions	Product	Overall Yield	Reference
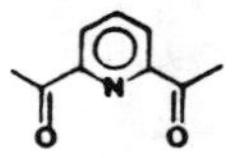 (not necessary)	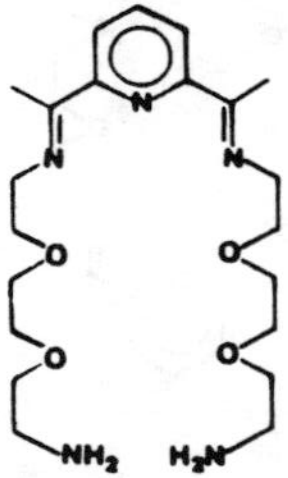	Ba^{2+}	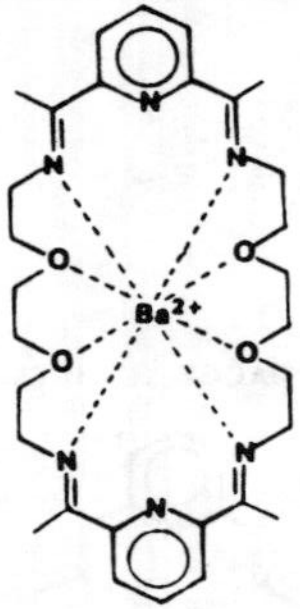	30%	46
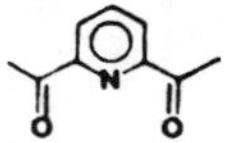	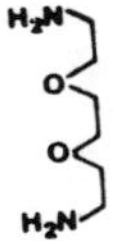	Pb^{2+}	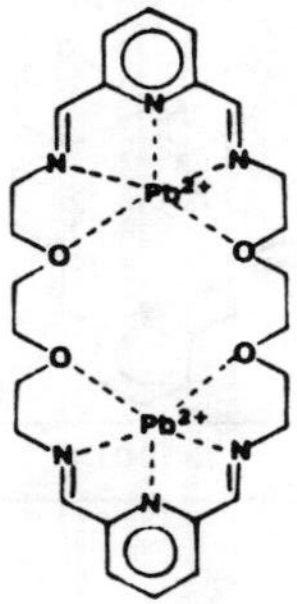	80%	47

Table 7.3. (continued)

Aldehyde or Ketone	Polyamine	Metal Ions	Product	Overall Yield	Reference
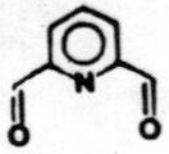		Ba^{2+}	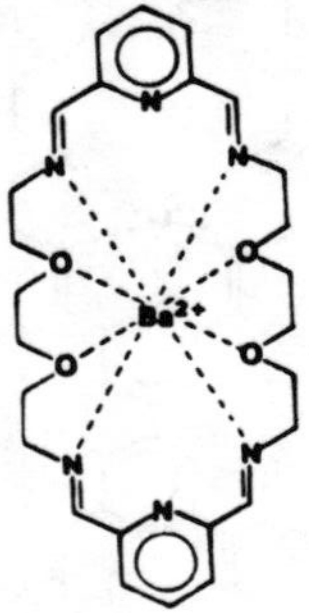	35%	46
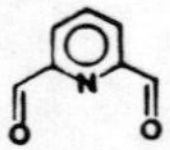	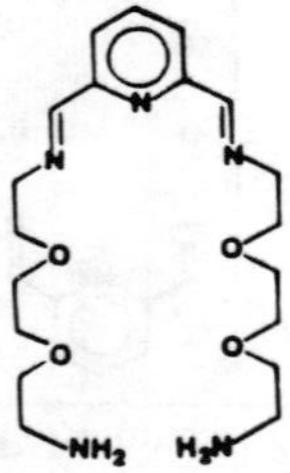	Ba^{2+}	same as above	35%	46

Table 7.3. *(continued)*

Aldehyde or Ketone	Polyamine	Metal Ions	Product	Overall Yield	Reference
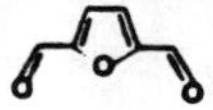	same as above	Ba^{2+}	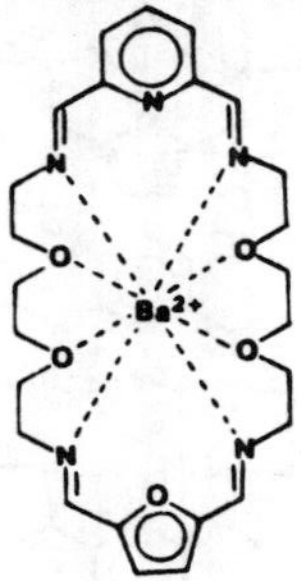	40%	46
		Ba^{2+}	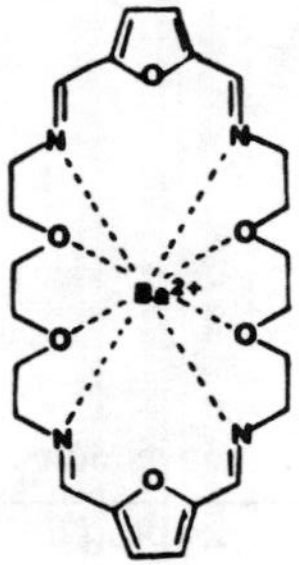	45%	46

Table 7.3. *(continued)*

Aldehyde or Ketone	Polyamine	Metal Ions	Product	Overall Yield	Reference
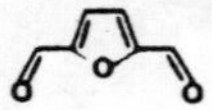	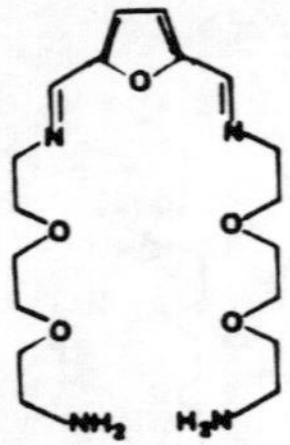	Ba^{2+}	same as above	60%	46
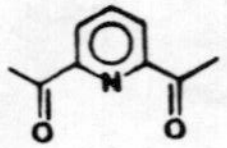	2 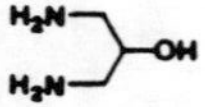	Ba^{2+} then Ag^+	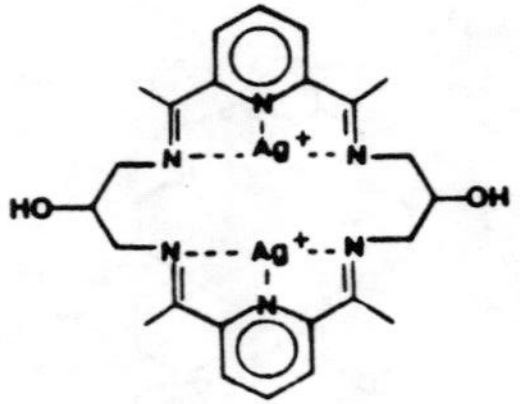	63%	48
2 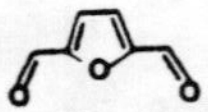	2 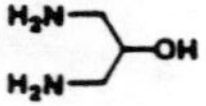	Ba^{2+}	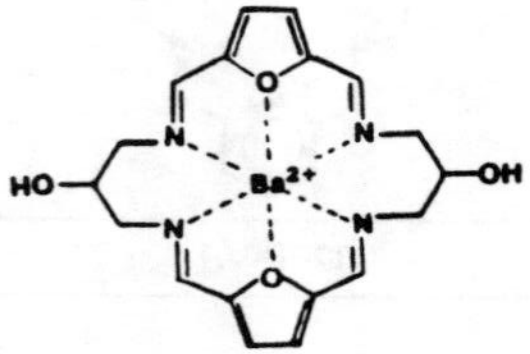	55%	48

Table 7.3. *(continued)*

Aldehyde or Ketone	Polyamine	Metal Ions	Product	Overall Yield	Reference
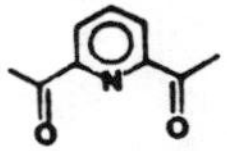	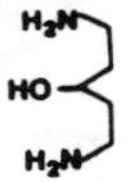	Pb^{2+}	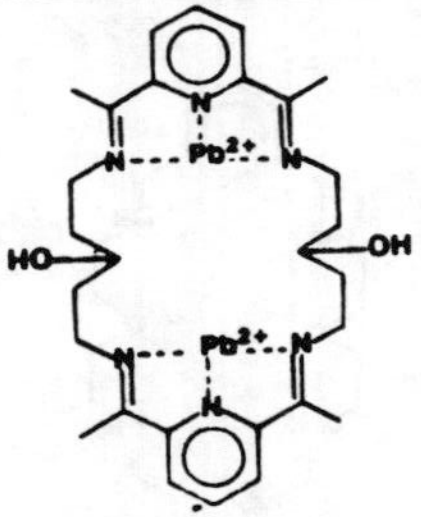	55%	48
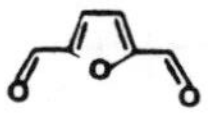	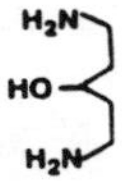	Ba^{2+}	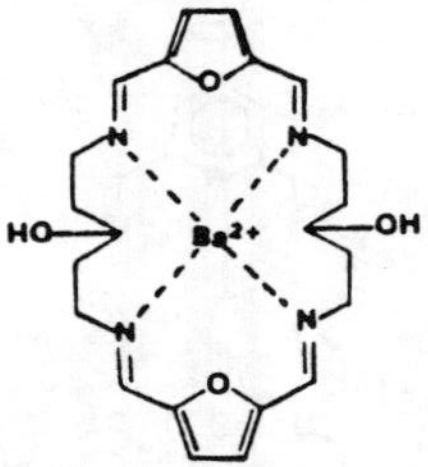	55%	48
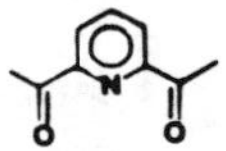	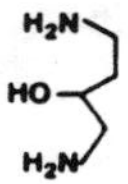	Pb^{2+}	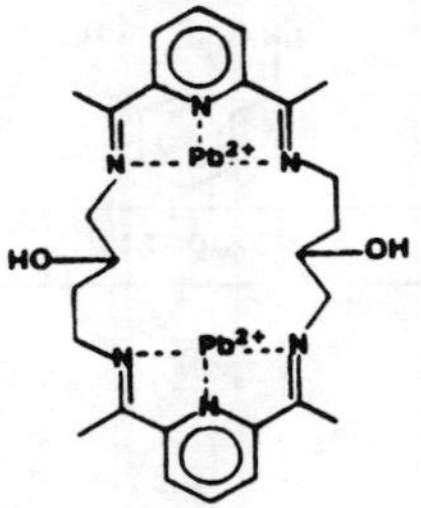	75%	48

Table 7.3. (continued)

Aldehyde or Ketone	Polyamine	Metal Ions	Product	Overall Yield	Reference
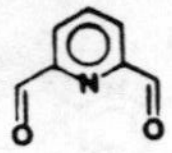		Pb²⁺ (for Schiff base formation)	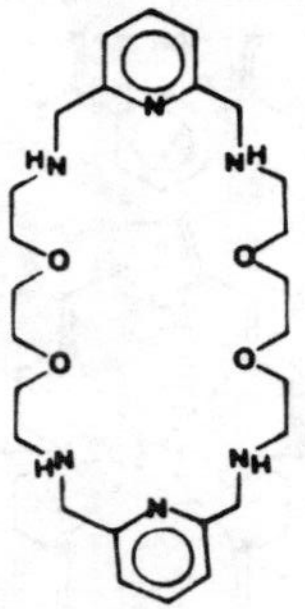	85%	8, 41, 49
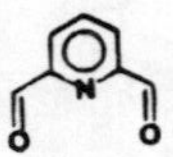	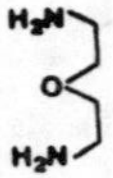	Pb²⁺ (for Schiff base formation)	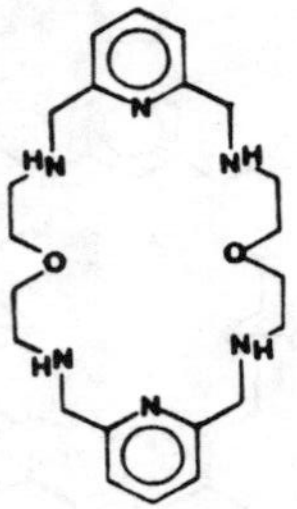	67%	9, 41

Table 7.3. (continued)

Aldehyde or Ketone	Polyamine	Metal Ions	Product	Overall Yield	Reference
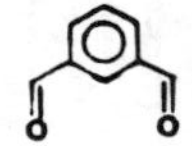	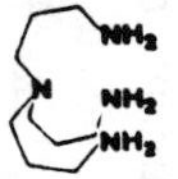	AgNO$_3$, then Cu$^+$	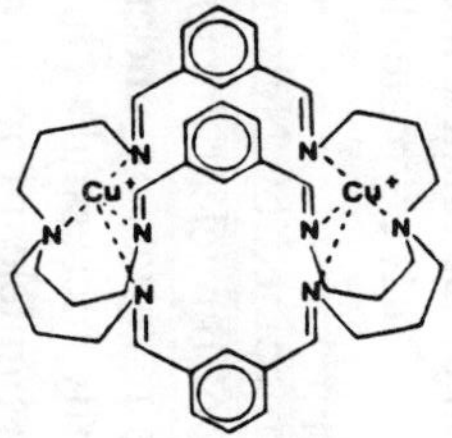	75%	23
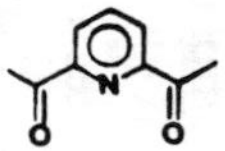	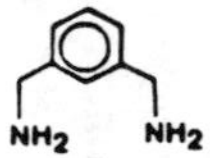	M = Cu^{2+}, Ni^{2+} then Na$_4$EDTA	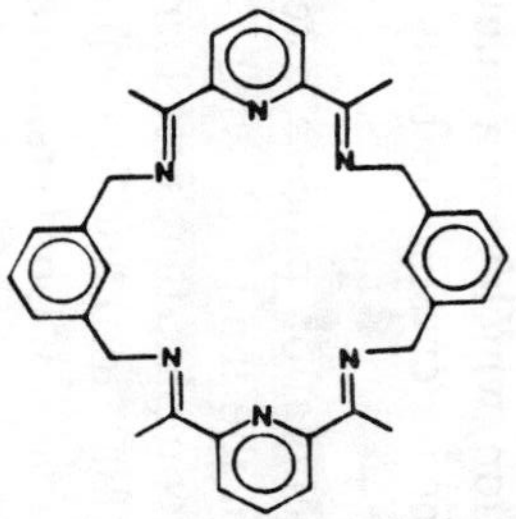	~60%	50

until later that progress was made on the nontemplate synthesis of these ligands. Many of the template-based syntheses did not work for the non-template procedure, which seems to be much more sensitive to the nature of the starting materials. At the present time the template procedure offers an alternative to the nontemplate synthetic methods and often succeeds in producing the desired ligand when the nontemplate method gives only mixtures of products. Perhaps one of the most unusual cryptand complexes to be synthesized thus far is the cryptate derived from isophthaldehyde and tris(aminopropyl)amine in a 3+2 tripodal condensation to give the corresponding Schiff base. The ligand could not be made by the nontemplate synthesis, even though many solvents and reaction conditions were attempted. The synthesis was finally accomplished by preparing the Ag(I) cryptate, which was converted to the corresponding Cu(I) cryptate by trans-metallation.[23] The bis copper(I) complex was also prepared by a Cu(II) template reaction, which gave a solution of the binuclear Cu(II) chelate that gradually formed a crystalline binuclear Cu(I) cryptate illustrated in Table 7.3.

The transmetallation process, a term coined by Nelson and co-workers, is important in the template synthesis of binuclear complexes. Frequently, the binuclear complex that first forms involves an undesirable metal as a template (e.g., Ba^{2+} ion) and is frequently successful as in the case of many formulas in Table 7.3. Conversion to a binuclear copper or other metal ion macrocyclic or cryptate structure depends on replacing the metal used in the synthesis with the metal of interest: hence the term *transmetallation*. This is accomplished by treating the complex having the undesirable metal with an excess of the soluble salt of the desired metal. The examples cited thus far for transmetallation usually involve an organic solvent.

7.4 Future Research

7.4.1 Synthetic Methods

Methods of synthesis are dealt with first because it is necessary to synthesize the macrocycles of cryptate ligands before experiments can be carried out with them. Virtually any conceivable synthetic problem can be solved by the use of the methods of Lehn and co-workers,[27] involving the stepwise synthesis in which the polyamines are protected by tosyl groups. Frequently a shortcut is possible in which an acid chloride is condensed with a polyamine by a multipodal synthetic step at high dilution. A general procedure for such methods has been described.[17,27] The most convenient methods of synthesis, however, are the template and nontemplate procedures outlined in this paper, because they involve simple one- or two-step condensations of polyamines with dialdehydes to form Schiff bases, followed by hydrogenation to the saturated macrocycle or cryptand ligands. The limitations of these synthetic methods and their possibilities for further extension will be described in the following.

The nontemplate condensation of a dialdehyde with a poly primary amine to form 22 Schiff bases and their reduction production product is outlined in Table 7.2. Apparently these reactions require a rigid dialdehyde and a flexible polyamine ligand such as TREN or DIEN, and the resulting Schiff base is formed in relatively high yield for a one-step process. The dialdehydes listed in Table 7.2 are five in number and may be expanded further; only two polyamines are employed, however, and this number may also be greatly expanded. In fact, in cases where the method does not work, the main product may be a mixture of polymeric Schiff bases. The extent of polymerization can be greatly affected by the time factor and the use of alternative solvents. For example, when acetonitrile is used as a solvent, the condensation of pyrroledicarboxaldehyde and TREN produces only polymers. However, when methanol was used as a solvent, a good yield of the cyclic condensation product, formula *20* in Table 7.2, was obtained. Similarly glutaric acid dialdehyde and succinnic acid dialdehyde would not condense in a cyclic fashion with diethylenetriamine or TREN even though trials with a number of solvents were attempted. Apparently the aldehydes are too flexible; however, this does not indicate that a condensation would not be successful in a different solvent or under different conditions. In any case, the condensation of other rigid aldehydes such as propylenedialdehyde (methane dicarboxaldehyde) is not precluded, and various reaction conditions or solvents should be attempted with these compounds. It is noted that the two primary amines that were employed in the examples given in Table 7.2 involve bridging groups consisting of secondary and tertiary amino groups: they were not involved in the condensation with the Schiff base. Therefore, the structures of the primary amines used in these condensations could be greatly altered, so that the final macrocyclic or cryptand ligands would vary considerably. Thus far, no difficulties have been observed in the hydrogenation of the Schiff bases to form the saturated macrocycle or cryptand ligands, the yields being almost quantitative in nature.

The template method has been used for many types of reactions of starting materials to produce macrocyclic or macrobicyclic ligands, but the most convenient method recommended here is the condensation of a polyamine with a dialdehyde to produce, through 2+2 or 2+3 dipodal or tripodal condensation, the tetra or hexa Schiff bases that may form binuclear complexes. This type of reaction seems to be greatly aided by a central ion acting as an organizer for the Schiff base condensations. From the examples listed in Table 7.3, it seems that a large metal ion such as Ba(II) or Ag(I) is by far the more successful and that smaller metal ions such as Pb(II), which are still fairly large, may produce a larger amount of mononucealating ligand as well as the binucleating ligands illustrated in Table 7.3. The first-row transition metal ions and Cu(I) are not very successful in promoting the condensations that produce binucleating metal complexes of macrocyclic and macrobicyclic ligands. It should also be noted from the examples given that the desired compounds are frequently the binuclear complexes of the smaller metal ions, and these are best formed through transmetallation, starting with the mononuclear complexes of the larger metal ions. The difference in coordination number of the large versus the small metal ions makes the difference in whether mononuclear or binuclear complexes will be formed. It seems as though a number of solvents should be used in attempts to produce good yields of the cyclic condensation products even when the metal template is employed. All of the procedures involving the nontemplate process, which give poor yields, should instead be attempted by the template method to produce the barium(II) or silver(I) complexes of the desired ligands; here a change in solvent would be desirable in most cases. Another strategy that has not yet been employed is the use of preassembled binuclear metal complexes of the type that produce Fe(III) hemocyanin mimics, extensively investigated by Armstrong and Lippard,[51] Wieghardt and co-workers,[52] and others. It seems that a Fe(III) core, held together by oxo and carboxylato groups, is a stable structure, with three or four coordination sites unsatisfied. However, it can combine with the coordinating groups of a macrocycle or cryptand ligand to give an even more stable structure. The first example of this method was applied to the binuclear copper imidazolate core by Salata et al.[50] This is a new method for synthesizing macrocyclic or macrobicyclic ligands; the metal can be extracted at the end of the process by the use of EDTA, cyanide, or some other appropriate reagent.

7.4.2 Macrocyclic Complexes

The analogs of BISDIEN shown in Tables 7.2 and 7.3, and further extensions of these structures obtained through development of the synthetic methods involved, provide a wide variation in the structures of the binucleating macrocyclic ligands available for investigation. Various metal ions, particularly those of the first-row transition metal series that includes zinc and copper(I), offer a wide variety of binuclear complexes available for investigation. The number and variety of bridging ligands available for this large number of binuclear complexes is almost limitless. A study of the structures of the binuclear complexes with or without bridging ligands—or failing the crystallization process and X-ray analysis, CPK space-filling models of these complexes—would demonstrate the most effective size of the bridging ligand. Sizes will vary greatly with the flexibility or rigidity of the separating groups provided by the dialdehyde involved. The coordinating groups of both the Schiff bases and their reduction products provided by the polyamines employed will also result in a variation of the metal ions by the donor groups provided on each end of the binucleating ligand.

The BISDIEN dioxygen complex of cobalt is a very stable structure but an additional bridging group, which is also a reductant, shows facile oxidation and reduction of the bridging dioxygen to water. This reaction may involve two- or four-electron reduction, depending on whether the end result is Co(III) or a Co(II) complex. In the case of the latter, the reaction may be catalytic in nature. There may be other substrates that react with dioxygen encapsulated by a macrocyclic complex in this manner. A review of the literature readily indicates several bifunctional molecules that may act both as bridging ligands and reducing agents, such as hydroxamic acids, ethylenediamine, glycine, glyoxal, phosphinic acid, hydroxymethylphosphonic acid, dimethylglyoxime, and the like. This is only a partial list; a more complete list could contain 20 or 30 ligands. The more stable of these would probably not involve a redox reaction but could still bridge between the two cobalt(II) centers. However, the more reactive reducing agents, such as hydroxamic acids or formylphosphinic acid, would certainly be expected to undergo redox reactions with coordinated dioxygen. It must be remembered that the constraints on the host molecule are considerable, causing it to recognize only certain bifunctional ligands as guests. Not only are the hosts restricted by the macrocyclic structure of the primary ligand but also by the dioxygen that forms a bridging peroxo group between the two metal centers. Thus far the bridging groups that have been successful in spanning the two metal centers as well as undergoing redox reactions are oxalate, mesoxalate, and catecholate, but the list could be greatly expanded, given the large number of potential candidates for this type of reaction. Thus far the only macrocyclic ligand employed has been BISDIEN, but Tables 7.2 and 7.3 list a large number of other candidates for this type of oxygen complex formation.

The formation of successful models for oxyhemocyanin and oxyhemerythrin remains an elusive, though probably achievable, objective of bioinorganic chemistry. There is no problem in making cobalt–dioxygen complexes, and BISDIEN is a good example. However, the copper(I)–dioxygen complexes obtained thus far are stable mainly at low temperatures and warming them to ambient temperature results in rapid degradation. Thus far the Cu(I) complexes reported by Kida et al.[15] and by Martell et al.[14] seem to be the most stable, but they have a limited lifetime at 5.0 °C and an even more limited lifetime at room temperature. Thus, the formation of a stable oxygen complex with stability comparable to that observed for oxyhemerythryn and oxyhemocyanin remains as a yet-to-be achieved objective.

Among the Cu(I) complexes that form adducts with dioxygen, formula 13 of Table 7.2 forms a binuclear Cu(II)–dioxygen adduct that deserves further investigation with respect to its oxidation of various substrates. Thus, at 5.0 °C over the few hours of the complex's lifetime, a considerable amount of substrate may be oxidized, especially if the latter reaction takes place much more rapidly than the dioxygen complex's natural degradation to the Cu(II) species. The example cited is only one dioxygen adduct of the Cu(I) complexes that may be investigated in this regard; the discovery of additional complexes would greatly expand the field. Similarly, the synthesis of iron–dioxygen complexes might some day be based on a variation of the primary donor groups of the macrocycles that complex the iron and of the bridging groups involved. The catalytic nature of such a dioxygen complex in oxidizing other substrates would be of great interest. Thus far, catalytic oxygen complexes of iron have been described by Sheu et al.[53] and Barton and co-workers,[54] but the catalytic species involved are simple complexes of pyridine carboxylic acid or pyridine dicarboxylic acid in nonaqueous solvents. Iron complexes of macrocyclic and macrobicyclic ligands have yet to be explored.

7.4.3 Macrobicyclic (Cryptand) Complexes

The cryptands described in Table 7.2 indicate a wide variety of ligands that may be investigated for recognition of various bifunctional bridging ligands between the metal centers. In addition, a variety of metal ions is available for investigation, but the number of ligands recognized by these metals will vary not only with the metal but with the metal ions' distance of separation and also by the nature of the connecting groups. When the connecting groups are long or rigid enough, the binucleating ligand donor groups are so far apart that dioxygen cannot be complexed. However, a number of longer bifunctional ligands may be complexed in such binucleating structures, depending on whether or not the groups connecting the coordinating sites are compatible with the functional bridging groups being accommodated by the bimetallic centers. Thus is follows that the isoelectronic hydroxide and fluoride ligands are accommodated in a different way by O-BISTREN and C-BISTREN. Similar effects will be observed with other bifunctional groups. Lehn[28,30,55,56] has reported a few of these effects mainly with multifunctional ligands, but the vast majority of bridging ligands, especially those involving different coordinating groups in the cryptand, remain to be investigated.

The success of O-BISTREN as an oxygen carrier has been described earlier.[10] The discovery of an appropriate substitute of O-BISTREN remains the elusive objective of research programs designed to achieve successful separation of oxygen from air. The type of dioxygen complex generally sought in these investigations involves a 1:1 complex, because that is much less stable than the 2:1 complex formed in binuclear peroxo-bridged structures. Thus the dry caves of Busch and coworkers[57,58] are of great interest and would be successful in achieving the separation process if they did not undergo such rapid degradation. The lack of degradation in the peroxo bridged O-BISTREN–dicobalt complex makes it the most successful oxygen carrier formed to date. However, the ligand itself is much too expensive and hard to synthesize for it to be of practical value as a successful oxygen separation catalyst. Therefore, the hunt is on for a substitute for this remarkable complex.

Acknowledgment

This research was supported by the Office of Naval Research.

References

1. Martell, A. E.; Motekaitis, R. J.; Lecomte, J. P.; Lehn, J. M. *Inorg. Chem.* **1983**, *22*, 609.
2. Coughlin, P. K.; Lippard, S. J.; Martin, A. E.; Bulkowski, J. R. *J. Am. Chem. Soc.* **1980**, *22*, 609.
3. Lehn, J. M. *Pure Appld. Chem.* **1980**, *52*, 2441.
4. Coughlin, P. K.; Dewan, J. C.; Lippard, S. J.; Watanabe, E.; Lehn, J. M. *J. Am. Chem. Soc.* **1979**, *101*, 265.
5. Coughlin, P. K.; Martin, A. E.; Dewan, J. C.; Watanabe, E.; Bulkowski, J. R.; Lehn, J. M.; Lippard, S. J. *Inorg. Chem.* **1984**, *23*, 1004.
6. Coughlin, P. K.; Lippard, S. J. *J. Am. Chem. Soc.* **1984**, *103*, 2328.
7. Coughlin, P. K.; Lippard, S. J. *J. Am. Chem. Soc.* **1981**, *106*, 3228.
8. Basallote, M. G.; Martell, A. E. *Inorg. Chem.* **1988**, *27*, 4219.
9. Menif, R.; Chen, D.; Martell, A. E. *Inorg. Chem.* **1989**, *28*, 4633.
10. (a) Motekaitis, R. J.; Martell, A. E. *J. Chem. Soc., Chem. Commun.* **1988**, 915.
    (b) Motekaitis, R. J.; Martell, A. E. *J. Am. Chem. Soc.* **1988**, *110*, 8059.
11. Szpoganicz, B.; Motekaitis, R. J.; Martell, A. E. *Inorg. Chem.* **1990**, *29*, 1467.
12. Motekaitis, R. J.; Martell, A. E. *Inorg. Chem.* **1991**, *30*, 1396.
13. Menif, R.; Martell, A. E. *J. Chem. Soc., Chem. Commun.* **1989**, 1521.
14. Ngwenya, M. P.; Chen, D.; Martell, A. E.; Reibenspies, J. *Inorg. Chem.* **1991**, *30*, 2732.
15. Nishida, Y.; Takahashi, K.; Kuramoto, H.; Kida, S. *Inorg. Chim. Acta* **1981**, *54*, L103.
16. Motekaitis, R. J.; Martell, A. E.; Lehn, J. M.; Watanabe, E. *Inorg. Chem.* **1982**, *21*, 4253.
17. Motekaitis, R. J.; Martell, A. E.; Murase, I. *Inorg. Chem.* **1986**, *25*, 938.
18. Motekaitis, R. J.; Martell, A. E.; Murase, I.; Lehn, J. M.; Hosseini, M. W. *Inorg. Chem.* **1988**, *27*, 3630.
19. Motekaitis, R. J.; Martell, A. E.; Rudolf, P.; Clearfield, A. *Inorg. Chem.* **1989**, *28*, 112.
20. Motekaitis, R. J.; Martell, A. E. *J. Chem. Soc., Chem. Commun.* **1988**, 1020.
21. Motekaitis, R. J.; Martell, A. E. *J. Am. Chem. Soc.* **1988**, *110*, 7715.
22. Motekaitis, R. J.; Chen, D.; Martell, A. E., unpublished results.
23. Ngwenya, M. P.; Martell, A. E.; Reibenspies, J. *J. Chem. Soc., Chem. Commun.* **1990**, 1207.
24. Menif, R.; Martell, A. E.; Squattrito, P. J.; Clearfield, A. *Inorg. Chem.* **1990**, *29*, 4723.
25. Menif, R.; Martell, A. E., Reibenspies, J. *Inorg. Chem.* **1991**, *30*, 3446.
26. Chen, D.; Martell, A. E., *Tetrahedron* **1991** *47*, 6895.
27. Dietrich, F.; Hosseini, M. W.; Lehn, J. M.; Sessions, R. B. *Helv. Chim. Acta* **1985**, *68*, 289.
28. Jazwinski, J.; Lehn, J. M.; Meric, R.; Vigneron, J. P.; Cesarrio, M.; Juilhem, J.; Pascard, C. *Tetrahedron Lett.* **1987**, *28*, 3489.
29. Alcock, W.; Kingston, R. G.; Moore, P.; Pierpoint, C. J. *J. Chem. Soc., Dalton Trans.* **1984**, 1937.
30. Jazwinski, J.; Lehn, J. M.; Lilienbaum, D.; Ziessel, R.; Guilhem, J.; Pascard, C. *J. Chem. Soc., Chem. Commun.* **1987**, 1691.
31. McDowell, D.; Nelson, J. *Tetrahedron Lett.* **1988**, *29*, 385.
32. Drew, M. G. B.; McDowell, D.; Nelson, J. *Polyhedron,* **1988**, *7*, 22.
33. McKee, V.; Robinson, W. T.; McDowell, D.; Nelson, J. *Tetrahedron Lett.* **1989**, *30*, 7453.
34. Curtis, N. F. *J. Chem. Soc.* **1960**, 4409.

35. Curtis, N. F.; House, D. A. *Chem. & Ind.* **1961,** *42,* 1708.

36. (a) Thompson, M. C.; Busch, D. H.; Burke, J. A., Jr.; Jica, D. C.; Morris, M. I. *Adv. Chem Ser.* **1963,** *37,* 125.
    (b) Thompson, M. C.; Busch, D. H.; Burke, J. A., Jr.; Jica, D. C.; Morris, M. I. *Chem. & Eng. News* **1962,** *39,* 57.

37. Nelson, S. M.; Esho, F. S.; Drew, M. G. B. *J. Chem. Soc., Dalton Trans.* **1982,** 407.

38. Drew, M. G. B.; Nelson, J.; Esho, F. S.; McKee, V.; Nelson, S. M. *J. Chem. Soc., Dalton Trans.* **1982,** 1837.

39. Tadokoro, M.; Sakiyama, H.; Matsumoto, N.; Okawa, H.; Kida, S. *Bull. Chem. Soc. Jpn* **1990,** *63,* 3337.

40. Nelson, M. S.; Esho, F. S.; Drew, M. G. B. *J. Chem. Soc., Chem. Commun.* **1981,** 388.

41. Nelson, M. S. *Inorg. Chim. Acta* **1982,** *62,* 39.

42. Nelson, S. M.; Esho, F. S.; Lavery, A.: Drew, M. G. B. *J. Am. Chem. Soc.* **1983,** *105,* 5693.

43. Drew, M. G. B.; Yates, P. C.; Trocha-Grimshaw, J.; Lavery, A.; McKillop, K. P.; Nelson, S. M. *J. Chem. Soc., Dalton Trans.* **1988,** 347.

44. Kol'chinskii, A. G.; Yatsimirskii, K. B. *Theor. Exp. Chem.* **1984,** *20,* 90.

45. Drew, M. G. B.; Nelson, J.; Nelson, S. M. *J. Chem. Soc., Dalton Trans.* **1981,** 1678.

46. Nelson, M. S.; Knox, C. V.; McCann, M. *J. Chem. Soc., Dalton Trans.* **1981,** 1669.

47. Cook, D. H.; Fenton, D. E.; Rodgers, A.; McCann, M.; Nelson, S. M. *J. Chem. Soc., Dalton Trans.* **1979,** 414.

48. Adams, H.; Bailey, N. A.; Fenton, D. E.; Good, R. J.; Moody, R.; Rodriguez de Barbarin, C. O. *J. Chem. Soc., Dalton Trans.* **1987,** 207.

49. Burnett, M. G.; McKee, V.; Nelson, S. M.; Drew, M. G. B. *J. Chem. Soc., Chem. Commun.* **1980,** 829.

50. Salata, C. A.; Youinou, M. T.; Burrows, C. J. *J. Am. Chem. Soc.* **1989,** *111,* 9278.

51. Armstrong, W. H.; Lippard, S. J. *J. Am. Chem. Soc.* **1983,** *105,* 4837.

52. Wieghardt, K.; Pohl, K.; Gebert, W. *Angew. Chem.* **1983,** *95,* 729.

53. Sheu, C.; Sobkowiak, A.; Jeon, S.; Sawyer, D. T. *J. Am. Chem. Soc.* **1990,** *112,* 879.

54. Barton, D. H. R.,; Csuhai, E.; Coller, D.; Ozbalik, N.; Balavoine, G. *Proc. Natl. Acad. Sci. USA* **1990,** *87,* 3401.

55. Lehn, J. M. *Science* **1985,** *227,* 849.

56. Lehn, J. M. *Ann. NY Acad. Sci.* **1986,** *471,* 41.

57. Delgado, R.; Glogowski, M. W.; Busch, D. H. *J. Am. Chem. Soc.* **1987,** *109,* 6855.

58. Lance, K. A.; Goldsby, K. A.; Busch, D. H. *Inorg. Chem.* **1990,** *29,* 4573.

CHAPTER 8

Developments in the Field of Functionalized Tetraazamacrocycles

Thomas A. Kaden
Institute of Inorganic Chemistry
Spitalstrasse 51
Basel CH-4056
Switzerland

8.1 Introduction

The chemistry of classical macrocycles such as the porphyrins, the phthalocyanines, and the corrins has been known for a long time. In the 1960s, interest in new macrocyclic compounds started from several different points. The discovery of valinomycin,[1] the synthesis of the crown ethers,[2] the study of template reactions,[3] and the preparation of cryptands[4] mark some of the milestones of this first exciting period of macrocyclic chemistry.

The potential of these new compounds became apparent very quickly, and several research groups commenced studies of their fundamental properties, such as stability constants, complexation selectivities, host–guest interactions, new coordination geometries of metal complexes, and kinetic behaviour.

One of the many developments in this field of research was the introduction of side chains containing functional groups. The purpose of side chain introduction was manifold. Solubility change became possible through lipophilic rests; additional donors were made available to increase the stability and/or selectivity in the complexation of metal ions; and side chains could either be polymerized or used to covalently attach the macrocycle to a polymeric support or to label proteins. This field has gone through a rapid expansion, and a large number of such compounds bearing side chains have been synthesized. Because several reviews have appeared,[5,6] we shall concentrate on one type of macrocycles, the tetraazacycloalkanes, using these compounds as models and examples and discussing their salient points.

8.2 Functionalized Tetraazamacrocycles

Although functionalization of N_4-macrocycles is given most easily at the secondary amines, there are also several examples of functionalization at the carbon backbone. N-functionalization allows the introduction of side chains, which can be designed so that additional five- or six-membered chelate rings are formed when the donor group of the side chain coordinates to the metal ion. C-functionalization has the great advantage of not modifying the donor properties of the nitrogens and thus allows the introduction of chains, which can be used as anchoring groups.

8.2.1 C-Functionalized Derivatives

The synthesis of 13- and 14-membered tetraazamacrocycles, as performed by Tabushi et al.,[7] by reaction of open-chain tetraamines with substituted malonic ester derivatives (both of which are easily accessible) allows one to prepare C-functionalized compounds (*1* and *2*, illustrated in Figure 8.1.)

The yields of this cyclization procedure are generally not very good, and often the products must be purified by chromatographic techniques. However, in some cases the cyclic amides precipitate from the reaction mixture in relatively pure form and can be used directly for further steps. The same reaction has also been used to introduce a 2'-pyridylmethyl side chain into 13- to 15-membered cyclic amides (*1*, R = $CH_2-C_5H_4N$, *n,m,p* 2 or 3).[8] Compounds of this type have been studied by Fabbrizzi et al. to perform selective metal ion transport (*1*, R = $C_{16}H_{32}$, *m* = *p* = 2, *n* = 3)[9] or to

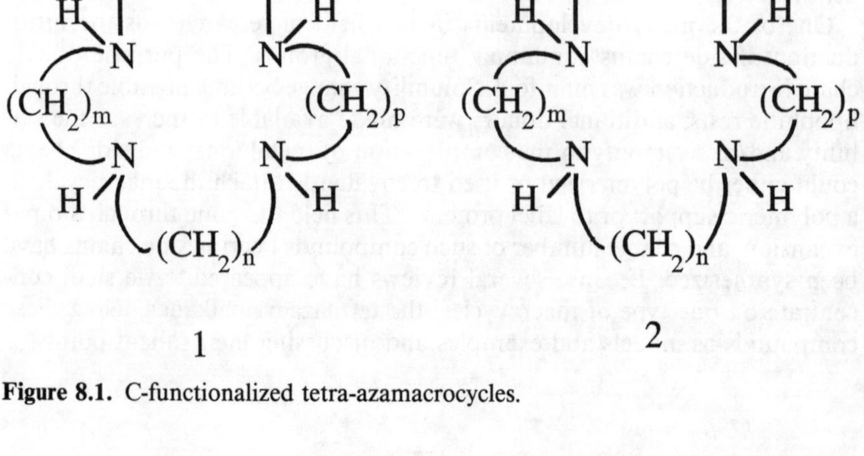

Figure 8.1. C-functionalized tetra-azamacrocycles.

transport electrons (2, R = $C_{16}H_{32}$, $m = p = 2$, $n = 3$)[10] through a lipophilic phase.

A further synthetic approach to C-functionalized tetraazamacrocycles stems from the work of Sargeson et al.,[11] who have used the condensation of formaldehyde and nitroethane to prepare the Co^{3+} sarcophagines. If other metal ions such as Ni^{2+} or Cu^{2+} are taken as templates instead of Co^{3+}, 13- to 16-membered monocyclic compounds, such as 3, are obtained.[12] The pendant nitro group coordinates to the metal ion only in the larger and more flexible 15- and 16-macrocycles, whereas for the 13- and 14-membered derivatives no such interaction takes place.

After reduction of the nitro group to an amine (compound 4), one obtains an additional donor, which can bind to the metal ion or be used as a starting point for further modifications. (See Figure 8.2.)

A very interesting synthetic route is the reaction of acryl esters or acryl lactams with open-chain tetraamines and the subsequent reduction of the amide to give α-substituted macrocycles such as 5, shown in Figure 8.3. Kimura et al. have prepared several derivatives of this type, in which R is a phenol,[13] an imidazole,[14] or a pyridine.[15] All of these groups can bind in the axial position of the metal ion, thus giving penta- or hexacoordinated species. It is worth pointing out that in these cases only a small change in the metal/macrocyclic moiety takes place when the side chain coordinates, in contrast to the more drastic changes in the analogous N-derivatives.[16]

Although it is relatively easy to obtain 13- to 16-membered C-function-alized tetraazamacrocycles by the methods previously mentioned, it is much more difficult to prepare 12-membered derivatives. Meares et al.[17] have proposed an elegant synthesis of a C-functionalized tetraazacyclododecane

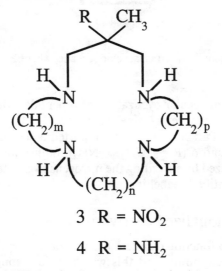

$$3 \quad R = NO_2$$

$$4 \quad R = NH_2$$

Figure 8.2. NO_2- and NH_2-bearing tetraazamacrocycles (see text).

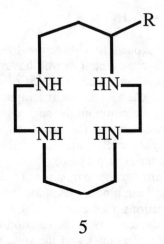

5

Figure 8.3. α-substituted tetraazamacrocycle, where R can include a phenol, imidazole or pyridine binding group.

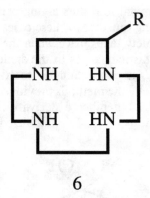

6

Figure 8.4. C-functionalized tetraazacyclododecane, where R = $CH_2C_6H_4NO_2$.

by using the intramolecular cyclization of a peptide, which then has been reduced to the cyclic tetraamine **6**, illustrated in Figure 8.4 (R = $CH_2\_C_6H_4-NO_2$). Parker et al.[18] have reported a second preparation that also gives compound **6** (R = $(CH_2)_4-NH_2$). Both macrocycles have been further functionalized by alkylating the nitrogens to give tetraacetates, which can be used for antibody labelling.

8.2.2 N-Functionalized Derivatives

The easiest way to functionalize tetraazamacrocycles is the tetra alkylation of the four nitrogen atoms, and this was first done long ago. These tetra derivatives are excellent ligands, but their coordination chemistry is complicated because of the many possible ways that they can bind metal ions. From the coordination chemists' point of view, the mono-N-derivatives are simpler to understand, but their synthesis is more demanding. Di- and trialkylation have also been performed,[19] but the few examples will not be discussed here.

8.2.2.1 Tetra-N-Functionalized Macrocycles

The four secondary nitrogens of tetraazamacrocycles have been substituted with a large number of different side chains bearing functional groups such as $-COOH$,[20] $-OH$,[21] $-NH_2$,[22] $-CN$,[23] $-PO_3H_2$,[24] and heterocycles.[25] The coordination chemistry of compound 7, displayed in Figure 8.5, is very rich and interesting. The X-ray structures of their metal complexes show a large variety of coordination modes (see Figure 8.6). Examples of metal ions encompassed by the macrocycle with additional coordination of two side-chain donor groups in a transoctahedral arrangement have been observed in the 14-membered tetraacetates.[26] For the smaller, 12-membered derivatives, however, the metal ion cannot fit into the cavity, resulting in a cis-octahedral arrangement with a folded macrocycle.[27] Only in lanthanides, which exhibit large coordination numbers, are all of the donor groups involved in the coordination of the metal ion.[28]

Beside metal complexes with 1:1 stoichiometry, 2:1 species have also been observed. In the case of the tetraacetates, complicated sheet or chain structures result,[29] whereas for the aminoalkyl derivatives discrete 2:1 species are formed,[30] in which the two metal ions are outside of the macrocyclic cavity and are coordinated by two nitrogen atoms of the ring and two of the side-chain amino groups. In several cases, an additional bridging between the two metal centers through an exogenous ligand also takes place.[31]

Several papers have dealt with the determination of the stability constants of the tetraacetates' ligands, which are extremely good chelates for all types of metal ions, starting from alkali ions to transition metal ions and up to

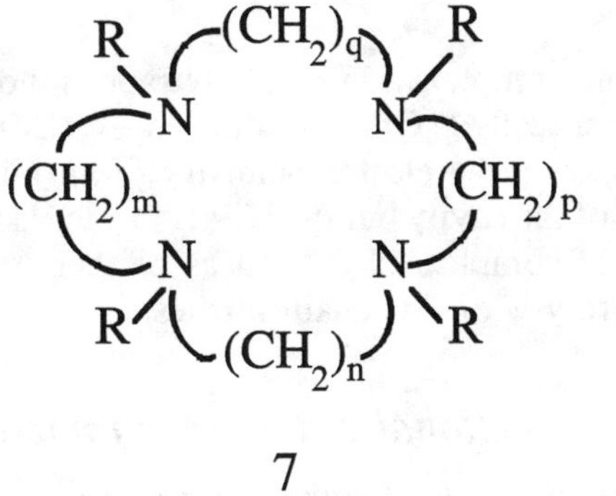

7

Figure 8.5. Tetra-N-functionalized tetraazamacrocycles.

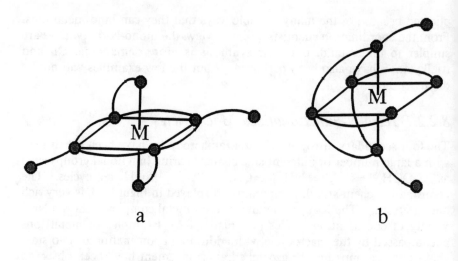

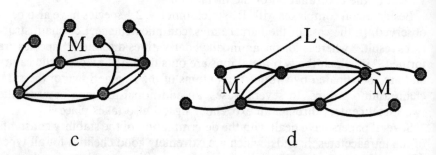

Figure 8.6. Structures of metal complexes with tetra-N-substituted macrocycles: a,b,c 1:1 species; d 2:1 species.

lanthanides.[20,32] The complexation of the tetraphosphonates has also been studied potentiometrically.[24] Other studies cover NMR-measurements on lanthanide complexes,[33] the electrochemistry of an Ni^{2+} species, in which the Ni^{2+} is outside of the cavity but the Ni^{3+} is inside,[34] kinetic investigations of the formation of complexes[35] as well as of their rearrangement;[36] and studies of the hydrolysis of side-chain nitriles.[37]

8.2.2.2 Mono-N-Functionalized Tetraazamacrocycles

The different synthetic routes for the preparation of mono-N-substituted tetraazamacrocycles have been previously discussed.[5] The trend in recent years has been to simplify the relatively demanding synthesis of these products. The use of the tritosyl derivative *8* (R = tosyl, R' = H), which is accessible in one step with relatively good yield, has opened the possibility of preparing such monosubstituted products in an easier way.[38] (See Figure 8.7.) Also, selective monoalkylation using an excess of the macrocycle has made the synthesis of a series of monoderivatives of *8* possible (R = H, R' = $(CH_2)_n$−COOH, n = 1–3), carrying a carboxylic group in the side chain.[39] The same idea was also taken up to prepare macrocycles with a lipophilic side chain.[40] In both of these syntheses, the side chain imparts to the monoderivative solubility or extraction properties distinctly different from those of the unsubstituted parent compound, so that isolation and purification become possible.

The trimethylated species of *8* (R = CH_3, R' = H) is also a convenient starting compound. Since there is only one nitrogen that can be alkylated, a series of macrocycles with one pendant group can be prepared.[41] The spectral properties of a series of Cu^{2+} complexes as well as the structures of two Cu^{2+} complexes of this type[42] clearly indicate that the macrocycle has the trans-I configuration, which is typical for macrocycles with tertiary amines. In several cases, the side-chain donor group is coordinated to the metal ion, giving pentacoordinated species.

Mono-N-substituted tetraazamacrocycles exhibit an interesting reaction involving the side chain (see Figure 8.8). In several examples, the on/off equilibrium could be determined. It was shown that the equilibrium constant depends on the nature of the functional group (especially its basicity), on the nature of the metal ion, and on the length of the side chain.[39,43,44]

This reaction is a consequence of the different reactivity of the macrocyclic unit, which gives a kinetically stable complex, and that of the donor group of the side chain, which is kinetically labile. Studies of structures of such

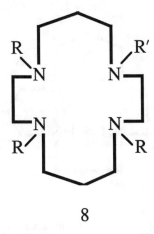

8

Figure 8.7. Mono-N-functionalized tetraazamacrocycle.

complexes indicate that the geometrical and structural changes associated with the coordination of the side chain can be very different.[16] In certain cases, no change at all is observed; in other cases, the metal ion is displaced from its position in the direction of the axial donor group. For example, a drastic rearrangement in the metal-macrocyclic moiety was observed in the copper complex of macrocycle *9* (R = $(CH_2)_3-NH_2$),[44] associated with a different configuration of the nitrogen atoms for the complex, between a complex in which the side chain is coordinated and one in which it is not. Macrocycle *9* is illustrated in Figure 8.9.

In addition to these examples with a ligating group in the side chain, there are monofunctionalized tetraazamacrocycles in which the side-chain functional group fulfils other purposes. In two Cu^{2+} complexes with a carbonic ester (*8*, R = $(CH_2)_n-COOEt$) and a phosphonate ester (*8*, R = $(CH_2)_2-PO(OEt)_2$), respectively, the kinetics of the hydrolysis in their side chains was studied.[42] Although in these cases no large reactivity enhancement was observed, in the analogous Cu^{2+} complex of *8* (R = CH_2-CN) the nitrile hydrolysis is extremely fast ($\tau_{1/2} \approx 50$ ms at pH = 12).[45]

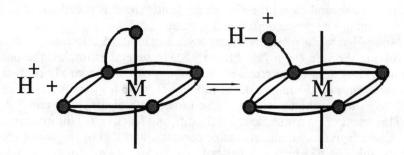

Figure 8.8. On/off equilibrium of the side chain in mono-N-substituted macrocyclic metal complexes.

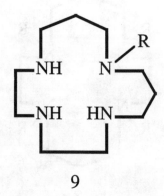

9

Figure 8.9. A mono-N-substituted tetraazamacrocycle; here R = $(CH_2)_3$-NH_2 (see text).

Other monofunctionalized tetraazamacrocycles have side-chain function-alities that can be used to attach the macrocyclic metal complex to a protein or an antibody.[46]

8.3 Conclusions and Perspectives

The examples discussed earlier allow one to extrapolate what can be done in the future and what kind of new results can be expected. Continuing to attach one side chain after another to macrocycles will not provide many more innovations unless those side chains introduce some interesting prop-erties that control the geometry or the stereochemistry of the complex or that induce new properties into the system. Due to the rigidity of the ma-crocyclic metal unit, it is conceivable that by cleverly designing the side chain this can be achieved.

Little has been done yet with reactive side chain groups. Imagine the applications of side chains that orient the reactive group so that an optimal geometric arrangement results for an interaction with the metal ion, or that bring the reactive group close to another reactant so to enhance reaction rates, or that fix the stereochemistry so that selective and stereospecific reactions can take place. Because there are many metal-promoted reactions, this field is quite large, and the possibilities of designing such systems seem worth the effort. Once we have learned how to control the geometry and the stereochemistry in the coordination sphere of the metal/macrocyclic unit and how to build more rigid, structurally demanding macrocycles, we should encounter new and unexpected results.

Metal-induced reactions, in which the reacting group is incorporated in the side chain, are of course stoichiometric processes, but by modifying such systems so that a substrate could bind to the metal ion, metal ion catalysis might be achieved. In addition to hydrolytic reactions, one can also postulate redox reactions that will be metal ion induced or catalyzed.

Also with respect to catalysis, one could design systems that would allow the performance of organometallic chemistry in the coordination sphere of a metal ion coordinated to a suitable macrocycle. Perhaps one could even perform organometallic chemistry in aqueous solution, as nature does with corrin macrocycles.

Another aspect that will be of great importance in the future is the ap-plication of macrocyclic metal complexes in medicine. In the field of nuclear magnetic imaging we are only just beginning to use Gd(DOTA)$^-$ as a contrast agent,[47] and other metal complexes exist that could be used for the same purpose. The large number of patents granted in this field over the past few years shows the high level of interest in this kind of medical application.[48]

Functionalized macrocycles have also a great potential in the diagnostic and therapeutic application of nuclear medicine. Examples of such macro-cyclic complexes, which have been covalently attached to antibodies, have been published.[46] But here again we are only at the beginning of a new development.

These few points clearly show that although this field has taken about 30 years to reach maturity, it is yet an area of great potential in which new ideas can still be found.

Acknowledgments

The support of this work by the Swiss National Science Foundation is gratefully acknowledged.

References

1. Pressman, B. C. *Fed. Proc. Fed. Am. Soc. Exp. Biol.* **1968**, *27*, 1283.

2. Pedersen, C. J. *J. Am. Chem. Soc.* **1967**, *89*, 7017.

3. (a) Curtis, N. F. *Coord. Chem. Rev.* **1968**, *3*, 3.
   (b) Lindoy, L. F.; Busch, D. H. *Prep. Inorg. Reactions* **1971**, *6*, 1.

4. Dietrich, B.; Lehn, J. M.; Sauvage, J. P. *Tetrahedron Lett.* **1969**, 2885 and 2889.

5. Kaden, T. A. *Topics Curr. Chem.* **1984**, *121*, 157.

6. (a) Hay, R. In *Current Topics in Macrocyclic Chemistry in Japan*; Kimura, E., Ed.; Hiroshima School of Medicine; 1987; p 56.
   (b) Bernhardt, P.; Lawrance, G. A. *Coord. Chem. Rev.* **1990**, *104*, 297.

7. Tabushi, I.; Taniguchi, Y.; Kato, H. *Tetrahedron Lett.* **1977**, 1049.

8. Kimura, E.; Koike, T.; Machida, R.; Nagai, R.; Kodama, M. *Inorg. Chem.* **1984**, *23*, 4181.

9. DiCasa, M.; Fabbrizzi, L.; Perotti, A.; Poggi, A.; Tundo, P. *Inorg. Chem.* **1985**, *24*, 1610.

10. Desantis, G.; Fabbrizzi, L.; Poggi, A.; Seghi, B. *J. Chem. Soc., Dalton Trans.* **1990**, 2729.

11. Geue, R. J.; Hambley, T. W.; Harrowfield, J. M.; Sargeson, A. M.; Snow, M. R. *J. Am. Chem. Soc.* **1984**, *106*, 5478.

12. (a) Comba, P.; Curtis, N. F.; Lawrance, G. A.; O'Leary, M. A.; Skelton, B. W.; White, A. H. *J. Chem. Soc., Dalton Trans.* **1988**, 497 and 2145.
    (b) Lawrance, G. A.; Rossignoli, M.; Skelton, B. W.; White, A. H. *Austr. J. Chem.* **1988**, *41*, 1533.
    (c) Lawrance, G. A., O'Leary, M. A. *Polyhedron* **1987**, *6*, 1291.

13. Kimura, E.; Koike, T.; Takahashi, M. *J. Chem. Soc., Chem. Commun.* **1985**, 385.

14. Kimura, E.; Shionoya, M.; Mita, T.; Iitaka, Y. *J. Chem. Soc., Chem. Commun.* **1987**, 1712.

15. Kimura, E.; Koike, T.; Nada, H.; Iitaka, Y. *J. Chem. Soc., Chem. Commun.* **1986**, 1322.

16. Kaden, T. A. *Comments Inorg. Chem.* **1990**, *10*, 25.

17. Moi, M.; Meares, C. F.; DeNardo, S. J. *J. Am. Chem. Soc.* **1988**, *110*, 6266.

18. Cox, J. P. L.; Jankowski, K. J.; Kataki, P.; Parker, D.; Beeley, N. R. A.; Boyce, B. A.; Eaton, M. A. W.; Millar, K.; Millican, A. T.; Harrison, A.; Walker, C. *J. Chem. Soc., Chem. Commun.* **1989**, 797.

19. (a) Alcock, J. W.; Moore, P.; Omar, H. A. A. *J. Chem. Soc., Chem. Commun.* **1985**, 1058.
    (b) Helps, I. M.; Parker, D.; Morphy, J. R.; Chapman, J. *Tetrahedron* **1989**, *45*, 219.
    (c) Riesen, A.; Kaden, T. A.; Ritter, W.; Mäcke, H. R. *J. Chem. Soc., Chem. Commun.* **1989**, 460.

20. (a) Stetter, H.; Frank, W. *Angew. Chem.* **1976**, *88*, 76.
    (b) Stetter, H.; Frank, W.; Mertens, R. *Tetrahedron* **1981**, *37*, 767.

21. (a) Buoen, S.; Dale, J.; Groth, P.; Krane, J. *J. Chem. Soc., Chem. Commun.* **1982,** 1172.
    (b) Madeyksi, C. M.; Michael, J. P.; Hancock, R. D. *Inorg Chem.* **1984,** *23,* 1487.

22. (a) Wainwright, K. P. *J. Chem. Soc., Dalton Trans.* **1983,** 1149.
    (b) Kida, S.; Murase, I.; Harada, C.; Daizeng, L.; Mikuriya, M. *Bull. Chem. Soc. Jpn.* **1986,** *59,* 2595.
    (c) Murase, I.; Ueda, I.; Marabayashi, N.; Kida, S.; Matsumoto, N.; Kudo, M.; Toyahara, M.; Hiate, K.; Mikuriya, M. *J. Chem. Soc., Dalton Trans.* **1990,** 2763.

23. Wainwright, K. P. *J. Chem. Soc., Dalton Trans.* **1980,** 2117.

24. (a) Kabachnik, M. J.; Polykarpov, J. M. *Zh. Obshch. Khim. SSSR* **1988,** *58,* 1937.
    (b) Delgado, R.; Kaden, T. A. *Helv. Chim. Acta* **1990,** *73,* 140.

25. (a) Alcock, N. W.; Balakrishnan, K. P.; Moore, P. *J. Chem. Soc., Dalton Trans.* **1986,** 1743.
    (b) McLaren, F.; Moore, P.; Wynn, A. M. *J. Chem. Soc., Chem. Commun.* **1989,** 798.
    (c) Norante, G. D.; Divaira, M.; Mani, F.; Mazzi, S.; Stoppioni, P. *Inorg. Chem.* **1990,** *29,* 2822.
    (d) Pietraszkiewicz, M.; Karpiuk, J.; Bilewski, R.; Kasprzyk, S. P. *J. Coord. Chem.* **1990,** *21,* 75.

26. (a) Moi, M. K.; Yanuck, M.; Deshpande, S. V.; Hope, H.; DeNardo, S. J.; Meares, C. F. *Inorg. Chem.* **1987,** *26,* 3458.
    (b) Riesen, A.; Zehnder, M.; Kaden, T. A. *Acta Cryst.* **1988,** *C44,* 1740.

27. Riesen, A.; Zehnder, M.; Kaden, T. A. *Helv. Chim. Acta* **1986,** *69,* 2067.

28. Spirlet, M. R.; Rebizant, J.; Desreux, J. F.; Loncin, M. F. *Inorg. Chem.* **1984,** *23,* 359.

29. Riesen, A.; Zehnder, M.; Kaden, T. A. *Helv. Chim. Acta* **1986,** *69,* 2074.

30. Mikriya, M.; Kida, S.; Murase, I. *J. Chem. Soc., Dalton Trans.* **1987,** 1261.

31. Mikuriya, M.; Kida, S.; Murase, I. *Bull. Chem. Soc. Jpn.* **1987,** *60,* 1355 and 1681.

32. (a) Delgado, R.; Frausto DaSilva, J. J. *Talanta* **1982,** *29,* 815.
    (b) Loncin, M. F.; Desreux, J. F.; Mercigny, E. *Inorg. Chem.* **1986,** *25,* 2646.

33. Desreux, J. F.; Loncin, M. F. *Inorg. Chem.* **1986,** *25,* 69.

34. Kimura, E.; Koike, T.; Yamaoka, M.; Kodama, M. *J. Chem. Soc., Chem. Commun.* **1985,** 1341.

35. Kasprzyk, S. P.; Wilkins, R. G. *Inorg. Chem.* **1982,** *21,* 3349.

36. Clarke, P.; Henslow, A. M.; Keough, R. A.; Lincoln, S. F. *Inorg. Chem.* **1990,** *29,* 1793.

37. Freeman, G. M.; Barefield, E. K.; Van Derveer, D. G. *Inorg. Chem.* **1984,** *23,* 3092.

38. Pallavicini, P. S.; Perotti, A.; Poggi, A.; Fabbrizzi, L. *J. Am. Chem., Soc.* **1987,** *109,* 5137.

39. Studer, M.; Kaden, T. A. *Helv. Chim. Acta* **1986,** *69,* 2081.

40. Blain, S.; Appriou, P.; Chaumeil, H.; Handel, H. *Anal. Chim. Acta* **1990,** *232,* 331.

41. (a) Barefield, E. K.; Foster, K. A.; Freeman, G. M.; Hodges, K. D. *Inorg. Chem.* **1986,** *25,* 4463.
    (b) Tschudin, D.; Basak, A. K.; Kaden, T. A. *Helv. Chim. Acta* **1988,** *71,* 100.

42. Tschudin, D.; Riesen, A.; Kaden, T. A. *Helv. Chim. Acta* **1989,** *72,* 737.

43. (a) Lotz, T. J.; Kaden, T. A. *Helv. Chim. Acta* **1978,** *61,* 1376.
    (b) Hediger, M.; Kaden, T. A. *Helv. Chim. Acta* **1983,** *66,* 861.
    (c) Alcock, N. W.; Kingstone, R. G.; Moore, P.; Pierpoint, C. *J. Chem. Soc., Dalton Trans.* **1984,** 1937.
    (d) Pallavicini, P. S.; Perotti, A.; Poggi, A.; Seghi, B.; Fabbrizzi, L. *J. Am. Chem. Soc.* **1987,** *109,* 5139.

44. Schiegg, A.; Kaden, T. A. *Helv. Chim. Acta* **1990,** *73,* 716.

45. Schibler, W.; Kaden, T. A. *J. Chem. Soc., Chem. Commun.* **1981,** 603.

46. (a) Kaden, T. A. *Nachr. Chem. Techn. Lab.* **1990,** *38,* 728.
    (b) Parker, D. *Chem. Br.* **1990,** *26,* 942.

47. (a) Doucet, D.; Meyer, D.; Bonnemain, B.; Doyon, D.; Caille, J. M. In *Enhanced Magnetic Resonance Imaging*; Runge, V., Ed.; Mosky; St. Louis, 1989.
    (b) Meyer, D.; Schaefer, M.; Bonnemain, B. *Invest. Radiol.* **1988,** *23,* 232.

48. FR Pat. 2 539 996 (1984); EP 0 292 689 (1988); EP 0 305 320 (1988); EP 0 287 465 (1988).

CHAPTER 9

Working on the Cyclam Framework to Take Advantage of the Redox Activity of the Bound Metal Center

Giancarlo De Santis, Luigi Fabbrizzi, Maurizio Licchelli, and Piersandro Pallavicini
Dipartimento di Chimica Generale
Università di Pavia
Pavia 27100
Italy

9.1 Introduction

Transition metal centers display a versatile multi-electron redox activity, which is related to the special nature of d orbitals. Access to the different oxidation states of the metal is regulated by the magnitude of the successive ionization potentials (I_n): these quantities refer to processes taking place in the gaseous phase. Coordination of the metal center by a ligating set and, to a minor extent, dissolution in a given medium substantially reduce the ionization energy values. Moreover, the coordinating features of the ligand system (i.e., the nature of the donor atoms and stereochemical arrangement) may drastically alter the relative stabilities of adjacent oxidation states or favor access to uncommon and otherwise unstable states. One hundred years of coordination chemistry has produced a number of ligands that allow us to promote and control the redox activity of the bound metal center.

One of these ligands is undoubtedly cyclam (1,4,8,11-tetraaza-cyclotetra-decane, *1*, illustrated in Figure 9.1). Cyclam is the most popular represent-ative of the cyclic polyamines (also known as *saturated macrocycles,* whereas the most classical *unsaturated macrocycles* are porphyrins and phthalocy-anines)[1] and can be considered as a rather sophisticated product of the evolution of one of most ancient ligands: ammonia. First, linking of am-

147

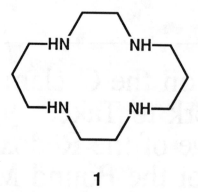

1

Figure 9.1. Structure of cyclam (1,4,8,11-tetraazacyclotetradecane).

monia molecules by aliphatic chains in an open mode produced multidentate ligands (ethylenediamine, triethylenetetramine, etc.) that are able to
form complexes of enhanced solution stability (the so-called *chelate effect*).
Later on, closure of open-chain polyamines produced macrocycles, very
special ligands whose complexes exhibit unprecedented properties, defined
as *macrocyclic properties*.[1] One of these properties is related to the versatile
redox activity of the encircled metal center, as mentioned earlier. Another
one refers to the kinetic stability of macrocyclic complexes; the metal center
cannot be extruded from the macrocyclic ring except in extremely drastic
conditions. For instance, in spite of the affinity of the amine group towards
the proton, metal complexes of cyclic polyamines are stable in strongly acidic
solutions.[2] This has opened the way to the investigation of the reactivity
(e.g., redox behavior) of amine complexes in very acidic media.[3]

Polyamine macrocycles represent a broad class of ligands that vary in
number of donor atoms and in size of the ring. They are conventionally
indicated by the general formula: $[x]$-aneN$_y$, where x is the number of atoms
(C and N) forming the ring and y is the number of amine nitrogen atoms
(e.g. cyclam = [14]-aneN$_4$). The so-called macrocyclic properties apply to
almost all the members of the class but are especially evident in the cyclam
complexes. To make just a couple of examples, the most exothermic complexation of divalent 3d metal ions is observed with cyclam,[4] and the most
kinetically stable complexes are still those of cyclam.[5]

Cyclam was first obtained in extremely low yield as a side product in the
Van Alphen[6] reaction of 1,2-dibromoethane with excess 1,3-propane diamine to prepare the linear tetramine 1,4,8,11-tetrazaundecane (2.3.2-tet); in
this type of reaction, the formation of the cyclic molecule is contrasted by
a combination of unfavorable entropy and enthalpy effects. However, the
first investigations of cyclam complexes and their new properties[7] commenced in 1965 and were carried out using the ligand obtained through a
method based on the rather chancy procedure described by Van Alphen. In

the same years, Busch[8] and Curtis[9] independently developed new high-yield methods for the preparation of tetra-aza macrocycles, in which a 3d metal center [mainly Ni(II)] assisted the combination of the reacting fragments by acting as a template. Finally, in 1972, Barefield[10] reported a very convenient template synthesis of [Ni(II) cyclam](ClO$_4$)$_2$ (yield: 60–70%). This synthesis was based on the Schiff condensation of glyoxal with the appropriate open-chain tetramine (1,5,8,12-tetraazadodecane, 3.2.3-tet), which was preorganized by the square-planar metal center, followed by hydrogenation with BH$_4^-$ of the two C=N double bonds.[11] The procedure is very simple, and multigram amounts of the [Ni(II)cyclam](ClO$_4$)$_2$ salt can be obtained even by inexperienced students after a few hours of extremely safe work. Eventually, the metal ion can be extruded from the cyclam ring by treating the complex with a boiling alkaline solution containing excess cyanide; work-up of the solution gives the pure ligand, a very stable, flaky solid, with a melting point of 190 °C

It is worth noting that the structural features of the macrocycle (number of donor atoms and ring size) are very critical to the success of the synthesis; as a matter of fact, the metal template Schiff base–condensation works well only with cyclam and does not work when the number of nitrogen atoms is increased or the atomicity of the tetra-aza ring is varied from 14. Therefore, tedious and multistep nonmetal-assisted procedures are required to prepare other saturated (and unsubstituted) poly-aza macrocycles than cyclam.[12] Thus, it is a very lucky situation that macrocyclic properties are associated with the complexes of the "cheapest" ligand among macrocycles. However, the truth is that the success of metal-assisted synthesis derives from the same factors that are responsible for the enhanced macrocyclic properties of the cyclam complexes. In particular, the cyclam molecule is tailor-made to encircle a di- or trivalent 3d metal ion; when the macrocycle is relaxed to its minimum energy configuration, the four nitrogen atoms are in the most favorable positions (i.e., in the corners of a square) to transfer electronic charge to the d orbitals of the metal center according to a σ-mode that establishes very strong coordinative interactions. The formation of strong M–N interactions more than compensates for the energy loss associated with the cyclization (the template effect), makes the folding of the ligand (which precedes demetallation) quite difficult (the kinetic macrocyclic effect), and raises the energy of the antibonding level, which is essentially metallic in character and from which the electron is extracted during oxidation (the stabilization of high oxidation states of the encircled metal).

As a consequence, metal cyclam complexes are unique as reagents for the design of new redox reactions. In particular, cyclam's metal center ensures a fast and reversible one-electron redox change; it is protected from the surroundings by a rigid aliphatic belt, and the electron transfer is conveyed along the z axis through the axially bound unidentate ligands (e.g., X$^-$ anions); and finally, coordinated cyclam is a convenient framework for appending further subunities and inserting new functionalities. Because of the latter feature and especially because of its very moderate cost, cyclam should be strongly considered for the design of multicenter devices involving the transfer of electrons. A few examples in this vein are described in the following sections.

9.2 Appending an Alkylammonium Side Chain at the Cyclam Ring

Nickel(II) cyclam complexes are by far the most investigated of the metal azamacrocycle systems. One explanation for this could be that the very efficient metal-assisted synthesis is based on the Ni(II) template, producing the [Ni(II) cyclam](ClO$_4$)$_2$ salt. But perhaps a more valid reason is that the [Ni(II)cyclam]$^{2+}$ species is prone to a fast and reversible one-electron oxidation process, giving a stable authentic trivalent complex.[13] This Ni(II)/Ni(III) oxidation process takes place within the complexes of nearly all polyaza macrocycles of different denticity and ring size, at least in the appropriate medium, but it is much easier in the case of cyclam.[14]

This is clearly illustrated by the diagram in Figure 9.2, in which the redox potentials associated with the Ni(III)/Ni(II) couple, measured in acetonitrile (MeCN) solution, are plotted against the ligand's atomicity x by means of cyclic voltammetry experiments performed on a series of complexes with macrocycles having the general formula [x]-aneN$_y$. The ligands examined in Figure 9.2 are quadridentate ($y = 4$), quinquedentate ($y = 5$), and sexidentate ($y = 6$). As the potential becomes more positive, access to the Ni(III) state becomes more difficult. The [Ni(III)cyclam]$^{3+}$ complex is formed at the lowest potential and presents the highest thermodynamic stability. The species could be more correctly described by the formula [Ni(III)cyclamX_2]$^{n+}$, where X can be a solvent molecule ($n = 3$) or a mononegative anion ($n = 1$). The complex, which typically exhibits a green or green-yellow color, displays an elongated octahedral stereochemistry, as shown by X-ray investigations on crystalline products.[15] The Ni(III) cation is d^7 low-spin, and the existence of an axial distortion is accounted for in terms of the Jahn-Teller effect.[16]

Ni(III)cyclam complexes can be obtained as solution-stable complexes in a number of solvents by using a platinum anode or a conventional oxidizing agent. In nonaqueous solvents (e.g., MeCN), NOBF$_4$ is very efficient; in water, a very rapid oxidation can be performed by S$_2$O$_8^{2-}$. The axial ligand X plays a substantial role in the stabilization of trivalent nickel; the more coordinating the anion, the less positive the Ni(III)/Ni(II) redox potential will be. A special stabilizing effect is exerted by the chloride ion: E (1 mol HCl) = 0.73 V versus SHE. That means that [Ni(II)cyclam]$^{2+}$ is a stronger reducing agent than the conventionally used Fe(II) ion ($E = 0.77$ V versus SHE); for this reason, in the presence of air, [Ni(II)cyclam]$^{2+}$ in aqueous HCl is oxidized by dioxygen to the indefinitely stable [Ni(III)cyclamCl$_2$]$^+$ species.

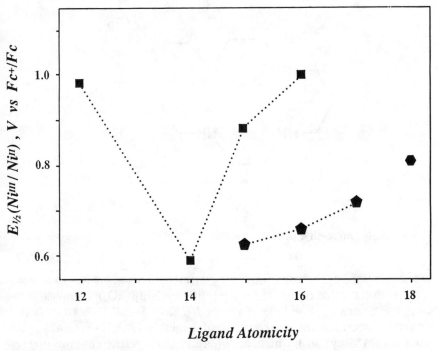

Figure 9.2. Half-wave potentials for the Ni(III)/Ni(II) redox change in acetonitrile solution, made 0.1 mol dm^{-3} in Bu$_4$NClO$_4$, for the complexes of a series of polyamine macrocycles of varying *denticity* and *atomicity*, as expressed by the general formula: x-aneN$_y$. The denticity y (i.e., the number of nitrogen atoms in the ring) is defined by these symbols: (■) $y = 4$; (⬠) $y = 5$; (⬢) $y = 6$. The atomicity x (i.e., the number of carbon and nitrogen atoms) that makes the ring varies 12 (the quadridentate macrocycle, 12-ane-N$_4$) to 18 (the sexidentate ligand, 18-aneN$_6$). Potentials are given versus the Fc$^+$/Fc internal reference. The lower the Ni(III)/Ni(II) potential, the easier the attainment of the trivalent state.

The potential of the Ni(III)/Ni(II) redox couple inside the cyclam ring can be controlled in several ways: for instance, by means of a proximate electrically charged center. This has been observed in the case of the complex of the ligand 1-aminoethyl-1,4,8,10-tetraaza-cyclotetradecane (*N*-(aminoethyl)cyclam, *2*, illustrated in Figure 9.3). The aminoethyl side chain is able to "sting" the in-plane chelated metal center with its "tail." For this reason, taking inspiration from zoology, this ligand and other similar macrocycles with an appended and potentially coordinating chain have been called *scorpiands*.[17] When the metal *scorpiate* complex is dissolved in a fairly acidic solution, the primary amine group of the tail is protonated. Thus, redox changes on the metal center take place under the influence of a positive charge (from the ammonium group) at a fixed distance. In the case of the {Ni(II)[*N*-(aminoethyl)cyclam]}(ClO$_4$)$_2$ complex, oxidation to the trivalent state takes place at a potential of 1.19 *V* versus *SHE* in 1 mol HClO$_4$ solution,

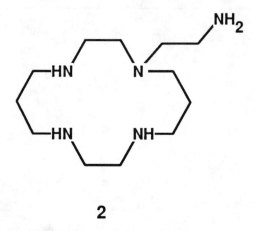

2

Figure 9.3. N-aminoethyl pendant cyclam.

as shown by the cyclic voltammetry experiment. This value has to be compared with the value observed for the [Ni(II)cyclam](ClO$_4$)$_2$ complex in the same conditions ($E_{1/2}$ = 0.98 V versus *NHE*, in 1 mol HClO$_4$) or better, with that measured for the complex of 1-methyl-1,4,8,10-tetra aza-cyclotetradecane (Mecyclam), which, as *N*-(aminoethyl)cyclam, contains one tertiary and three secondary nitrogen atoms ($E_{1/2}$ = 1.11 V versus *SHE*, in 1 mol HClO$_4$). The higher potential value observed with the *N*-(aminoethyl)cyclam complex reflects the additional work necessary when the charge of the metal center is increased from +2 to +3 in the electrostatic field generated by the $-NH_3^+$ charge at a distance of 5.31 Å.

The {Ni(II)[*N*-(aminoethyl)cyclam]}(ClO$_4$)$_2$ complex is an interesting example of redox activation. If dissolved in a solution containing concentrated perchloric acid (3 mol dm^{-3} and over), the solution turns an intense greenyellow color, and dichlorine develops. The color is that of the Ni(III) complex, whose formation is confirmed by the ESR experiment on the frozen solution. A redox process occurs in which the oxidizing agent is the ClO$_4^-$ ion and the reducing agent is the [Ni(II)(HL*)]$^{3+}$ complex [L* = *N*-(aminoethyl)cyclam]. In acidic solution, the perchlorate ion is a strong oxidizing agent ($E°$ = 1.39 V versus *SHE*). In particular, it should be strong enough to oxidize [Ni(II)(HL*)]$^{3+}$ to [Ni(III)(HL*)]$^{4+}$. However, nobody seems to care about the intense oxidizing power of perchlorate and perchloric acid; most redox chemists use them as background electrolytes without much concern. As a matter of fact, ClO$_4^-$ is an extremely inert reagent and remains intact in solution indefinitely, despite its thermodynamic instability, even in the presence of strong reducing agents such as dithionite or Cr(II). Thus, the Ni(II) scorpiate complex has to play some role in perchlorate activation. It is possible that at a 3 mol dm^{-3} and over concentration of HClO$_4$, ClO$_4^-$ ions might be more or less tightly bound to the Ni(II) center along the z axis. However, this should happen also with the $[Ni(II)(cyclam)]^{2+}$ complex, which does not react with concentrated perchloric acid. Thus, the simple interaction with a reducing metal center, such as Ni(II), is not the determining factor for the activation of ClO_4^-. It is probable that some hydrogen-bonding interaction exists between a perchlorate oxygen atom and the ammonium group of the side chain; this interaction impoverishes the negative charge of the perchlorate ion and promotes the uptake of one electron from the proximate reducing center, Ni(II). Molecular models show that two oxygen atoms of the perchlorate ion have the requisite "sting" to encompass the metal center (at the coordinative bond distance) and the ammonium group of the side chain (at the hydrogen-bonding distance). The existence of an electron transfer in this crucial step of the activation process is demonstrated by the fact that the $[Cu(II)(HL^*)]^{3+}$ complex, which is a much weaker reducing agent than $[Ni(II)(HL^*)]^{3+}$, does not give any reaction under the same conditions.

Using the fashionable language of *supramolecular chemistry*,[18] one could say that $[Ni(II)(HL^*)]^{2+}$ is an efficient *molecular device*, designed to defuse the quiet but powerful perchlorate ion. This device has been fashioned by assembling two fragments displaying different properties: coordination and one-electron release (in the metal center) and hydrogen bonding (in the appended $-NH_3^+$ group). Combination of these two properties according to the correct *topology* generates the function (redox activation).

The example given is rather chancy and perhaps not too significant. However, the principles outlined in this section could be of very general application in the interesting area of the redox activation of inorganic anions.

9.3 Appending a Further Redox-Active Subunit at the Cyclam Ring

The metallocyclam moiety can be conjugated by means of a covalent bond or a bridge to any other kind of redox-active fragment. This gives rise to a multi-electron, multisite redox agent. The potential at which any single electron is exchanged should be very close to that observed for the unconjugated fragment. Moreover, mutual effects of an electrostatic or electronic nature may exist between the two redox centers, which can alter the values of the potentials of the different couples.

An example is given by the metal complexes of the molecule 1-anthraquinonyl-1,4,8,11-tetraaza-cyclotetradecane (*3*, L-RO$_2$, shown in Figure 9.4).[19] L-RO$_2$ reacts with Ni(II)X_2 to give complexes of formula $[X_2Ni(II)L$-$RO_2]$, in which the metal is coordinated by the coplanar tetra-aza ring and by two axial X^- anions. Due to their rather lipophilic nature, the complexes $[X_2Ni(II)L$-$RO_2]$ dissolve in poorly polar solvents such as dichloromethane (dcm). In voltammetric investigation of a CH_2Cl_2 solution containing $[(ClO_4)_2Ni(II)L$-$RO_2]_2$ and made 0.1 mol dm^{-3} in Bu_4NClO_4, a reversible one-electron peak is observed on both the oxidation scan and the reduction

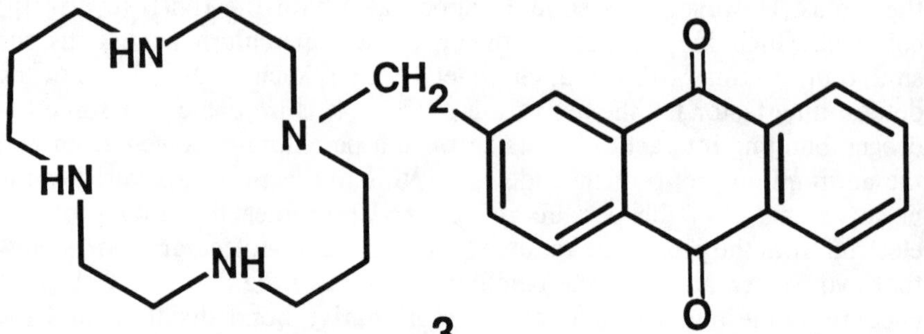

Figure 9.4. N-1-anthraquinonylmethyl pendant cyclam.

scan. The anodic peak should refer to the oxidation of the metal center [Ni(II)/Ni(III)], and the cathodic peak should correspond to the reduction of the quinone subunit to the semiquinone form, as described by the two following half-reactions:

$$[(ClO_4)_2Ni(III)L\text{-}RO_2]^+(ClO_4)^- + e^-$$
$$= [(ClO_4)_2Ni(II)L\text{-}RO_2] + (ClO_4)^- \quad (9.1)$$

$$[(ClO_4)_2Ni(II)L\text{-}RO_2] + e^- + (Bu_4N)^+$$
$$= [(ClO_4)_2Ni(II)L\text{-}RO_2^-](Bu_4N)^+ \quad (9.2)$$

Values of the potential associated to the half-reactions (9.1) and (9.2) are reported in the unidimensional diagram illustrated in Figure 9.5.

For comparison, the diagram also reports the potential values for the Ni(II)/Ni(III) redox couple of a complex of 1-hexadecyl-1,4,8,11-tetraaza-cyclotetradecane (*N*-cetylcyclam, *4*, L), a lipophilic version of cyclam obtained by appending a C_{16} aliphatic chain on one of the amine nitrogen atoms of the 14-membered tetra-aza macrocycle, as well as the potential values for the one-electron reduction of 1-methyl-anthraquinone, MeAQ ($RO_2 + e^- = RO_2^-$).[20] Figure 9.5 shows that the potentials associated with the redox changes taking place in the conjugate system are distinctly different from those observed with the individual fragments. In particular, the oxidation of the metal center in the $[(ClO_4)_2Ni(II)L\text{-}RO_2]$ complex takes place at a more positive potential than that observed with the $[Ni(II)L(ClO_4)_2]$ species. This can be considered a substituent effect. The anthraquinonyl subunit is an electron-withdrawing group and reduces the donating properties of the macrocycle; it has been mentioned before how strictly donor tendencies of the ligand and attainment of higher oxidation states of the bound metal are related. On the other hand, uptake of the first electron by the anthraquinone fragment is made easier in the presence of an adjacent Ni(II)–cyclam moiety, as indicated by comparing the less negative value observed for the reduction of the $[(ClO_4)_2Ni(II)L\text{-}RO_2]$ conjugate species to

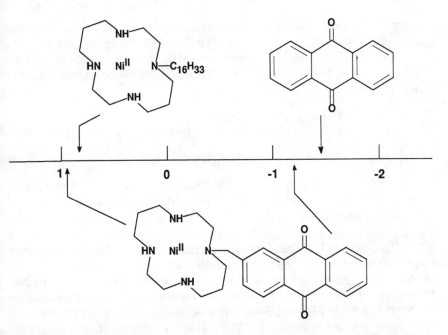

Figure 9.5. Oxidation and reduction behavior of the nickel(II) complex of the cy-clam/anthraquinone conjugate system, $[(ClO_4)_2Ni(II)L-RO_2]$, in a CH_2Cl_2 solution made 0.1 mol dm^{-3} in Bu_4NClO_4. In the lower portion of the diagram, the most positive potential corresponds to the oxidation of the metal center [Ni(II) to Ni(III)]; the more cathodic potential is associated with the one-electron reduction of the quinone ring to the semiquinone form. In the upper portion, the potentials associated with the redox behavior of reference systems under the same conditions are reported: the Ni(II) to Ni(III) oxidation in the Ni(II)L(ClO$_4$)$_2$ complex (L = 1-hexadecyl-1,4,8,11-tetraaza-cyclotetradecane) and the one-electron reduction of 2-methyl-an-thraquinone.

that of MeAQ. This stabilizing effect can be ascribed to the electrostatic interaction between the negatively charged semiquinone subunit and the neighboring dipositive metal center.

Interestingly, complexes of the $[X_2Ni(II)L-RO_2]$ type combine the lipo-philic nature of the organic subunit and the hydrophilic nature of the me-tallic moiety. However, the lipophilic nature predominates. As a matter of fact, the $[Cl_2Ni(II)L-RO_2]$ complex, when treated with a two-phase system containing equal volumes of CH_2Cl_2 and water (1 mol dm^{-3} HCl), dissolves completely in the organic phase; the intense absorption band of the $-RO_2$ chromophore is observed in the spectrum of the CH_2Cl_2 layer and is totally absent in the spectrum of the aqueous layer. However, if the aqueous phase is made 0.1 mol in Cr(II) and the two layers are equilibrated by vigorous shaking, the quinone band disappears from the spectrum of the organic phase, whereas quite intense bands at 350–450 nm are observed in the spectrum of the aqueous layer. These bands are typical of the hydroquinone chromophore. Thus, the following processes should occur: the quinone sub-unit in the $[Cl_2Ni(II)L-RO_2]$ complex is reduced to the hydroquinone form at the CH_2Cl_2/H_2O interface by aqueous Cr(II) in the presence of H^+ ions (eq. 9.3), and the $[Cl_2Ni(II)L-R(OH)_2]$ complex leaves the organic layer to water (eq. 9.4).

$$[Cl_2Ni(II)L-RO_2]_{(dcm)} + 2Cr(II)_{(aq)} + 2H^+_{(aq)}$$
$$= [Cl_2Ni(II)L-R(OH)_2]_{(dcm)} + 2Cr(III)_{(aq)} \quad (9.3)$$

$$[Cl_2Ni(II)L-R(OH)_2]_{(dcm)} = [Cl_2Ni(II)L-R(OH)_2]_{(aq)} \quad (9.4)$$

It should be noted that the reference system MeAQ, indicated as RO_2, undergoes the same two-electron reduction under two-phase conditions, but the hydroquinone form $R(OH)_2$ does not partition between the two layers and stays entirely in the organic phase. The special hydrophilicity of the $[Cl_2Ni(II)L-R(OH)_2]$ species, which is responsible for the transfer from CH_2Cl_2 to water, results from the combined effects of the Ni(II) center and of the two $-OH$ groups in the 9,10 positions of the anthracene moiety. Moreover, if the two-phase system containing the reduced species $[Cl_2Ni(II)L-R(OH)_2]$ is treated with an oxidizing agent (e.g., H_2O_2), the quinone form is found again in the CH_2Cl_2 layer. Therefore, H_2O_2 oxidizes the hydroquinone moiety to the quinone form, which goes into the CH_2Cl_2 layer. The process can be described in terms of the following stepwise equilibria:

$$[Cl_2Ni(II)L-R(OH)_2]_{(aq)} + H_2O_{2(aq)}$$
$$= [Cl_2Ni(II)L-RO_2]_{(aq)} + 2H_2O_{(aq)} \quad (9.5)$$

$$[Cl_2Ni(II)L-RO_2]_{(aq)} = [Cl_2Ni(II)L-RO_2]_{(dcm)} \quad (9.6)$$

These experiments illustrate a further case in which a molecular system can be transferred back and forth from an organic phase to an aqueous phase by varying the redox potential of the water layer. In this case, the Ni(II)–cyclam fragment does not play any active role in the redox event that drives the two-phase transfer process. It simply imparts a weighted hydrophilicity to the system, whose relative affinities toward water and CH_2Cl_2 are balanced as a result.

The example described here has illustrated how a multi-electron redox system can be obtained by linking two redox fragments of different natures through a methylene group. In such systems, electrons are exchanged at substantially different potentials that are reminiscent of those observed for the separated fragments. On the other hand, a redox system able to exchange two or more electrons at the same potential (or in a single shot) can be generated by assembling redox fragments of the same type. The metallo-cyclam subunit may be a convenient fragment for this purpose. In this connection, a series of biscyclam molecules has been obtained by linking two tetra-aza rings by a carbon chain of varying length through the nitrogen atoms (biscyclams, 5–9 in Figure 9.6).[21] Ligands of this family behave as

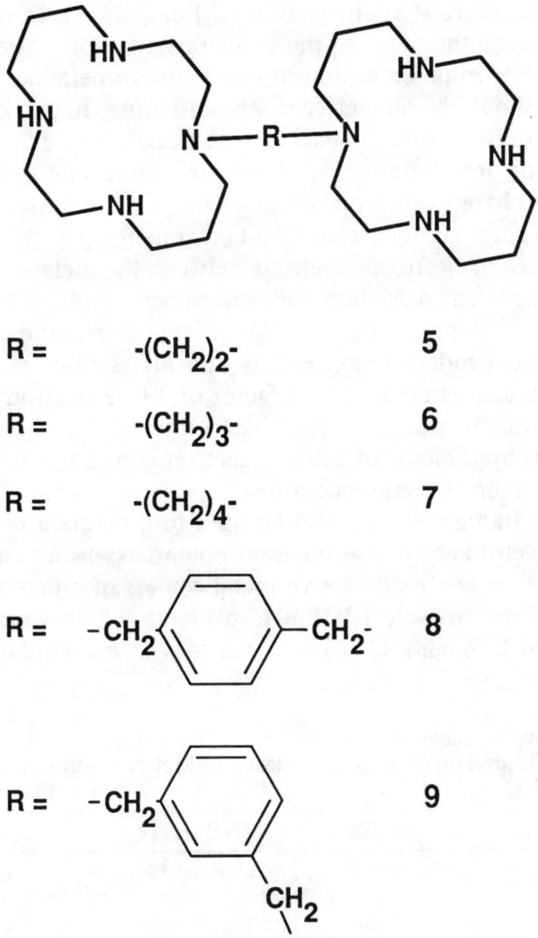

R =	$-(CH_2)_2-$	5
R =	$-(CH_2)_3-$	6
R =	$-(CH_2)_4-$	7

Figure 9.6. Bis(cyclam) molecules linked through the N atoms.

ditopic receptors for transition metal ions and may incorporate metal centers of the same or of different types.

Complexes of $[Ni(II)_2(biscyclam)](ClO_4)_2$ in MeCN solution undergo a two-electron oxidation process at the platinum electrode. The oxidation process occurs according to two reversible one-electron steps, whose electrode potentials are separated by the quantity $\Delta E = E_2 - E_1$.

$$E_1: [Ni(II)_2(biscyclam)]^{4+} = [Ni(II)Ni(III)(biscyclam)]^{5+} + e^- \quad (9.7)$$

$$E_2: [Ni(II)Ni(III)(biscyclam)]^{5+} = [Ni(III)_2(biscyclam)]^{6+} + e^- \quad (9.8)$$

Table 9.1 shows that the ΔE parameter varies with the type of the carbon chain joining the two cyclam rings. In particular, the potential difference is largest with the shortest chain system (C_2) and decreases with increasing separation between the rings. As the aliphatic nature of the ligand backbone excludes any electronic communication, mutual interactions between the metal centers should be only electrostatic in nature. In particular, the electrostatic effect derives from repulsive interactions between the metal centers; these repulsions are obviously greater in the dimetallic system holding a higher positive charge, which makes the second oxidation step more difficult. When the two rings are separated by a C_4 segment (*7*) or by one of the two xylyl spacers (*8, 9*), electrostatic effects between the metal centers in either divalent and trivalent oxidation states disappear, and ΔE approaches the limiting value of 36 mV. This potential difference is that expected for two one-electron independent redox events and arises from purely statistical effects; the first step is favored by a factor of 2 (the electron can come out from two equivalent sites and can go back on just one), whereas the second step is favored by a factor of ½ (the electron comes out from one center and can go back on two equivalent ones).

A biscyclam framework can also be used to generate a redox agent displaying two-electron activity at different potentials; in this case the ditopic ligand should host two redox-active metal centers of different natures. For instance, the bismacrocycle *10*, shown in Figure 9.7, in which two cyclam rings are linked through a C–C bond, reacts with one equivalent of Ni(II)

Table 9.1. Electrochemical parameters for the two-electron stepwise oxidation of [Ni$_2$(biscyclam)](ClO$_4$)$_4$ complexes in MeCN solution 0.1 M in Et$_4$NBF$_4$ at 25°C.

biscyclam	$\Delta E_{1/2}$ (mV)	$E_{1/2}(1)$ (V)	$E_{1/2}(2)$ (V)
5	100	0.755	0.855
6	66	0.697	0.763
7	54	0.687	0.732
8	50	0.735	0.785
9	50	0.720	0.770

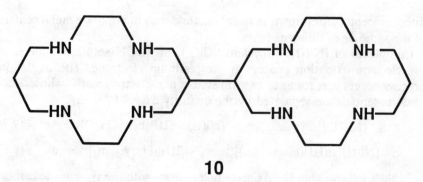

10

Figure 9.7. Bis(cyclam) linked through carbon atoms by a C-C single bond.

and one equivalent of Cu(II) to give as a major product the mixed metal complex [Ni(II)Cu(II)*10*](ClO$_4$)$_4$.[22] The homodimetallic species [Ni(II)$_2$*10*](ClO$_4$)$_4$ and [Cu(II)$_2$*10*](ClO$_4$)$_4$ are obtained in smaller amounts, as one could expect on statistical bases. In the very special solvent HClO$_4$ (concentration 70%), [Ni(II)Cu(II)*10*]$^{4+}$ undergoes a two-electron oxidation process at the platinum electrode, consisting of two one-electron steps, separated by a potential difference of 200 mV. This quantity is larger than the difference between the potential values, measured under the same conditions, for the [Cu(II,III)cyclam]$^{2+,3+}$ and the [Ni(II,III)cyclam]$^{2+,3+}$ couples ($\Delta E = 130$ mV) and again reflects the existence of mutual electrostatic repulsion effects between the adjacent metal centers.

These examples indicate that the cyclam ring may serve as a convenient and versatile building block in the production of multi-electron redox systems. Further redox fragments coming from organic chemistry (e.g., quinones, triazoles), from organometallic chemistry (e.g., ferrocene), or from classical coordination chemistry (e.g. metal complexes of 2,2'-bipyridine, phenanthroline) can be appended on it to generate new reagents able to exchange two or more electrons at different potentials and through a different electron transfer mechanism.

9.4 Lipophilic Metal Cyclam Complexes as Carriers of Electrons Across Liquid Membranes

Metal complexes of a lipophilic version of cyclam may behave as carriers in experimental transport of electrons across a *liquid membrane*. A liquid membrane is a layer of water-immiscible solvent that separates two aqueous layers (for instance, in a U-shaped glass tube).[23] To be a membrane, the interfacing layer must permit the transit, in a more or less selective way, of chemical substances from one aqueous phase to the other. The transit of a given species is possible if an appropriate molecular system (the carrier) is dissolved in the membrane; the carrier extracts the species of interest from the aqueous source phase, diffuses across the membrane layer, and releases it to the aqueous receiving phase.

Most experiments with liquid membranes have involved the carrier-mediated transport of cations,[24] including protons.[25] Transporting electrons across a membrane would seem to be an intriguing prospect for research, but only a limited number of electron transport experiments have been designed.[26] To carry out an electron transport experiment, an aqueous phase containing a reducing agent, Red (the so-called Electron Source Phase, ESP), and an aqueous phase containing an oxidizing agent, Ox (the Electron Receiving Phase, ERP), is interfaced by a layer of water-immiscible liquid in which a redox-active molecular system C has been dissolved. The redox system C may be prone to one-electron oxidation to the C$^+$ form. The mechanism of the three-phase redox process by which the electron transport experiment is carried out is illustrated by the scheme sketched in Figure 9.8.

The lipophilic redox system C releases an electron to the aqueous oxidizing agent Ox at the membrane/ERP interface, giving C^+. To maintain electroneutrality, C^+ extracts an X^- anion from the oxidizing aqueous phase, and the C^+X^- ion pair diffuses across the membrane to the other interface. Here, C^+ takes an electron from the aqueous reducing agent Red, again forming C, and X^- leaves the membrane to ESP. C is now available for the next redox cycle. After each turnover, an electron has been transferred from ESP to ERP, whereas an X^- ion has been transferred in the opposite direction.

The success of the three-phase process is contingent upon a thermodynamic requirement. The potential associated with the C^+/C redox couple must be intermediate between that of the Ox/Ox^- couple and that of the Red^+/Red couple:

$$E_{Ox} > E_C > E_{Red} \qquad (9.9)$$

The comparison is not straightforward because E_{Ox} and E_{Red} belong to the classical electrochemical scale in water (V versus SHE), whereas E_C refers to a very special scale used to measure potential in the water-immiscible solvent that serves as a membrane. As the value of the junction potential between the two liquids is not easily predictable, the juxtaposition of the

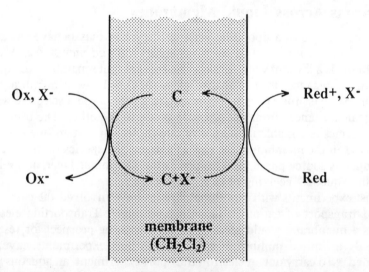

Figure 9.8. Scheme for the experimental transport of electrons from an aqueous layer containing a reducing agent, Red (the Electron Source Phase, ESP), to an aqueous layer containing an oxidizing agent, Ox (the Electron Receiving Phase, ERP), across an interfacing layer of a water-immiscible liquid (the membrane, e.g., CH_2Cl_2). The liquid membrane should contain a lipophilic redox-active molecular system C that is prone to the oxidation to C^+ in a catalytic amount. C transports electrons from ESP to ERP and C^+ transports anions X^- from ERP to ESP.

two electrochemical series must be empirically based. A redox system C/C$^+$ should satisfy these requirements to act as an efficient electron carrier: the redox system has to be confined in the membrane (i.e., both C and C$^+$ must be soluble in the membrane liquid and completely insoluble in water), and the one-electron C/C$^+$ redox change should be reversible, fast, and uncomplicated from a kinetic point of view. This latter requirement would suggest that transition metal complexes, in particular cyclam complexes, should be considered as candidates for the role of electron carrier. However, complexes of unsubstituted cyclam do not satisfy the first requirement, as they are soluble in water and poorly inclined to leave water for a water-immiscible liquid layer. However, cyclam and its complexes can easily be made definitively lipophilic by appending a long aliphatic chain on the ligand framework. For instance, a C_{16} aliphatic chain can be appended on one of the nitrogen atoms to give *N*-cetylcyclam (*4*, L), as mentioned in the preceding section. Metal complexes of L (e.g., those of the formula NiLX_2) are soluble in apolar or poorly polar solvents such as C_6H_6, $CHCl_3$, and CH_2Cl_2 and insoluble in water.[27]

In the transport experiments to be described in this section, the water-immiscible solvent chosen as a bulk liquid membrane was CH_2Cl_2. As a matter of fact, CH_2Cl_2 is polar enough to allow conventional electrochemical investigation and determination of the E_C values. In particular, cyclic voltammetric investigation reveals that the Ni(II)LX_2 complex in a CH_2Cl_2 solution containing 0.1 mol dm^{-3} Bu$_4$NCl undergoes a reversible one-electron oxidation process at the platinum electrode. The electrode process can be described by the following half-reaction:

$$[Ni(II)LCl_2] + Cl^- = [Ni(III)LCl_2]Cl + e^- \tag{9.10}$$

The corresponding $E_{1/2}$ value is 0.175 V, more positive than the potential of the redox couple used as an internal standard: ferrocene/ferrocenium (Fc/Fc$^+$). However, redox reactivity of the [Ni(II)LX_2] species under two-phase conditions (water/dichloromethane) cannot be predicted on the basis of the redox potential value but must instead be tested empirically. For instance, the [Ni(II)LCl$_2$] complex in the CH_2Cl_2 layer, pale violet in color, is oxidized to the [Ni(III)LCl$_2$]Cl species, bright yellow in color, through equilibration (read: vigorous shaking) with an aqueous layer containing the strong oxidizing agent $S_2O_8^{2-}$ and NaCl; the presence of chloride is essential to the formation of the [Ni(III)LCl$_2$]$^+$Cl$^-$ ion pair. The occurrence of this two-phase redox process indicates that the potential associated with the [Ni(III)LCl$_2$]Cl + e$^-$ = [Ni(II)LCl$_2$] + Cl$^-$ half-reaction (in CH_2Cl_2 solution) is lower than that associated with the half-reaction $S_2O_8^{2-} + 2e^- = 2SO_4^{2-}$ in water (E° = 1.99 V versus SHE). On the other hand, the [Ni(III)LCl$_2$]Cl species in the CH_2Cl_2 layer can be reduced to the [Ni(II)LCl$_2$] form by treatment with an aqueous layer containing the mild reducing agent Fe(II). Thus, the potential for the [Ni(III)LCl$_2$]Cl/[Ni(II)LCl$_2$] couple should be higher than that of the Fe(III)/Fe(II) couple (E° = 0.77 V versus SHE). These two experiments suggest that the two electrochemical scales for the media of interest (water and CH_2Cl_2) should be juxtaposed in such a way that the potential value for the [Ni(II)LCl$_2$]Cl/[Ni(II)LCl$_2$] redox change (in CH_2Cl_2 solution) lies between the potentials associated with the aqueous couples $S_2O_8^{2-}/SO_4^{2-}$ and Fe(III)/Fe(II) (see Figure 9.9).

If the tentative juxtaposition shown in Figure 9.9 is correct, all aqueous reducing agents with potentials lower than 0.77 V versus SHE should be able to reduce [Ni(III)LCl$_2$]Cl to [Ni(II)LCl$_2$] under two-phase conditions. Indeed, when the yellow-green CH_2Cl_2 layer of the [Ni(III)LCl$_2$]Cl complex is equilibrated with an aqueous layer containing Ti(III) {E°[Ti(IV)/Ti(III)] = 0.16 V versus SHE}, Cr(II) {E°[Cr(III)/Cr(II)] = −0.43 versus SHE}, or [Co(II)(diamsarH$_2$)]$^{4+}$ (E°{[Co(III)(diamsarH$_2$)]$^{5+}$/[Co(II)(diamsarH$_2$)]$^{4+}$} =

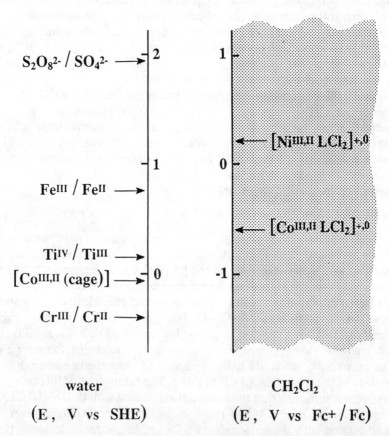

Figure 9.9. Tentative juxtaposition of the electrochemical scale in water and in CH_2Cl_2 solution. Such a juxtaposition accounts for all the investigated two-phase redox processes; a [M(II)LX_2] complex [M(III)/M(II) potential = E_C] will be oxidized to [M(III)LX_2]$^+$ by any aqueous oxidizing agent, whose potential E_{ox} is higher than E_C. On the other hand, a [M(III)LX_2]$^+$ complex will be reduced to [M(II)LX_2] by any aqueous reducing agent having a potential E_{red} lower than E_C.

−0.05 V versus SHE),[28] the pale violet color of the Ni(II) complex is restored. The two-phase redox processes involving the oxidation of [Ni(II)LCl$_2$] and the reduction of [Ni(III)LCl$_2$]Cl can be "assembled" to produce the three-phase (electron transport) process whose mechanism is illustrated in Figure 9.10.

Experiments have been performed[27] in which, using a V-shaped tube, a CH$_2$Cl$_2$ layer 0.001 mol dm^{-3} in [Ni(II)LCl$_2$] separates an aqueous layer 0.1 mol dm^{-3} in Na$_2$S$_2$O$_8$ and 1 mol dm^{-3} in NaCl (ERP) from an aqueous layer .01 mol dm^{-3} in one of the reducing agents (Red) mentioned earlier and 1 mol dm^{-3} in HCl (ESP). Progress of the three-phase redox process was followed by decreasing of the concentration of Red in ESP or, better, by monitoring the increasing concentration of Red$^+$. The transport of electrons from ESP to ERP (and the simultaneous countertransport of Cl$^-$ anions from ERP to ESP) occurs with all the mentioned reducing agents. The time required for complete oxidation of Red to Red$^+$, which involves 10 travels back and forth across the membrane layer of the macrocyclic carrier, varies dramatically with the nature of the reducing agent: from 10 minutes [Ti(III)] to 12 hours [Co(II)(diamsarH$_2$)]$^{4+}$. In particular, the electron transport rate decreases according to the following sequence:

$$TI(III) > Cr(II) > Fe(II) > [Co(II)(diamsarH^2)]^{4+} \qquad (9.11)$$

This sequence, which does not correlate with the redox potentials, should reflect the rate of the electron transfer process from Red to [Ni(III)LCl$_2$]$^+$,

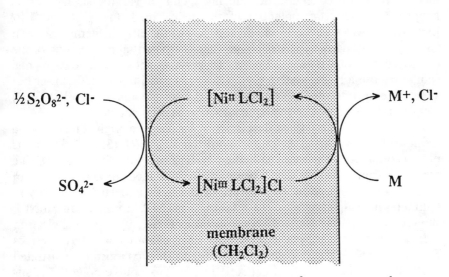

Figure 9.10. Scheme for the electron transport process from an aqueous layer containing a metal-centered reducing agent M (M = Cr(II), Fe(II), Ti(III), [Co(II) (diamsarH$_2$)]$^{4+}$) to an aqueous layer containing S$_2$O$_8^{2-}$, mediated by the [Ni(III)LCl$_2$]Cl/[Ni(II)LCl$_2$] redox system (L = *N*-cetylcyclam, **4**).

which takes place at the water/dicloromethane interface. The metal-centered reducing agents that have been investigated should be present in ESP in the form of chlorocomplexes such as $(FeCl_4^{2-})$ (notice that ESP is 1 mol dm^{-3} in HCl), with the exception of the Co(II) cage complex. Thus, it is possible that the electron transfer between the reducing species and the Ni(III) tetra-aza-macrocyclic complex (which has a *trans*-dichloro-octahedral stereo-chemistry) takes place by means of a chloride ion bridging the two metal centers. This could account for the very low rate of electron transport in the experiment involving the $[Co(II)(diamsarH_2)]^{4+}$ complex as a reducing agent; in fact, in this species, the ligand cage prevents chloride bridging.

The experiments previously described illustrate a further way to perform oxidation and reduction reactions that keep the solutions of the reagents separated—the electrons and ions are made to flow through the interfacing liquid membrane. The other, more classical, way to perform redox reactions under multiphase conditions is represented by the *voltaic cell*. In this case, the solutions of the reducing agent and of the oxidizing agent are also kept separated; electrons flow from ESP to ERP through a metallic support and, to balance the electrical charges, ions of opposite charge flow and counter-flow across the salt bridge.

Compared to Volta's device the redox-active membrane cell presents a unique feature; it may allow selective oxidation and reduction reactions to be performed by varying the potential of the redox system used as an elec-tron carrier. For instance, in the experiments illustrated earlier, it was ob-served that a strong, typically nonselective oxidizing agent, peroxydisul-phate, oxidized a collection of reducing agents of varying strength. This happened because the potential of the redox mediator, $[Ni(III)LCl_2]^+/$ $[Ni(II)LCl_2]$, was situated in the highest part of the electrochemical scale in CH_2Cl_2 and, in particular, was higher than the potential of each one of the employed aqueous reducing agents. To perform selective oxidation, one should choose a redox system C^+/C whose potential is distinctly lower but situated in an intermediate range, below that of the aqueous Fe(III)/Fe(II) couple and above that of the aqueous Cr(III)/Cr(II) couple.

To verify this, we modified the carrier by keeping the original macrocyclic framework but replacing the rather anodic Ni(III)/Ni(II) (35.17 V in the gas phase) couple with the more cathodic Co(III)/Co(II) couple (33.50 V). The green, low-spin $[Co(III)LCl_2]Cl$ complex was obtained through reaction of $CoCl_2$ with *N*-cetylcyclam in presence of dioxygen and HCl.[29] Cyclic vol-tammetry investigation of a CH_2Cl_2 solution, made 0.1 mol dm^{-3} in Bu_4NCl, showed that the $[Co(III)LCl_2]Cl$ complex undergoes a reversible one-electron reduction process at the platinum electrode at a potential of -0.61 V versus Fc$^+$/Fc internal reference couple. According to the juxtaposition illustrated in Figure 9.9, the $[Co(II)LCl_2]^+/[Co(II)LCl_2]$ redox couple potential is sit-uated between those of the Fe(III)/Fe(II) and Cr(III)/Cr(II) aqueous couples. Indeed, the $[Co(III)LCl_2]Cl$ complex in CH_2Cl_2 solution is reduced under two-phase conditions to the pink $[Co(II)LCl_2]$ species by aqueous reducing

Cr(II), but not by aqueous Fe(II). Moreover, aqueous Fe(III), in HCl 1 mol dm^{-3}, oxidizes under two-phase conditions [Co(II)LCl$_2$], to restore the green color of the CH$_2$Cl$_2$ layer. As a consequence, in three-phase experiments carried out using the [Co(III)LCl$_2$]Cl/[Co(II)LCl$_2$] redox system as a carrier, S$_2$O$_8^{2-}$ can discriminate between Fe(II) and Cr(II), oxidizing Cr(II) and leaving Fe(II) intact.

The multiphase chemistry described in this section seems, at this stage of investigation, nothing more than a chemical curiosity or an attractive experiment for student exercises. However, it is possible that chemical culture originating from these types of investigations could be instrumental in the development of new applications and technologies; for example, new membrane-based electrodes, sensitive to redox-active species, could be designed. Moreover, redox-active membranes could be useful in the separation of metal ions. In particular, selective or specific separation of a mixture of metal ions may be made easier by changing the oxidation state and the electrical charge of specific metal centers in a controlled manner. In this way the solution of interest, before being processed with a conventional metal-extracting membrane device, should be pretreated with an appropriate redox-active liquid membrane. Varying the nature of the electron carrier (i.e., moving the redox potential of the carrier up and down along the membrane electrochemical scale) may allow for the preparation of cations in the desired oxidation state for extraction.

The carriers described above are rather primitive and have been obtained by simple alkylation of one of the amine nitrogen atoms of the unsubstituted cyclam framework. However, more sophisticated redox systems based on the cyclam unit can be built to carry out more complex functions by following the principles described in this chapter. As an example, multi-electron transport could be designed by using multisite redox systems; further elements of selectivity in this type of electron transport might be introduced by modulating the difference between the potentials associated with the different redox centers. On the other hand, attention can also be focused on the countertransported anion; three-phase experiments could be designed to carry out selective transport and separation of a given anion X^-, profiting from a gradient of redox potential. In this case, the molecular system used as a carrier in its oxidized form C$^+$ should have the appropriate electronic and structural features to interact with X^- in a selective way.

References and Notes

1. Cotton, F. A.; Wilkinson, G. *Advanced Inorganic Chemistry*; 5th ed.; Wiley Interscience; New York, 1988, p 344.

2. Lindoy, L. F. *The Chemistry of Macrocyclic Ligand Complexes*; Cambridge University; Cambridge, 1989; p 201.

3. Bisi Castellani, C.; Fabbrizzi, L.; Licchelli, M.; Perotti, A.; Poggi, A. *J. Chem. Soc., Chem. Comm.* **1984**, *12*, 806.

4. Anichini, A.; Fabbrizzi, L.; Paoletti, P.; Clay, R. M. *J. Chem. Soc., Chem. Comm.* **1977,** *8,* 244.

5. Busch, D. H. *Acc. Chem. Res.* **1978,** *11,* 392.

6. Van Alphen, J. *Rec. Trav. Chim. Pays-Bas* **1937,** *56,* 343.

7. (a) Bosnich, B.; Poon, C. K.; Tobe, M. L. *Inorg. Chem.* **1965,** *4,* 1102.
   (b) Bosnich, B.; Tobe, M. L.; Webb, G. A.; *Inorg. Chem.* **1965,** *4,* 1109.

8. Busch, D. H. *Helv. Chim. Acta* **1967,** *50,* 174.

9. Curtis, N. F. *Coord. Chem. Rev.* **1968,** *3,* 3.

10. Barefield, E. K. *Inorg. Chem.* **1972,** *11,* 2273.

11. Barefield, E. K.; Wagner, F.; Herlinger, A. W.; Dahl, A. R. *Inorg. Synth.* **1976,** *16,* 220.

12. Richman, J. E.; Atkins, T. J. *J. Am. Chem. Soc.* **1974,** *96,* 2268.

13. Fabbrizzi, L. *Comments Inorg.Chem.* **1985,** *4,* 33.

14. Buttafava, A.; Fabbrizzi, L.; Perotti, A.; Poggi, A.; Poli, G.; Seghi, B. *Inorg. Chem.* **1986,** *25,* 1456.

15. (a) Ito, T.; Sugimoto, K.; Toriumi, K.; Ito, H. *Chem. Letters* **1981,** 1477.
    (b) Yamashita, M.; Toriumi, K.; Ito, T. *Acta Crystallogr., Sect. C: Cryst. Struct. Commun.* **1985,** *41,* 1607.
    (c) Yamashita, M.; Miyamae, H. *Inorg. Chim. Acta* **1989,** *156,* 71.

16. Bencini, A.; Fabbrizzi, L.; Poggi, A. *Inorg. Chem.* **1981,** *20,* 2544.

17. Pallavicini, P.; Perotti, A.; Poggi, A.; Seghi, B.; Fabbrizzi, L. *J. Am. Chem. Soc.* **1987,** *109,* 5139.

18. Lehn, J.-M. *Angew. Chem., Int. Ed. Engl.* **1988,** *27,* 89.

19. De Santis, G.; Fabbrizzi, L.; Mangano, C.; Poggi, A.; Seghi, B. *Inorg. Chim. Acta* **1990,** *177,* 47.

20. Di Casa, M.; Fabbrizzi, L.; Mariani, M.; Seghi, B. *J. Chem. Soc., Dalton Trans.* **1990,** *1,* 55.

21. Ciampolini, M.; Fabbrizzi, L.; Perotti, A.; Poggi, A.; Seghi, B.; Zanobini, F. *Inorg. Chem.* **1987,** *26,* 3527.

22. (a) Fabbrizzi, L.; Montagna, L.; Poggi. A.; Kaden, T. A.; Siegfried, L. C. *Inorg. Chem.* **1986,** *25,* 2671.
    (b) Fabbrizzi, L.; Montagna, L.; Poggi, A.; Kaden, T. A.; Siegfried, L. *J. Chem. Soc., Dalton Trans.* **1987,** *11,* 2631.

23. Noble, R. D.; Way, J. D. *Liquid Membranes. Theory and Applications*; ACS Symposium Series 347; Washington, DC, 1987.

24. Behr, J.-P.; Kirch, M.; Lehn, J.-M. *J. Am. Chem. Soc.* **1985,** *107,* 241.

25. Danesi, P. R.; Cianetti, C.; Horwitz, E. P. *Solvent Extraction Ion Exch.* **1983,** *1,* 289.

26. De Santis, G.; Fabbrizzi, L.; Poggi, A.; Seghi, B. *J. Chem. Soc., Dalton Trans.* **1991,** *1,* and references therein.

27. De Santis, G.; Di Casa, M.; Mariani, M.; Seghi, B.; Fabbrizzi, L. *J. Am. Chem. Soc.* **1989,** *111,* 2422.

28. diamsar = 1,8-diamino-3,6,10,13,16,19-hexaazabicyclo-(6.6.6)icosane; diamsarH$_2^{2+}$ is the form in which the amino groups in 1 and 8 positions (top and bottom of the cage) are protonated. See Geue, R. J.; Hambley, T. W.; Harrowfield, J. M.; Sargeson, A. M.; Snow, M. R. *J. Am. Chem. Soc.* **1984,** *106,* 5478.

29. De Santis, G.; Fabbrizzi, L.; Poggi, A.; Seghi, B. *J. Chem. Soc., Dalton Trans.* **1990,** *9,* 2729.

CHAPTER 10

Toward More Preorganized Macrocycles

Robert D. Hancock

Centre for Molecular Design
Department of Chemistry
University of Witwatersrand
Wits 2050, Johannesburg
South Africa

A strong historical trend in the study and synthesis of ligands has been to focus on those that are ever more highly *preorganized.* The term preorganization, coined by Cram,[1] refers to the degree to which the arrangement of the donor atoms in the free ligand is similar to that required in the complex for coordination to the metal ion. In ligands of the highest degree of preorganization, solvent molecules are sterically limited in their ability to solvate the donor atoms of the ligand, and high thermodynamic stability of the complexes formed may result because the energy of solvation of the free ligand does not have to be overcome before a complex can be formed. The trend to higher levels of preorganization can be traced back over the last 50 years of ligand synthesis and study of complex equilibria in solution. Early work was concerned only with unidentate ligands, which, along with potentially coordinating solvent molecules, are on the lowest level of preorganization (Figure 10.1). The first step along the path to higher levels of preorganization was the discovery of the *chelate effect.*[2] Here the donor atoms of the ligand are tied together with connecting bridges, and we see from Figure 10.1 that this results in greater complex stability. This greater complex stability arises mainly from the drop in translational entropy of the free ligand, as compared to a number of unidentate ligands containing the same number of donor atoms. A smaller, but still significant, contribution also comes from the increased basicity[3] of the donor atoms, which are moved along the primary, secondary, and tertiary series when used as attachment points for the alkane bridges used to hold the chelating ligand together. Chelates may exhibit high or low levels of preorganization, as seen in the example of the ligands EDTA and CYDTA. There is an entropy-controlled increase[4] in the $\log K_1$ for the CYDTA complex relative to the

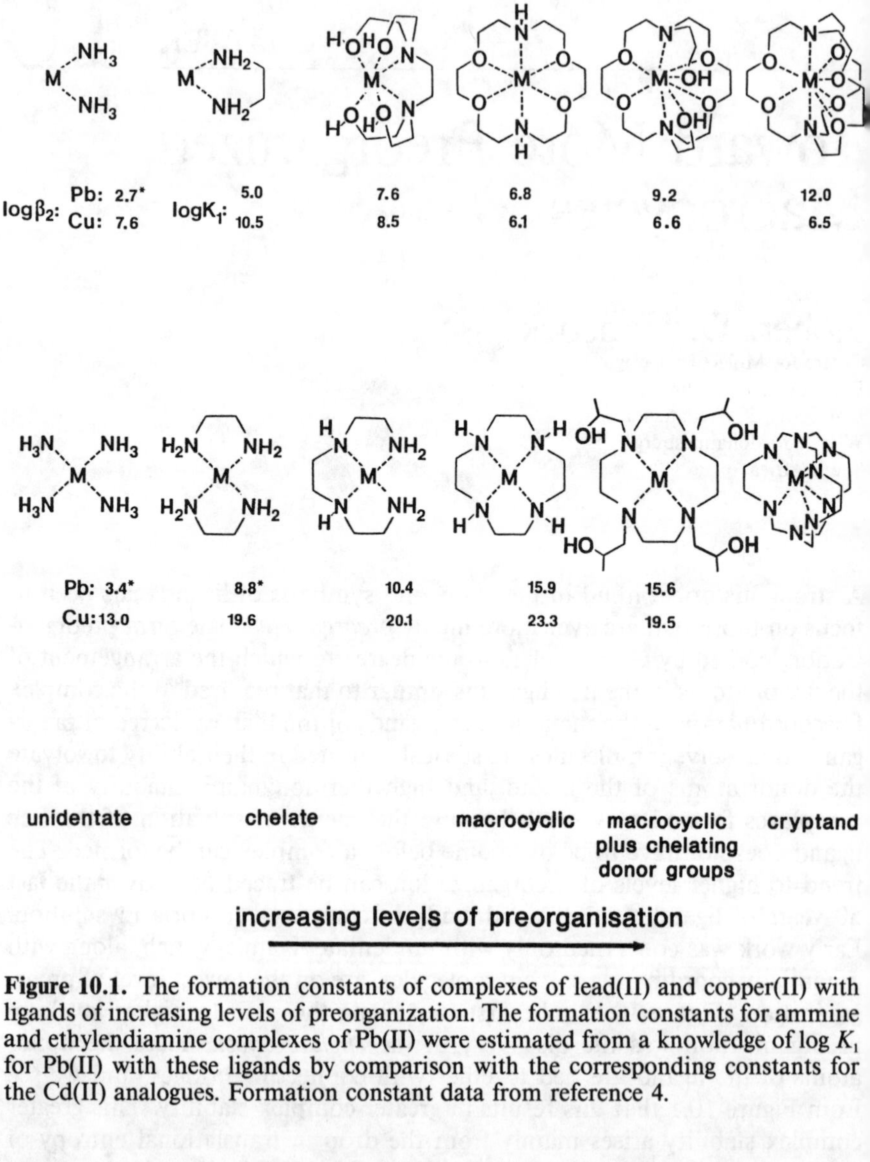

Figure 10.1. The formation constants of complexes of lead(II) and copper(II) with ligands of increasing levels of preorganization. The formation constants for ammine and ethylendiamine complexes of Pb(II) were estimated from a knowledge of log K_1 for Pb(II) with these ligands by comparison with the corresponding constants for the Cd(II) analogues. Formation constant data from reference 4.

EDTA complex brought about by the fact that the cyclohexane ring in CYDTA holds the donor atoms close to the conformation required for complex formation, whereas in the EDTA complex the donor atoms are free to move into the lower-energy *skew* conformation, which is not suitable for complex formation, as shown in Chart 1.

The next levels of preorganization were attained with the discovery of the macrocycles,[5] followed by the even more highly preorganized cryptands.[6] Here, as seen in Figure 10.1, there are increases in thermodynamic complex

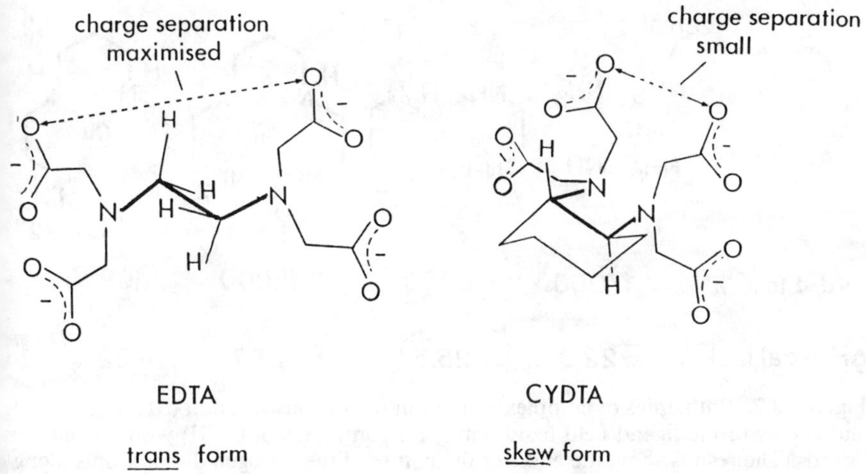

EDTA

trans form

CYDTA

skew form

Chart 1

stability (the macrocyclic and cryptate effects) but, just as importantly, there are large increases in selectivity. One should note in Figure 10.1 that the macrocycle with pendant chelating donor groups is more highly preorganized than a simple macrocycle, a point we shall return to in section 2. The effect of preorganization on selectivity is seen in Figure 10.1 for the large lead(II) ion relative to the small copper(II) ion. For the diammine complexes, the stability of the Cu(II) complex exceeds that of the Pb(II) complex by five log units, but by the time the 2,2,2-cryptand complexes are reached, the *selectivity* has reversed to be over five log units in favor of the Pb(II) ion. This selectivity of a ligand for Pb(II) over Cu(II) refers to the difference in $\log K_1$ for the two metal ions with that ligand. The macrocyclic and cryptate effects appear[3] to have several contributing factors:

1. Difficulty of solvation of the donor atoms within the confines of the ligand cavity[7]
2. Enforced electrostatic repulsion between the lone pairs of the donor atoms, which is relieved on complex formation[5,6]
3. The ligand is already in a conformation close to that required for complex formation[6-9]
4. The inductive effects of the alkane bridges used to connect the donor atoms together[3]

The latter effect is one which has not been widely considered in attempts to explain the coordinating properties of more highly preorganized ligands. The donor atoms in Figure 10.2 are seen to move from zeroth through primary to secondary in the series of ligands of increasing structural elaboration. Two distinct effects arise from this: a steady increase in the enthalpy change on complex formation as well as of the Ligand Field (LF) strength.

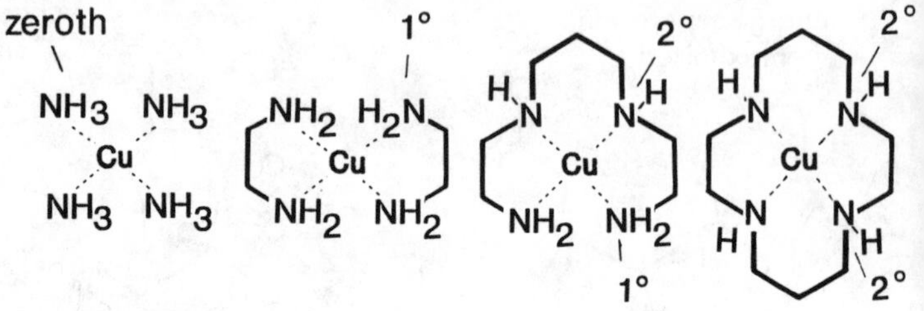

$\nu_{d\text{-}d}$ (cm^{-1}):	17000	18300	19000	19900
ΔH, kcal.mol^{-1}:	-22.0	-25.5	-27.7	-32.4

Figure 10.2. Enthalpies of complex formation in aqueous solution (ΔH, kcal mol^{-1}) and energy of the ligand field band (cm^{-1}) for complexes of Cu(II) with polyamine ligands. The results show the effect of the nature of the nitrogen donor atoms along the series primary < secondary < tertiary on ligand field strength and enthalpy of complex formation. Data from references 3 and 4.

Increases in LF strength are associated[8] with increase in overlap in the M–L bond. In the example in Figure 10.2, the increase in LF strength is brought about by the increased inductive effect of the added alkyl groups in passing from zeroth through to secondary nitrogens. However, as shown in Figure 10.3, the inductive effects of adding N-methyl groups to cyclam to give tetramethylcyclam (TMC) are outweighed by the adverse steric effects produced by the van der Waals repulsion between these methyl groups. The Ni–N bonds are stretched out from their strain-free values of 1.91 Å[9] to observed[10] values of 1.99 Å by the steric repulsion between the methyl groups in $[Ni(TMC)]^{2+}$. This results in a drop in log K_1 for Ni(II) from log K_1 = 20.1 for the cyclam complex[11] to 8.6 for the TMC complex[12] and a drop in the in-plane LF strength, Dq_{xy}, from 2043 cm^{-1} in the cyclam complex to 1782 cm^{-1} in the TMC complex. This highlights a very important aspect of ligand design; when alkyl groups are added to donor atoms, the potential inductive effects will only be realized if this is done in a sterically efficient manner. To achieve steric efficiency, the added alkyl group should be a bridging group leading to chelate ring formation, rather than a simply methyl or ethyl group. In the macrocycles, one is adding more alkyl groups than would be possible for a simple chelating ligand while still keeping all the added groups as bridging groups.

The historical trend we have noted here is one towards ever greater levels of preorganization in ligands, along the unidentate, chelate, macrocyclic, and cryptand series. This raises the question of whether even more highly preorganized ligands are possible. To answer this, it is first necessary to decide just how preorganized the macrocycles and cryptands really are. As has been pointed out,[1,3] the free ligands are not in the conformation required for complex formation, as seen in Figure 10.4, and so do not meet Cram's

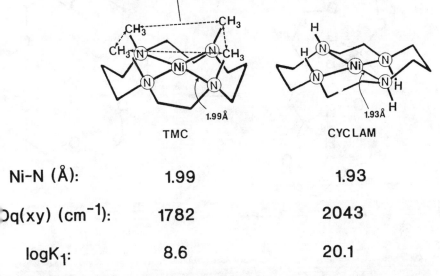

Ni–N (Å):	1.99	1.93
$Dq(xy)$ (cm^{-1}):	1782	2043
$logK_1$:	8.6	20.1

Figure 10.3. The effect of steric crowding of the C-methyl groups of tetramethylcyclam (TMC) on in-plane ligand field strength, Ni–N bond length, and formation constant of the nickel(II) complex tetramethylcyclam as compared to the less sterically crowded cyclam complex. Modified after reference 3.

criterion[1] for the highest levels of preorganization: the free ligand should have the same conformation as found in the complex with the target metal ion. Another important aspect of preorganization is that of specificity, which is to say that the ligand should be preorganized primarily for the target metal ion and display a high level of selectivity for it. An important idea in macrocyclic chemistry is that selectivity for metal ions is controlled by *size-match selectivity*, which is to say that the maximum complex stability is found where the match between the size of the cavity in the ligand and the size of the metal ion is closest. This idea has been examined for the tetraazamacrocycles,[13] and Figure 10.5 shows the variation in log K_1 for the tetraazamacrocycles 12-aneN$_4$ through 16-aneN$_4$. (See Figure 10.6 for key to ligand abbreviations). The results do not really accord with the idea of size-match selectivity. Extensive investigation of this problem by means of molecular mechanics (MM) calculation[14] shows that the tetraazamacrocycles are too flexible to show genuine size-match selectivity. In brief, apart from the trans-(III) conformer that might be expected to show size-match selectivity (Figure 10.7), other conformers such as the trans-I and cis-V conformers are adopted by metal ions that are too large for the macrocyclic cavity,[14] and the selectivity patterns of the tetraazamacrocycles are in reality not very different from those of open-chain ligands.

The factor controlling the selectivity patterns of open chain ligands (and also macrocycles) is the size of the chelate ring.[9,15,16] This control is achieved

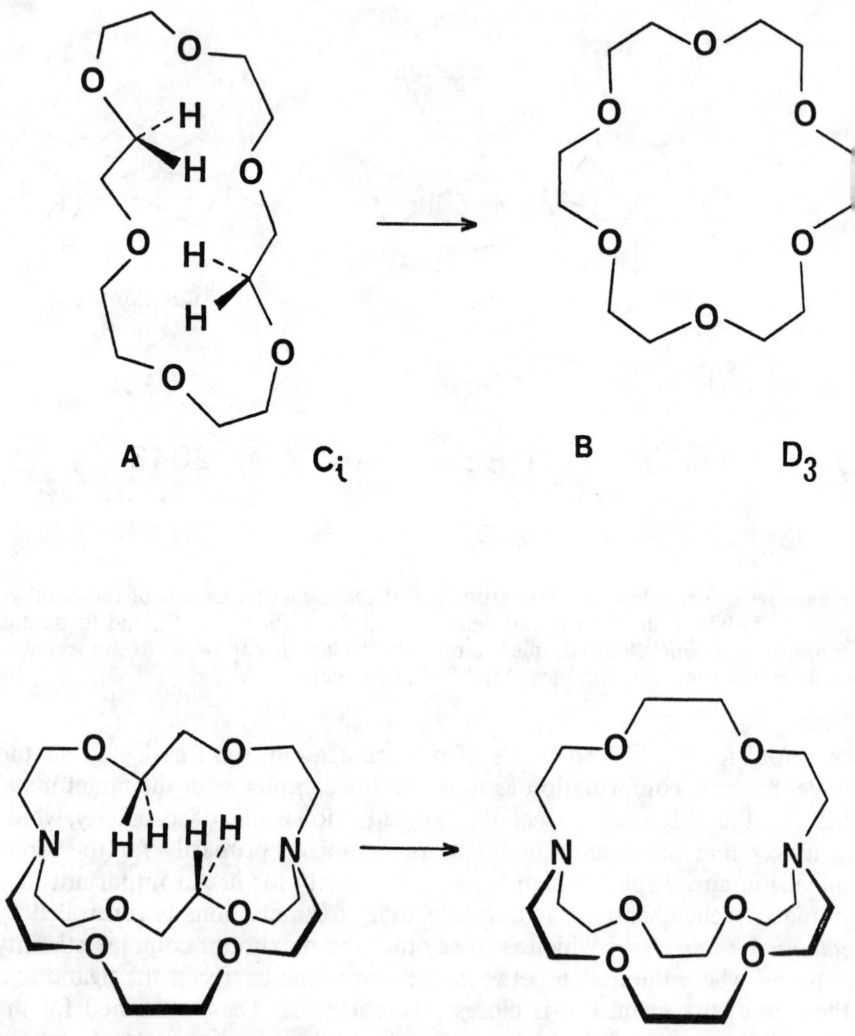

Figure 10.4. Conformers of 18-crown-6 and cryptand-2,2,2 as found (a) in the crystalline state and in solvents of low dielectric constant and (b) as required for formation of complexes with metal ions of about the size of K^+. Modified after reference 1.

in terms of how closely the metal ion can be placed at the focus of the lone pairs of the ligand in its minimum-strain energy conformation. This is summarized in Figure 10.8. For the best fit, the metal ion in a six-membered chelate ring should have a short strain-free M–L length and an L–M–L angle of about 109°, while in a five-membered chelate ring the metal ion should

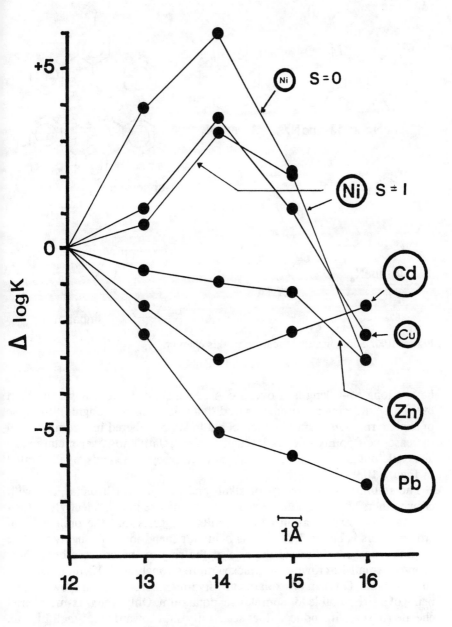

size of macrocyclic ring

Figure 10.5. Variation of formation constant for the X-aneN$_4$ tetraazamacrocycles as a function of macrocyclic ring size X. The variation has been expressed as $\Delta \log K$, which is the difference in log K_1 between that for the X-aneN$_4$ macrocycle and that for 12-aneN$_4$. The metal ions are shown inside circles proportional to their ionic radii. Modified after reference 9.

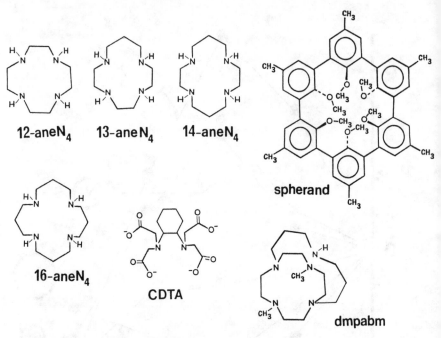

Figure 10.6. Some ligands discussed in this chapter.

have a long M–L length of over 2.5 Å and a small L–M–L angle of about 69^c. These differing requirements lead to a rule of ligand design:[17] "Increase of chelate ring size from five membered to six membered in a complex will increase the stability of complexes of smaller relative to larger metal ions." This rule holds for both open-chain and macrocyclic ligands, as illustrated in Figure 10.9

The stability of complexes of alkali and alkaline earth metal ions with crown ethers has been rationalized[5] in terms of size-match selectivity. However, a closer examination of the results[17] suggests that the preference of crown ethers is better described as being for metal ions of roughly the size of potassium, regardless of the size of the ring in the crown ether. Thus, even the small 12-crown-4 complexes more strongly with K$^+$ than with the small Li$^+$ ion. This can be explained by five-membered rings in crown ethers being of a size suitable for complexing potassium. Only when six-membered chelate rings are introduced does selectivity shift towards the small Li$^+$ ion, in accord with the previously mentioned rule.

The level of preorganization in cryptands appears to be considerably higher[16] than in simple macrocycles. The peaks in complex stability with alkali metal ions are very much sharper than is found for crown ethers. More importantly, the kinds of effects to be expected from high levels of preorganization are beginning to be manifested, such as the extraordinarily

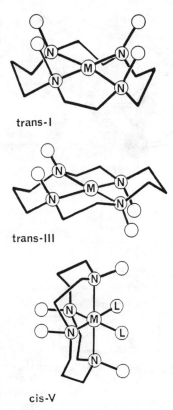

trans-I

trans-III

cis-V

Figure 10.7. Commonly found conformers of complexes of tetraazamacrocycles with metal ions, illustrated here for cyclam complexes. The open circles are the N–H hydrogens of the complex. L represents other coordinates ligands, such as water molecules.

high pKa of the cryptand-1,1,1,[18] as illustrated in Chart 2. One sees that as the cavity of the cryptand becomes too large for the proton, the pKa drops down to a value similar to that of an open-chain ligand such as triethanolamine. A similarly high protonation constant and affinity for metal ions is found[19] in the cryptandlike nitrogen donor ligand dmpabm in Figure 10.6.

It is just such effects that one is looking for in macrocyclic chemistry: effects that go way outside normal experience. The spherands of Cram[1] are examples of ligands that are much more highly preorganized than crown ethers and most cryptands. Smaller cryptands such as cryptand-1,1,1 appear to be more rigid and hence more highly preorganized than larger cryptands such as cryptand-2,2,2 and are possibly as highly preorganized as the spherands. The spherands show their high levels of preorganization through crystallographic investigation, which shows[1] the cavity of the free ligand to be about the same size as in the Li$^+$ complex and to be unsolvated, with the

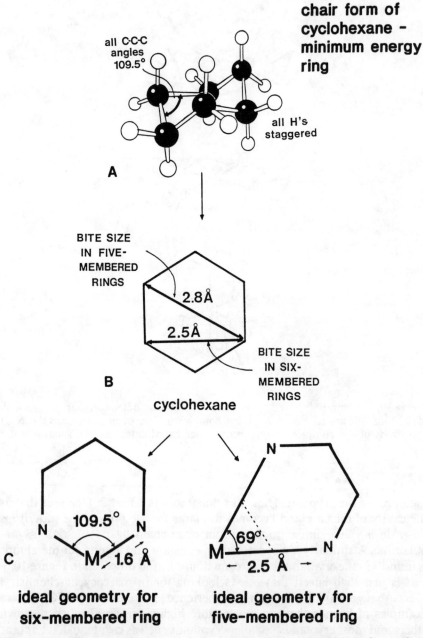

Figure 10.8. Diagram showing (a) the minimum-strain energy chair form of the cyclohexane ring and (b) how this leads to the minimum-energy bite sizes for chelate rings. This leads in turn to (c) the preference of six-membered chelate rings for coordination of small metal ions and of five-membered chelate rings for coordination of large metal ions.

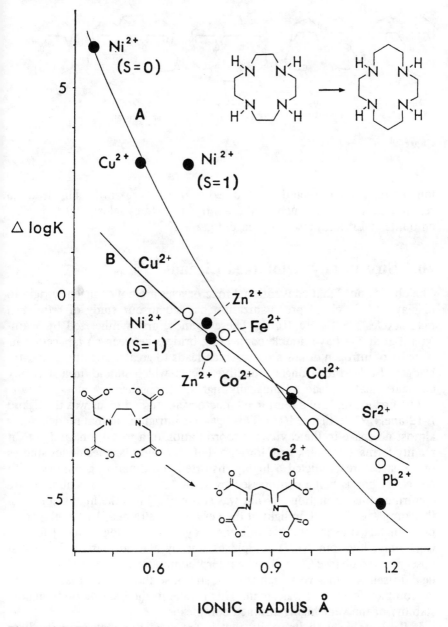

Figure 10.9. The change in formation constant, $\Delta \log K$, upon increase of chelate ring size from five membered to six membered in passing from EDTA to TMDTA and from 12-aneN$_4$ to cyclam complexes as a function of metal ion radius. The steeper slope for the 12-aneN$_4$/cyclam relationship as compared to the EDTA/ TMDTA relationship arises because in the 12-aneN$_4$/cyclam relationship two five-membered chelate rings are turned into six-membered rings, whereas in the EDTA/ TMDTA relationship only one five-membered chelate ring is turned into a six-membered ring. Data from reference 4.

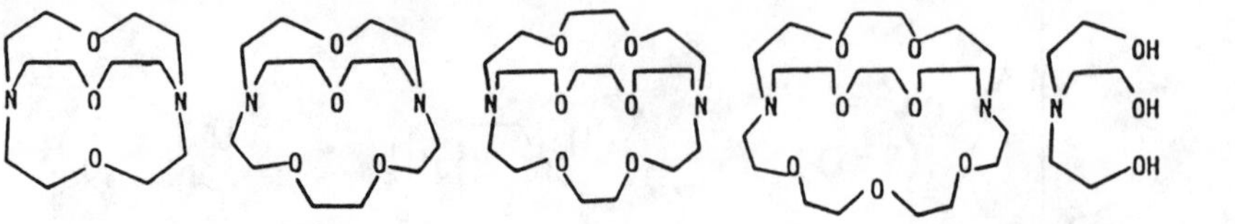

cryptand-111	cryptand-211	cryptand-222	cryptand-322	triethanolamine
$pK_1 \cong 17.8$	$pK_1 = 10.6$	$pK_1 = 9.60$	$pK_1 = 8.50$	$pK_1 = 7.8$

Chart 2

lone pairs on the oxygen donors forced into close proximity. This leads to remarkable properties such as the ability[1] to extract selectively the small amounts of Li$^+$ ion present in reagent grade KOH.

10.1 Structurally Reinforced Ligands

The observation[9] that conventional macrocycles are not sufficiently rigid to display high levels of preorganization has led to our study of *reinforced* macrocycles[20-24] (Figure 10.10). These ligands, first synthesized by Wainwright et al.,[25,26] have double connecting bridges between an adjacent pair or pairs of nitrogen donor atoms, which leads to greater rigidity. The extra bridges also lead to stronger inductive effects, which should, in a sterically efficient situation, lead to unusually high LF strengths.

The first example of a reinforced macrocycle studied by us was the ligand B-12-aneN$_4$[20,21,26] (Figure 10.6). This ligand is intriguing in that models show almost no space in the cavity for coordination of a metal ion, and yet[26] it readily forms a complex with low-spin Ni(II). Molecular mechanics studies on its structure predicted a highly distorted coordination geometry, with the Ni–N bonds being compressed by some 0.05 Å. The crystal structure confirmed this prediction.[20] What was surprising[20] was the high stability of the complexes formed, in light of the very highly strained B-12-aneN$_4$ complex. In fact, the thermodynamic macrocyclic effect [log K(mac)] for the Cu(II) and low-spin Ni(II) complexes was by a wide margin the largest observed to date (see Chart 3). The interpretation for this[20] was that the free ligand itself was in a very high state of strain, so that the increase in strain on complex formation was small. This results in the high thermodynamic stability of the very highly strained complexes.

If B-12-aneN$_4$, with its cavity that is too small for metal ions such as Cu(II) and Ni(II), can form complexes of high thermodynamic stability, then reinforced macrocycles with cavities that fit metal ions should be of considerable interest. In particular, the piperazine type of bridge does not (Figure 10.11) coordinate well to smaller metal ions. The homopiperazine group should coordinate distinctly better, and we have therefore synthesized[23] ligands such as NE-2,2-HP and AM-2,2-HP (Figure 10.11),

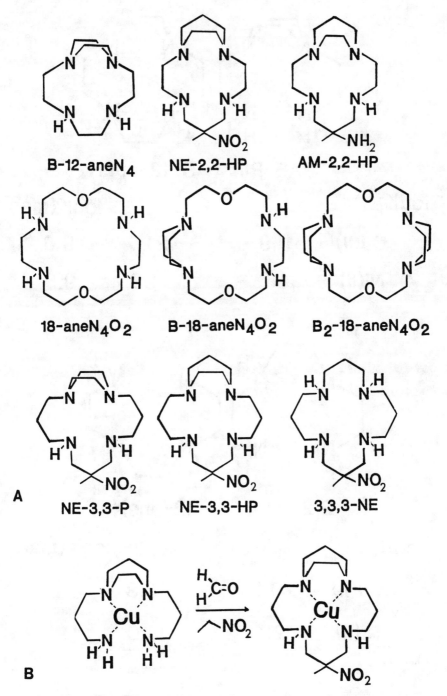

Figure 10.10. (a) Reinforced macrocycles with some reinforced analogues. (b) The template synthesis of Lawrance et al.[27] used to synthesize some of these ligands.

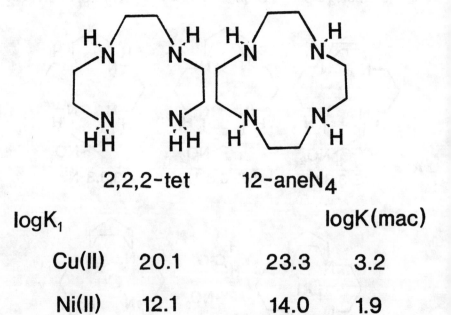

BAE-PIP   B-12-aneN$_4$

logK$_1$			logK(mac)
Cu(II)	11.9	21.5	9.6
Ni(II)	4.7	14.3	9.6

2,2,2-tet   12-aneN$_4$

logK$_1$			logK(mac)
Cu(II)	20.1	23.3	3.2
Ni(II)	12.1	14.0	1.9

Chart 3

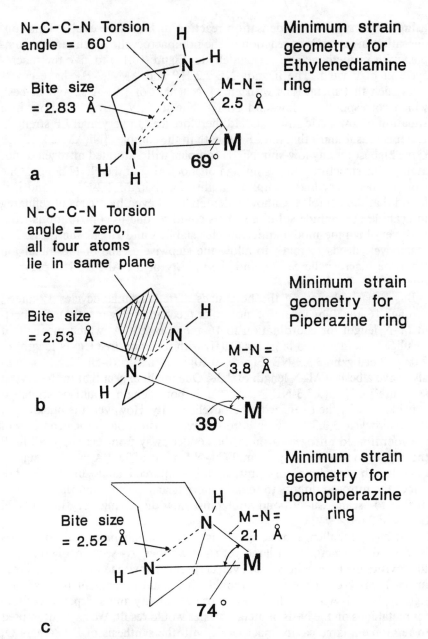

Figure 10.11. Minimum-strain geometries for chelate rings involving (a) ethylenediamine, (b) piperazine, and (c) homopiperazine.

using the convenient condensation reaction of Lawrance et al.[27] This condensation involves[27] the reaction of the nonmacrocyclic parent amine complex with nitroethane and formaldehyde (Figure 10.11) to give the macrocycle in high yield. The ligands NE-2,2-HP and AM-2,2-HP have cavity sizes such that metal ions with an M–N length of 1.91 Å should fit best, which corresponds to low-spin Ni(II). The tertiary nitrogen donors in a situation of low steric strain should therefore lead to very high LF strength, and it is thus found[23] that the LF strength in the complex [Ni(AM-2,2-HP)]²⁺ is the highest for any low-spin Ni(II) complex with saturated nitrogen donor atoms. The structure[23] shows no sign of shortening of the Ni–N bonds. One would expect very high complex stability for [Ni(AM-2,2-HP)]²⁺, and it is found that the complex cannot be demetallated even by means of refluxing in cyanide or sulphide solutions. This could be due to contributions from both very high thermodynamic complex stability and kinetic inertness. The macrocycle needs to fold[28] to allow the stepwise escape of the metal ion from the macrocyclic cavity, and the rigidity of the AM-2,2-HP prevents this.

Busch has calculated[29] the hole sizes of tetraazamacrocycles 12-aneN_4 through 16-aneN_4 by means of molecular mechanics (Figure 10.6). The best-fit M–N length for coordination to 16-aneN_4 is 2.38 Å, which means that this ligand is much too large for Cu(II) with its strain-free Cu–N length[9] of 2.03 Å. The ligand 3,3,3-NE (Figure 10.10) is similar to 16-aneN_4 and should also have a best-fit M–N length of 2.38. One might expect that in the crystal structure[27] of [Cu(3,3,3-NE)]²⁺, the Cu–N bond length would be stretched considerably by the oversized macrocyclic cavity. However, the buckling of the macrocycle 3,3,3-NE, with accompanying distortion of the geometry of the coordinated nitrogens around the copper away from planarity, allows[27] the copper to retain nearly normal Cu–N lengths of 2.02 Å, with no sign of bond-length stretching. In contrast, the reinforced analogue NE-3,3-HP forces the nitrogen donors to remain more nearly planar, and the copper is unable to escape Cu–N bond stretching, now displaying very long Cu–N bonds of 2.09 Å.[24]

The biggest challenge in the synthesis of more highly preorganized ligands is to produce ligands with large yet rigid cavities. As seen in Figure 10.5, the cavities in crown ethers (and even cryptands) have a tendency to collapse in on themselves in the free ligands. If this could be prevented by greater rigidity, ligands with much higher complex stability and sharper selectivity for metal ions on the basis of metal ion size would result. We have attempted to reinforce a large cavity macrocycle with the synthesis of B-18-aneN_2O_4 and B_2-18-aneN_2O_4 (Figure 10.10). The parent ligand 18-aneN_2O_4 forms complexes of considerable stability with metal ions such as Cu(II), which are much too small for the cavity in the ligand, because the ligand can buckle to coordinate to the undersized metal ion. As seen in Chart 4, the introduction of a single reinforcing bridge to 18-aneN_2O_4 to give B-18-aneN_2O_4 leads to a significant increase in selectivity for the large Pb(II) ion

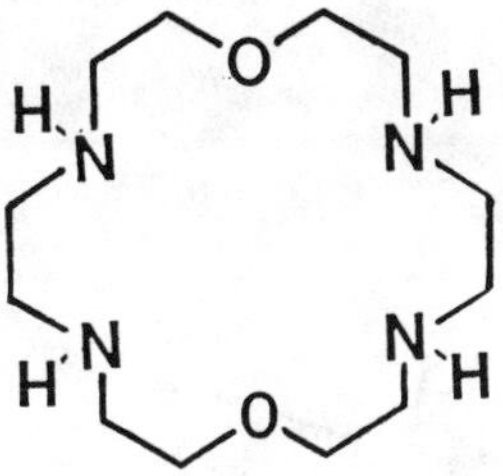

18-aneN$_4$O$_2$ 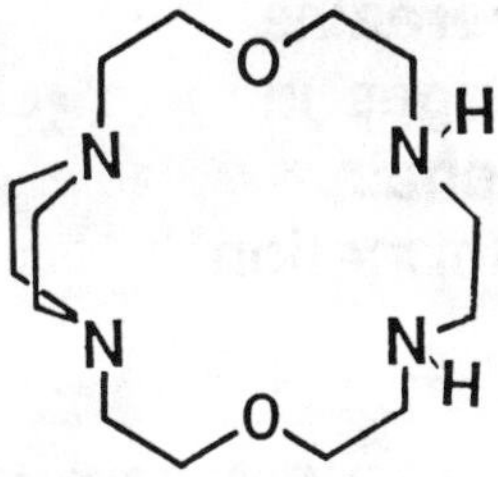   B-18-aneN$_4$O$_2$

ΔlogK

logK$_1$ Cu^{2+}	16.27		7.04	9.23
logK$_1$ Pb^{2+}	9.01		5.36	3.65

Chart 4

over the small Cu(II) ion. However, complex stability for both metal ions is lowered considerably. This relates to the fact that the piperazine groups in the free ligand can assume the lower-energy chair conformation, rather than the boat conformation required for coordination to a metal ion. This is illustrated for the free ligand B$_2$-18-aneN$_2$O$_4$ in Figure 10.12.

10.2 Macrocycles with Pendant Donor Groups

It has been pointed out that the macrocycle in Figure 10.1, which has pendant donor groups, is more highly preorganized than a simple macrocycle. The attraction of pendant hydroxyalkyl groups in particular is the ease with which they can be attached to amines. The effect they have on complex stability and selectivity is reminiscent of the coordinating properties of crown ethers[30-39] in that the complexes of large metal ions are stabilized by addition of hydroxyethyl groups to existing ligands, while those of small metal ions are destabilized. This is seen for the selectivity of the large Pb(II) ion relative to the small Cu(II) ion in Figure 10.1 where progressively more hydroxyethyl groups are added.

10.3 Future Trends

In this chapter the trend towards more highly preorganized ligands has been outlined. We now consider where this may yet lead and what applications might flow from this trend.

**piperazine
bridges in
chair
conformation**

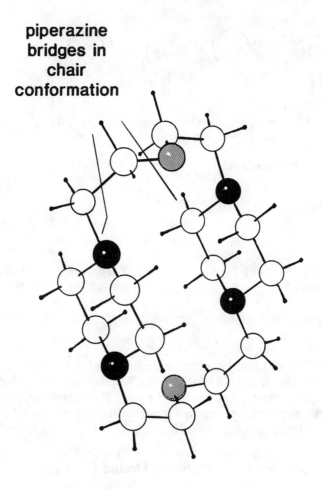

Figure 10.12. The structure of the free ligand B_2-18-aneN_4O_2, showing how the piperazine bridges are in the chair conformation rather than the boat conformation required for coordination to a metal ion. Redrawn after reference 22.

It is probably not possible at this stage to predict in detail what the more interesting properties of more highly preorganized macrocycles will be. Ligands such as bridged macrocycles give a hint as to what remarkable chemistry still may emerge. It has been shown that the stabilization of metal ions of unusually high oxidation states is promoted[40] by ligands of higher LF strength, and the high LF strength of reinforced macrocycles may well lead to the stabilization of metal ions in previously unknown oxidation states. Another possibility here is that the severe distortion of the coordination sphere of metal ions by more rigid ligands may lead via the *entatic* effect to useful catalytic properties, which may also involve the existence of unusual oxidation states.

In order to produce more highly preorganized ligands, especially ones with large cavities, more rigid building blocks will have to be used. Examples that spring to mind are structurally rigid units such as the cyclohexane, norbornane, or bispidine units, as well as homopiperazine and DACO units, in ligands such as those in Figure 10.13 (DACO = 1,4-diazacyclooctane). The ligand I, with its highly rigid bispidine units, would possibly achieve the very highest levels of preorganization. Ligand II would be highly preorganized, although the DACO group would not be as rigid as bispidine. Both bispidine and DACO produce tertiary nitrogens in an environment of low steric strain, and so ligand I and II should produce very high LF strengths. The norbornane unit in ligand III should hold the nitrogen donor atom in a very fixed orientation. The homopiperazine groups in ligand IV would be much better preorganized than would piperazine groups in B_2-18-aneN_4O_2 (see Figures 10.12 and 10.16). Extended aromaticity, as is already present in naturally occurring ligands such as porphyrins, could also be of considerable interest in producing enhanced rigidity. Depending on the intended application of the ligand, however, rigidity can also have a great drawback, which is that metallation and demetallation reactions can become very slow. This would be of great importance where metal ions must be rapidly sequestered, as in the use of ligands to complex toxic metal ions in the body or the environment. A solution to this problem is to combine more highly preorganized parts of the ligand with parts that are less preorganized. This has already been done with macrocycles with pendant coordinating groups.[29] As shown in Figure 10.1, the macrocycle with the chelating donor groups is clearly more highly preorganized than its unsubstituted macrocyclic analogue. The kinetic benefits of pendant donor groups have already been shown[30] in that tetraazamacrocycles with pendant hydroxyethyl groups have very rapid metallation reactions as compared with the unsubstituted analogues. As far as ligand design is concerned, two factors that we have identified as important are the greater levels of preorganization produced by more rigid structural units and the pendant donor groups, which enhance preorganization and increase lability with respect to metallation/demetallation reactions.

The highly preorganized ligands with slow metallation and demetallation reactions could be important in applications where metal exchange is undesirable, such as in the use of imaging agents in the body. However, a new area of applications could arise in the specific complexation of anions by

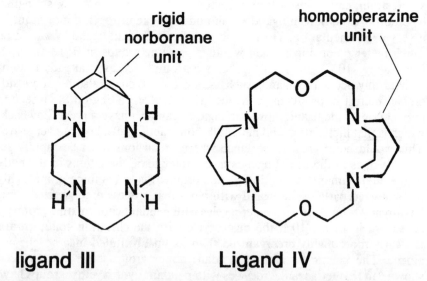

Figure 10.13. Some more highly preorganized ligands. The ligands I to IV are intended to show how rigid building blocks could be used to increase the level of preorganization of ligands. Ligand IV should be better preorganized than B_2-18-aneN$_4$O$_2$ in Figure 10.12, which in the free ligand has its bridging piperazine groups in the chair conformation as seen in Figure 10.16, which is unsuitable for coordination to metal ions.

metal ions held in rigid ligands attached to membranes or ion-exchange support materials. For example, complexes of copper(II) with macrocycles show strong and selective complexation[41] of anions such as cyanide on the axial coordination site. Without the resistance to demetallation conferred by high levels of preorganization, the metal ion would soon be lost. An intriguing aspect of the complexation of anions by metal ions held in macrocyclic ligands is seen in the remarkably strong complexation of anions by macrocyclic ligands coordinated to two metal ions.[42] In this situation,

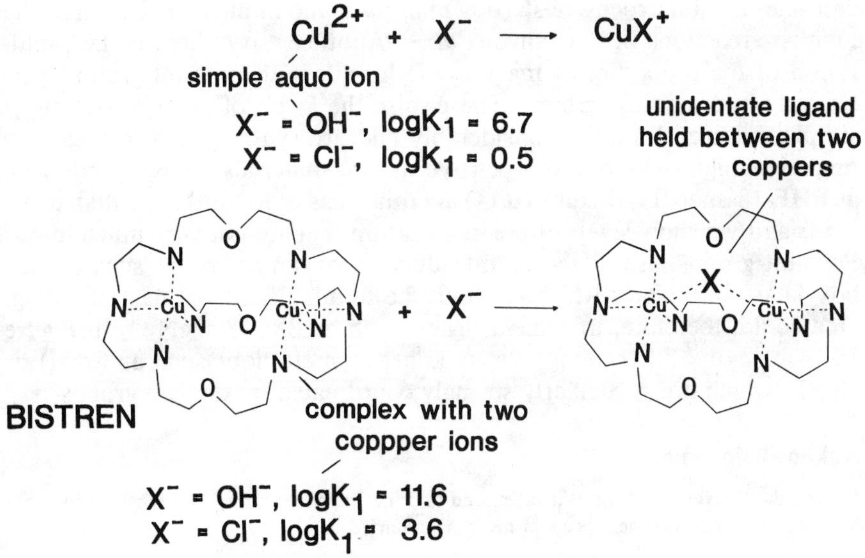

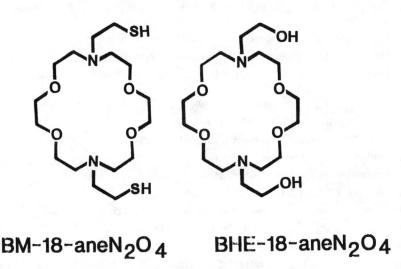

$$Cu^{2+} + X^- \longrightarrow CuX^+$$

simple aquo ion

$X^- = OH^-$, $\log K_1 = 6.7$

$X^- = Cl^-$, $\log K_1 = 0.5$

unidentate ligand
held between two
coppers

BISTREN

complex with two
coppper ions

$X^- = OH^-$, $\log K_1 = 11.6$

$X^- = Cl^-$, $\log K_1 = 3.6$

Chart 5

BM-18-aneN$_2$O$_4$ BHE-18-aneN$_2$O$_4$

Figure 10.14. Some macrocyclic ligands with pendent donor groups.

one has what is in effect a reverse chelate effect, with two metal ions holding a single ligand rather than two donor atoms coordinating to a single metal ion. The results on the strength of coordination are remarkable, as seen in the equilibria in Chart 5.

For macrocyclic ligands to achieve wide application, possibly the most important aspect to be addressed at this stage is bringing down their cost. To this end, procedures that look promising are syntheses that yield macrocycles from inexpensive starting materials, and in high yield, such as the template reactions of Lawrance et al. [27]. Another aspect here is the exploration of the properties of macrocyclic ligands with pendant groups, particularly hydroxyalkyl groups. These raise the levels of preorganization of simple macrocycles quite considerably and are synthetically very easy to prepare in high yield from inexpensive starting materials. Thus, ligands such as BHE-18-aneN_2O_4 (Figure 10.14) are much easier to synthesize than cryptands and yet show levels of preorganization that are not very much lower. Pendant groups also allow for introduction of donor groups such as carboxylates or thiolates which can only be terminal and not part of a ring. One could, therefore, imagine a ligand such as BM-18-aneN_2O_4 in Figure 10.14 being useful in removing toxic heavy metal ions such as Cd(II) or Hg(II), which are particularly strongly coordinated by thiolate groups.

Acknowledgments

I thank the University of the Witwatersrand and the Foundation for Research Development for generously funding the aspects of my work reported here.

References

1. Cram, D. J.; Kaneda, T.; Helgeson, R. C.; Brown, S. B.; Knobler, C. B.; Maverick, E.; Trueblood, K. N. *J. Am. Chem. Soc.* **1985,** *107,* 3645.

2. Schwarzenbach, G. *Helv. Chim. Acta* **1952,** *35,* 2344.

3. Hancock, R. D.; Martell, A. E. *Comments Inorg. Chem.* **1988,** *6,* 237.

4. Martell, A. E.; Smith, R. M. *Critical Stability Constants*; Plenum; New York, 1974–1988; vols. 1–6.

5. Pedersen, C. J. *J. Am. Chem. Soc.* **1967,** *89,* 2459.

6. Lehn, J-M. *Acc. Chem. Res.* **1978,** *11,* 49.

7. Hinz, D.; Margerum, D. W. *Inorg. Chem.* **1974,** *13,* 2941.

8. Burdett, J. K. *J. Chem. Soc., Dalton Trans.* **1976,** 1725.

9. Hancock, R. D. *Prog. Inorg. Chem.* **1989,** *37,* 187.

10. Hambley, T. W. *J. Chem. Soc., Dalton Trans.* **1986,** 565.

11. Evers, A.; Hancock, R. D. *Inorg. Chim. Acta* **1989,** *160,* 245.

12. Hancock, R. D.; Nakani, B. S. *S. Afr. J. Chem.* **1983,** *36,* 117.

13. Thom, V. J.; Hosken, G. D.; Hancock, R. D. *Inorg. Chem.* **1985,** *24,* 3378.

14. Thom, V. J.; Fox, C. C.; Boeyens, J. C. A.; Hancock, R. D. *J. Am. Chem. Soc.* **1984,** *106,* 5947.

15. Hancock, R. D.; Wade, P. W.; Ngwenya, M. P.; de Sousa, A. S.; Damu, K. V. *Inorg. Chem.* **1990,** *29,* 1968.

16. Hancock, R. D.; Martell, A. E. *Chem. Rev.* **1989,** *89,* 1875.

17. Hancock, R. D. *Acc. Chem. Res.* **1990,** *23,* 253.

18. Smith, P. B.; Dye, J. L.; Cheney, J.; Lehn, J. M. *J. Am. Chem. Soc.* **1981,** *103,* 6044.

19. Ciampolini, M.; Micheloni, M.; Orioli, P.; Vizza, F.; Mangani, S.; Zanobini, F. *Gazz. Chim. Ital.* **1986,** *116,* 189.

20. Hancock, R. D.; Dobson, S. M.; Evers, A.; Wade, P. W.; Ngwenya, M. P.; Boeyens, J. C. A.; Wainwright, K. P. *J. Am. Chem. Soc.* **1988,** *110,* 2788.

21. Hancock, R. D.; Evers, A.; Ngwenya, M. P.; Wade, P. W. *J. Chem. Soc., Chem. Commun.* **1987,** 1129.

22. Wade, P. W.; Hancock, R. D.; Boeyens, J. C. A.; Dobson, S. M. *J. Chem. Soc., Dalton Trans.* **1990,** 483.

23. Hancock, R. D.; Ngwenya, M. P.; Wade, P. W.; Boeyens, J. C. A.; Dobson, S. M. *Inorg. Chim. Acta* **1989,** *164,* 73.

24. Pattrick, G.; Hancock, R. D. *Inorg. Chem.* **1991,** *30,* 1419.

25. Wainwright, K. P. *Inorg. Chem.* **1980,** *19,* 1396.

26. Wainwright, K. P.; Ramasubbu, A. *J. Chem. Soc., Chem. Commun.* **1982,** 277.

27. (a) Comba, P.; Curtis, N. F.; Lawrance, G. A.; O'Leary, M. A.; Skelton, B. W.; White, A. H. *J. Chem. Soc., Dalton Trans.* **1988,** 2145.
    (b) Comba, P.; Curtis, N. F.; Lawrance, G. A.; Sargeson, A. M.; Skelton, B. W.; White, A. H. *Inorg. Chem.* **1986,** *25,* 4260.

28. (a) Lin, C. T.; Rorabacher, D. B.; Cley, G. R.; Margerum,D. W. *Inorg. Chem.* **1975,** *11,* 288.
    (b) Hertle, L.; Kaden, T. *Helv. Chim. Acta* **1981,** *64,* 33.

29. (a) Kaden, T. A. *Top. Curr. Chem.* **1984,** *121,* 157.
    (b) Gokel, G. W.; Goli, D. M.; Minganti, C.; Echegoyen, L. *J. Am. Chem. Soc.* **1983,** *105,* 6786.

30. Madeyski, C. M.; Michael, J. P.; Hancock, R. D. *Inorg. Chem.* **1984,** *23,* 1487.

31. Thom, V. J.; Hancock, R. D. *Inorg. Chim. Acta* **1985,** *96,* L43.

32. Hancock, R. D.; Bhavan, R.; Shaikjee, M. S.; Wade, P. W.; Hearn, A. *Inorg. Chim. Acta.* **1986,** *112,* L23.

33. Thom, V. J.; Shaikjee, M. S.; Hancock, R. D. *Inorg. Chem.* **1986,** *25,* 2992.

34. Damu, K.; Shaikjee, M. S.; Michael, J. P.; Howard, A. S.; Hancock, R. D. *Inorg. Chem.* **1986,** *25,* 3879.

35. Wade, P. W.; Hancock, R. D. *Inorg. Chim. Acta.* **1987,** *130,* 251.

36. Hancock, R. D.; Shaikjee, M. S.; Dobson, S. M.; Boeyens, J. C. A. *Inorg. Chim. Acta* **1989,** *154,* 229.

37. Hancock, R. D.; Bhavan, R.; Wade, P. W.; Boeyens, J. C. A.; Dobson, S. M. *Inorg. Chem.* **1989,** *28,* 187.

38. Bhavan, R.; Hancock, R. D.; Wade, P. W.; Boeyens, J. C. A.; Dobson, S. M. *Inorg. Chim. Acta* **1990,** *171,* 235.

39. Damu, K. V.; Hancock, R. D.; Wade, P. W.; Boeyens, J. C. A.; Billing, D. G.; Dobson, S. M. *J. Chem. Soc., Dalton Trans.* **1991,** 293.

40. Fabbrizzi, L. *Comments Inorg. Chem.* **1985,** *4,* 33.

41. Hancock, R. D.; Darling, E. A.; Hodgson, R. H.; Ganesh, K. *Inorg. Chim. Acta,* **1984,** *90,* L83.

42. (a) Motekaitis, R. J.; Martell, A. E.; Dietrich, B.; Lehn, J. M. *Inorg. Chem.* **1982,** *24,* 1588. (b) Evers, A.; Hancock, R. D.; Murase, I. *Inorg. Chem.* **1986,** *25,* 2160.

CHAPTER 11

Tailormade Large Molecular Cavities for the Selective Complexation of Organic and Inorganic Guests

Christian Seel and Fritz Vögtle

Institut für Organische Chemie und Biochemie
Universität Bonn
D-5300 Bonn 1
Federal Republic of Germany

11.1 Introduction

The principle of receptor/substrate binding is one of the foundations of biological chemistry. To imitate this phenomenon of molecular recognition by molecular complementarity and to reveal its driving forces is the object of intensive research in the field of supramolecular chemistry. The purpose is to study theoretically and experimentally the specific interactions between the matching host and guest molecules.

The properties, functions, and applications of molecular aggregates (supramolecules and superstructures) can be investigated by developing host components supplied with efficient binding functional groups and a fitting molecular framework that enables them to entangle guest molecules—inorganic or organic, ionic or neutral—from all sides. To encapsulate stereofunctionally complementary guests in the center of a large spherical cavity, macrooligocyclic phane host systems proved to be valuable,[1] especially when they can be built up according to a modular concept (building block concept), using easily available starting materials and simple structural subunits. By attaching appropriate intraannular donor groups and through variation of spacer units, host molecules containing large cavities can be tailormade, their cavity size and binding properties leading to an optimal fit with respect to the guests desired.

11.2 Cagelike Ligands Bearing Three Catechol Units

Many microorganisms produce substances that are capable of complexing iron, which is essential to life but forms hardly soluble hydroxides at physiological pH. A great variety of these siderophores (from the Greek: *iron carriers*) contain three catechol or related structural units. The prototype of the catechol siderophores is enterobactin (*1*)[2] (Fig. 11.1). The association constant for the sideroplex [Fe(III)·1]³⁻ was determined to be about 10^{52}.[3] This seems to be the highest value ever determined for any natural product. The strength of complexation of ferric iron by siderophores stimulated the search for new, efficient, synthetic iron-sequestering agents for medical applications. A variety of synthetic analogues of enterobactin (*1*) has recently been developed.[4] In all of these ligands, conformationally flexible sidearms bearing the catechol units are fixed to linear, branched, or monocyclic anchor groups. In MECAM (*2*) (Fig. 11.1) for example, the cyclic triester of *1*, being sensitive to hydrolysis, is replaced by a benzene ring.[5]

With the intention of obtaining a chelating agent supplied with an optimal preorganized donor geometry, we synthesized the macrobicyclic "siderand"[6] *3* in 1984.[7] In this ligand, new in principle, the three donor groups are integrated in a large cagelike molecular framework, so that the conformational flexibility is strongly restricted. In the hexaanion of *3*, an octahedral geometry of the donor oxygen atoms is preorganized, which leads to extremely strong metal binding properties (macrobicyclic effect). In the course of the synthesis of *3*, a tricarboxylic acid chloride is reacted with a triamine in the decisive cyclization step under high dilution conditions (path A in Figure 11.2). Because of the clearly higher yields, this strategy is more favorable than the "one-pot reaction" using two equivalents of triamine and three equivalents of diacid chloride (path B in Figure 11.2).

The ligand *3* exhibits the highest association constant ever determined for any iron(III) complex: $K_{ass} \approx 10^{59}$ M⁻¹. The complex is stable even at

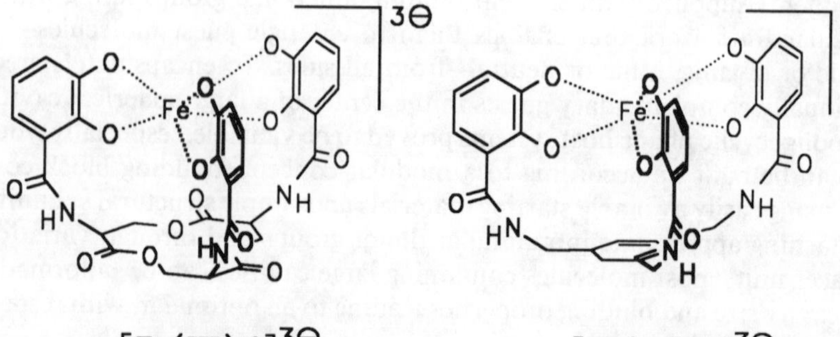

$$[\text{Fe(III)}\cdot\text{1}]^{3\ominus} \qquad\qquad [\text{Fe(III)}\cdot\text{2}]^{3\ominus}$$

Figure 11.1. The iron(III) complexes of enterobactin (*1*) and its synthetic analogue MECAM (*2*).

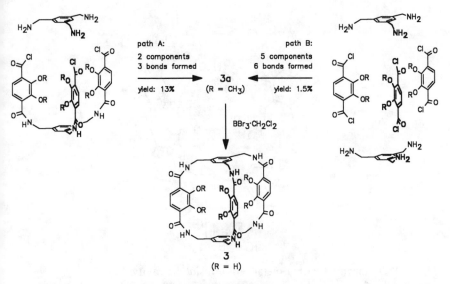

Figure 11.2. Two alternative synthetic strategies for the synthesis of the macrocyclic "siderand" *3*.

pH 3 and in competition experiments with open-chained reference com-pounds, *3* proves definitely to be superior.

The volume of the cavity formed by the molecular framework of *3* can be enlarged by introducing more extended spacer units (e.g., 1,3,5-triphen-ylbenzene and 1,1,1-triphenylethane) instead of the central benzene rings.[8] In ligand *4*, the six methoxy groups can be arranged in a convergent and cooperative manner inside the cavity if a metal cation is complexed in the center.[9] (Fig. 11.3). The stability of the alkali metal complexes formed by *4* are in the same order of magnitude as those of crown ether complexes. A certain selectivity for cesium is determined ($pK_{ass} \approx -5.3 \pm 0.5$); Cs$^+$ is bound six times more strongly than potassium.

11.3 Tris(bipyridine) Ligands Forming Molecular Cavities of Different Size and Shape

By clamping three bipyridine units to macrocyclic cagelike molecules, a series of new ligands has been obtained. Due to their high degree of preor-ganization,[10] they exhibit unique properties with regard to selective molec-ular recognition, complex stability, and photophysical processes.

11.3.1 Synthetic Strategies

In the course of the synthesis of triply bridged tris(bipyridine) ligands, the molecular frameworks are built up by forming amine or amide bonds. Sup-

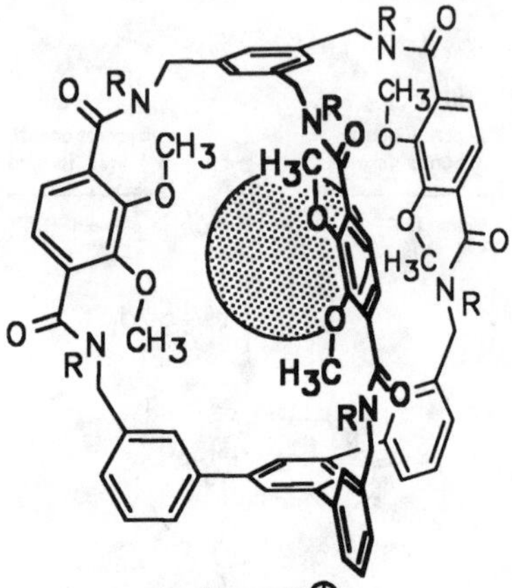

$$[M \cdot 4]^{\oplus}$$
R = Benzyl

Figure 11.3. Ligand *4* with a metal cation inside its cavity.

ported by metal template assistance, the host molecules can be prepared via one-pot reactions. The sodium complex of *7* thus is formed in 4% yield as a solution of the triamine *5* in acetonitrile is dropped into a suspension of 5,5′-bis(bromomethyl)-2,2′-bipyridine (*6*) and Na_2CO_3 in acetonitrile (Fig. 11.4).[11] The bicyclic compound *7* is not formed using Cs_2CO_3 alternatively. The larger cesium cation obviously hinders the bicyclic clamping of the components. According to an analogous strategy, the cage complex of the hexalactam *9* can be obtained by reacting *5* with the ruthenium(II) complex of 5,5′-bis(chlorocarbonyl)-2,2′-bipyridine.[12]

An important criterion for the complexation of many transition metals is the orbital geometry of the cation. Fe(II) and Ru(II), for example, demand a strictly octahedral ligand field. Hosts with insufficient conformational flexibility such as the cryptand *8* (Fig. 11.5),[13] built up of 6,6′-disubstituted bipyridine units, prove to be unsuitable.[14] The larger cavity of *9*, however, allows an octahedral coordination. Its Fe(II) complex shows remarkable stabilities towards strong oxidants like H_2O_2.[15]

11.3.2 Photochemical Properties

Ruthenium(II) polypyridine complexes exhibit chemoluminescence and are of some interest as sensitizers in photochemical processes. One problem in complexes like $[Ru(bipy)_3]^{2+}$ is the relatively quick radiation-free decay of its lowest excited state. This ³MLCT state (MLCT = *m*etal-to-*l*igand *c*harge *t*ransfer) determines the luminescence properties and the energy and electron transfer processes of these systems. Another disadvantage is the photodissociation, in the course of which a distorted ³MC state (MC = *m*etal-*c*en-

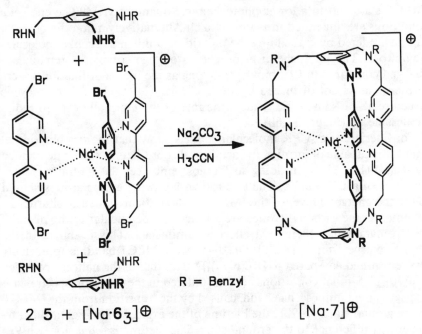

2 5 + [Na·6₃]⊕   [Na·7]⊕

Figure 11.4. Template-assisted synthesis of the sodium complex of the cagelike ligand 7.

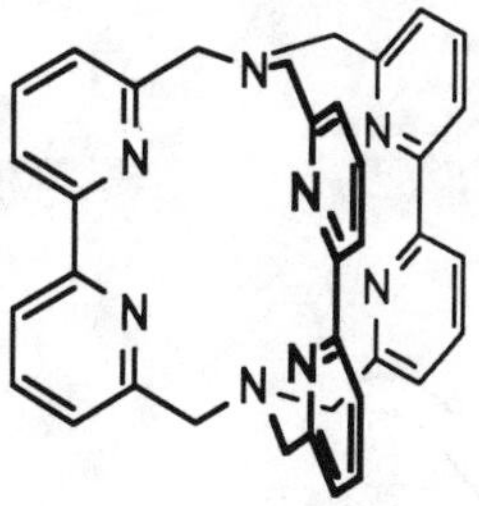

8

Figure 11.5. The tris(bipyridine) cryptand 8.

tered) is formed as an intermediate by (thermally activated) radiationless transition, accompanied by the cleavage of one of the six metal–nitrogen bonds. One possibility of increasing the photostability of such complexes is the enlargement of the energy gap between the ^{3}MLCT and the ^{3}MC state. As the energy level of the latter cannot be increased, the ^{3}MLCT energy has to be lowered by appropriate derivatization of the bipyridine units. Yet this leads to enforced (temperature-independent) radiation-free transition to the ground state and consequently to a dramatically shortened lifetime of the

³MLCT state and to a lower photochemical quantum yield. The energetic conditions are illustrated in Figure 11.6a.[16] Alternatively, photodissociation can be hindered by embedding the bipyridine units in a cagelike molecular framework. The photochemical properties of cage complexes remain unchanged compared to [Ru(bipy)₃]²⁺, as long as the first coordination sphere of the metal is not disturbed by steric effects. The rigidity of the bicyclic structure at the same time reduces the rate of the thermally excited radiationless decay (Figure 11.6b).

Considerations of CPK molecular models show that the tris(bipyridine) ligand 9 is capable of forming the octahedral coordination sphere demanded by Ru(II) (Fig. 11.7). In doing so the host embraces the central cation in the form of a triple helix. The electron-withdrawing amide groups should lower the energetic level of the ³MLCT state and consequently shorten its lifetime. As the carbonyl groups are arranged perpendicular to the plane of the aromatic rings, their −M effect is diminished. This is shown in the absorption spectrum of [Ru·9]²⁺; its d→π* band (MLCT band) is appreciably less red-shifted compared to [Ru(bipy)₃]²⁺ than that of the halfcage complex [Ru·10]²⁺.[17] Similar conditions are encountered in the emission spectra with respect to the luminescence band, caused by the transition from the ³MLCT to the ground state. At 90K the lifetime of the excited state is controlled by the energy difference to the ground state. The lifetime is 4.8 μs for [Ru·9]²⁺ as well as for [Ru(bipy)₃]²⁺ (see Table 11.1). At room temperature, in contrast, the emission state of the cage complex is more than twice as long lived

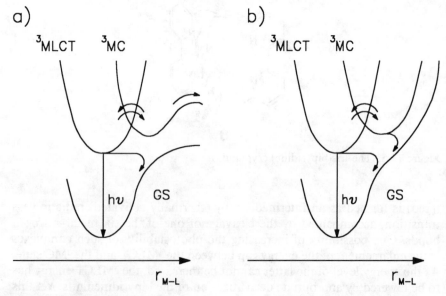

Figure 11.6. Schematic representation of the potential energy curves of (a) [Ru(bipy)₃]²⁺ and (b) [Ru·9]²⁺.

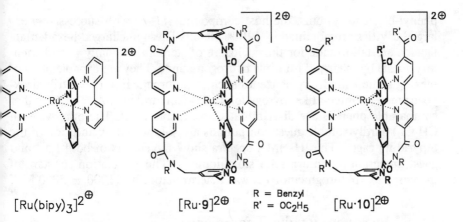

$[Ru(bipy)_3]^{2\oplus}$ $[Ru\cdot9]^{2\oplus}$

R = Benzyl
R' = OC_2H_5 $[Ru\cdot10]^{2\oplus}$

Figure 11.7. The Ru(II) complexes of the ligands 2,2'-bipyridine, the cage 9, and the halfcage 10.

Table 11.1. Some photophysical data of the different tris(bipyridine) complexes.

Complex	Absorption	Emission					Quantum Yield
	298 K	298 K		90 K			298 K
	λ_{max} [nm] (ϵ)	λ_{max} [nm]	τ [μs]	λ_{max} [nm]	τ [μs]		ϕ_p
$[Ru(bipy)_3]^{2+}$	452 (13000)	615	0.8	582	4.8		$1.7\cdot10^{-2}$
$[Ru\cdot9]^{2+}$	455 (10400)	612	1.7	597	4.8		$< 10^{-6}$
$[Ru\cdot10]^{2+}$	477 (9500)	640	0.45	620	1.9		$< 10^{-5}$

compared to the bipyridine complex (1.7 and 0.8 μs, respectively). Here the lifetime is determined by the thermally activated transition to the distorted ^{3}MC state. Because the metal cation cannot easily escape the cage, this transition is not followed by dissociation of the complex, and so the quantum yield of the photodissociation is at least four orders of magnitude smaller than in $[Ru(bipy)_3]^{2+}$. Caused by the shielding of the relevant orbitals of the metal, the three-dimensional encapsulation of the central cation effects an oxygen quenching of the ^{3}MLCT state that is five times slower compared to $[Ru(bipy)_3]^{2+}$.

Summing up, it can be stated that the photochemical properties of the tris(bipyridine) complexes can be improved significantly if the donor centers are integrated into suitable macrobicyclic molecular frameworks. While the absorption and emission characteristics remain comparable, an obviously enlarged photostability and a remarkably higher lifetime of the luminescence state results.

11.3.3 Complexation of Organic Guests

By replacing the bridging 1,3,5-trisubstituted benzene spacers in the bicyclic host 9 by larger building blocks like 3,3',3''- or 4,4',4''-trisubstituted triphenyl-benzene systems, the host compounds *11–13* with successively en-larged cavities are obtained in 7–14% yield.[18] These uncharged hexadentate ligands are tailormade for the enclosure of stereofunctionally complemen-tary guests by means of multiple hydrogen bonds. They are capable of com-plexing some trihydroxy benzenes such as phloroglucinol (1,3,5-trihydrox-ybenzene). Isomers like pyrogallol (1,2,3-trihydroxybenzene) and other comparable phenols are discriminated (see Table 11.2). The otherwise (in CH_2Cl_2) hardly soluble guest compounds are solubilized by addition of the bipyridine cages. The ^{1}H-NMR spectra show the signals of both host and guest; integration conforms 1:1 stoichiometry. The association constant of the complex [phloroglucinol·*12*] was determined to be 11000 ± 2000 M^{-1}.

11.4 Macropentacyclic Triscrowns

The discovery of the capability of crown ethers to complex metal cations marked the beginning of the development of supramolecular chemistry. Even organic species such as some ammonium cations and neutral guests like acetonitrile are bound by crowns and cryptands. Incorporation of three crown ether building blocks into oligocyclic structures leads to host systems that are able to adjust their endo- and exopolarophilic properties to the character of the enclosed guest and the polarity of the solvent.[19] In water, nonpolar neutral guests should be encapsulated in the cavity, being hydro-phobic in this case whereas in organic solution, topologically complemen-tary polar or protic guests should be enclosed by tritopic receptor molecules

Table 11.2. Molecular recognition of trihydroxy benzenes by the hosts *11, 12,* and *13* in organic layer (CH_2Cl_2). (+): complexation; (−): no appreciable complexation.

Guests				
11	+	+	−	−
12	+	+	−	−
13	−	+	−	−

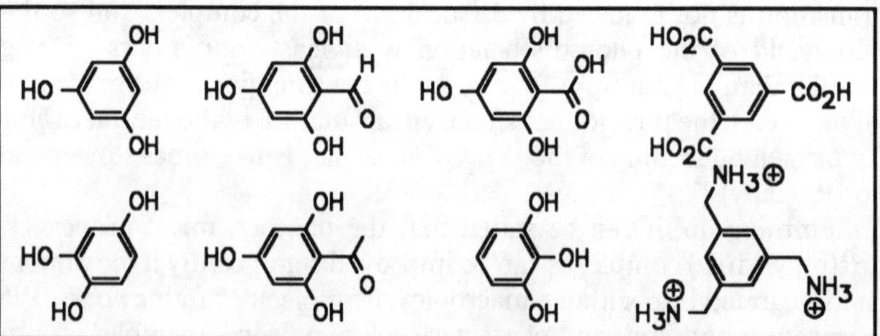

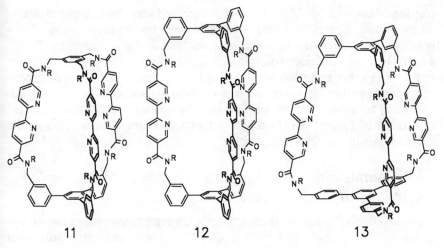

Figure 11.8. Large cagelike tris(bipyridine) ligands for the selective complexation of trihydroxy benzenes.

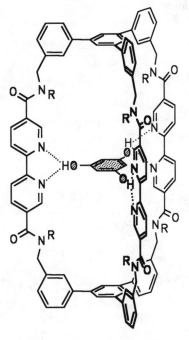

[phloroglucinol·12]

Figure 11.9. Assumed structure of the complex of host *12* and phloroglucinol.

like the triscrowns *14–17*.[20] These were obtained according to our modular donor–spacer concept (building block principle) for the preparation of macrobicyclic hexalactams, followed by reduction to the hexaamines (Figure 11.10). In acidic aqueous solution, *16* and *17* form complexes with β-naphthol and some naphthalenediols as guests. Phenol, di-, and trihydroxybenzenes as well as 1,3,5-benzene tris(methylammonium) salts are not complexed. The molecular recognition of the guests was determined (as illustrated in Figure 11.11) on account of significant upfield shifts of the guest signals in the ^{1}H-NMR spectra.

11.5 Complexation of Lipophilic Guests in Water by Ligands with Hydrophobic Cavity

Complexation of nonpolar organic guests in water is supported by the hydrophobic effect as driving force and is thus sometimes easier to obtain than complexation in organic phase. Apart from hydrophilic groups (amino

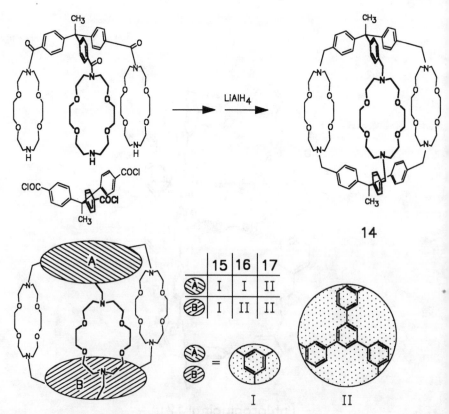

Figure 11.10. Synthesis of the macropentacyclic triscrown *14* and schematic representation of the hosts *15–17*.

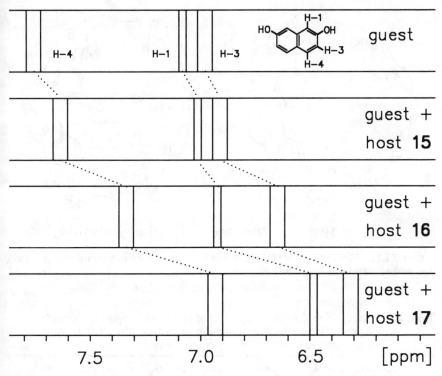

Figure 11.11. Chemical shifts of the signals of the 2,7-naphthalenediol protons in the presence of the triscrowns *15–17.*

or sulfonate groups) effecting the water solubility, appropriate host compounds require hydrophobic units to embrace and bind the guest molecules by dispersive interactions. A classical building block is Koga's diphenyl methane system.[21] By introducing three of these units in spheric macrobicyclic structures, some very effective hosts have been prepared.[22] The hexaamine *18,* which is fairly soluble in acidic aqueous solution, was obtained in a one-pot reaction under dilution conditions starting with trimesic acid chloride and the matching diamine, followed by reduction of the resulting hexalactame.[23] This host, with its large hydrophobic cavity, shows remarkable capabilities for the complexation of a series of condensed aromatic hydrocarbons with complementary size and shape, such as triphenylene, perylene, or pyrene, whereas partly hydrogenated derivatives of these systems are not accepted as guests (Fig. 11.12). Likewise, uncomplexed molecules, such as benzene, naphthalene, or coronene, remain too small or too large to fit inside the cavity. The discrimination of anthracene against phenanthrene was observed for the first time. A summary of the complexed and the discriminated compounds is given in Table 11.3. Localization of the guests inside the cavity is evident, proved by the highfield shifts of the guest

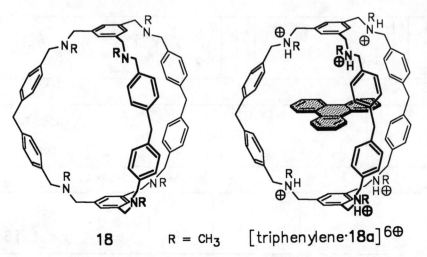

18   R = CH₃   [triphenylene·18a]⁶⊕

Figure 11.12. The spheric host *18* and the complex of its hexaammonium cation *18a* and triphenylene as guest.

Table 11.3. Selective molecular recognition of condensed arenes as guests by the ligand *18a* and the discrimination of partly hydrogenated derivatives and other noncomplementary compounds.

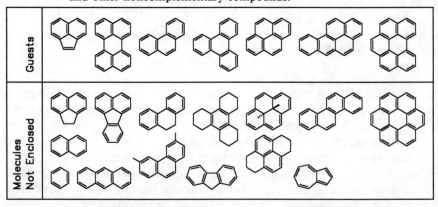

signals in the ¹H-NMR spectra and the sharpenings and shifts of the host signals. For pyrene as a guest, a binding constant of 10²·⁵ M⁻¹ was determined. Phase transfer experiments (liquid–liquid and liquid–solid), for example, lead to the separation of pyrene from decahydropyrene and triphenylene from dodecahydrotriphenylene. Applying the U-shaped-tube method, phenanthrene was extracted from its mixture with naphthalene in *n*-hexane.

The bicyclic host compounds *21* and *22* were synthesized by reaction of the tribromide *19* and the tritosyl amide *20* and subsequent detosylation²⁴ (Figure 11.13). The cyclization step yields both compounds in a 1:1 ratio:

79% in total (!). *21* and *22* are in/out isomers that cannot be interconverted by conformational alterations (configurational isomers!). The resulting difference in cavity size significantly influences the complexation behavior; although naphthalene as well as 2,6- and 2,7-dihydroxynaphthalene are complexed by both hosts in aqueous solution, only the larger out/out-isomer *21* is capable of enclosing adamantane and 1-adamantanethiol (Fig. 11.14) (see Table 11.4). This seems to be the very first complexation of a pure aliphatic hydrocarbon by any synthetic concave receptor.

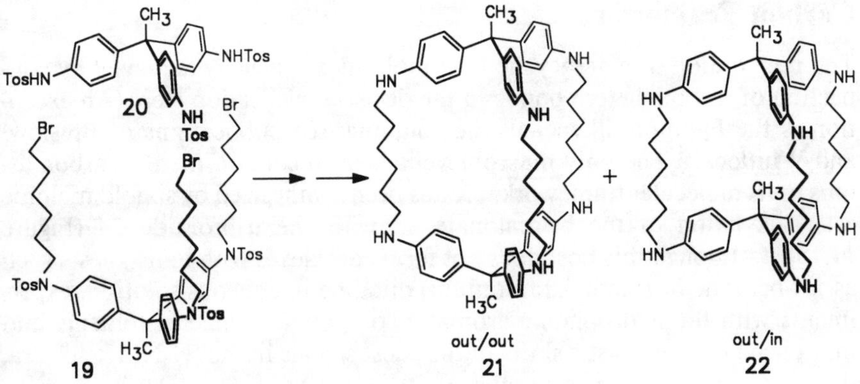

Figure 11.13. Synthesis of the in/out isomers *21* and *22*.

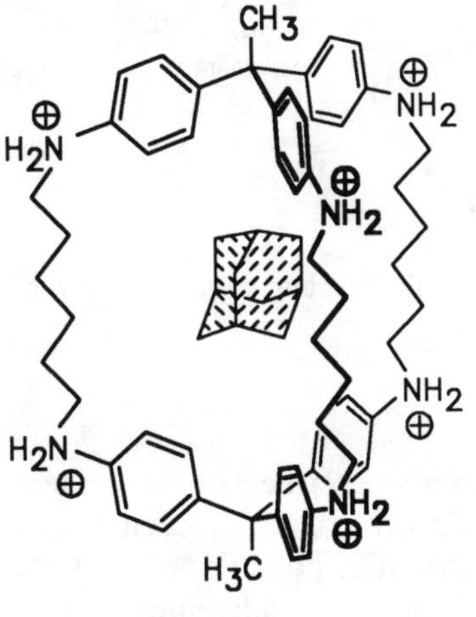

$$[\text{adamantane} \cdot 21a]^{6\oplus}$$

Figure 11.14. Assumed structure of the complex of host *21a* and adamantane.

Table 11.4. Complexation selectivities of the hosts *21* and *22* in acidic aqueous solution. (+): complexation; (−): no appreciable complexation.

Guests	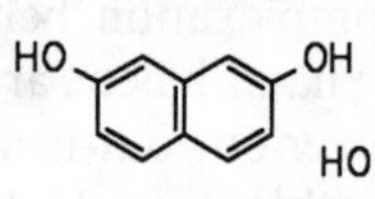 HO⟨naphthalene⟩OH	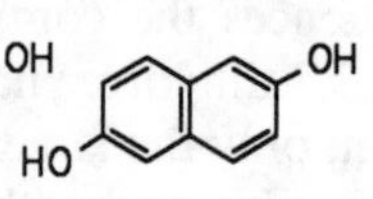 ⟨naphthalene⟩OH with HO	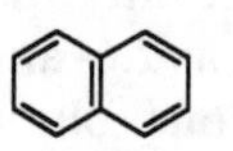 ⟨naphthalene⟩	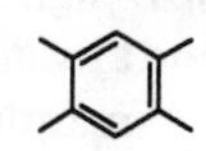 ⟨dimethylbenzene⟩	 ⟨adamantane⟩	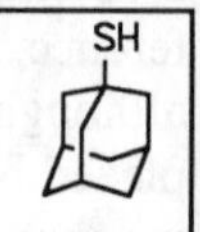 ⟨adamantane-SH⟩
21	+	+	+	−	+	+
22	+	+	+	−	−	−

11.6 A Macrobicyclic Dodecacarboxylic Acid with an All-Carbon Framework

The preparation of macrocyclic host molecules is usually achieved by connection of carbon–hetero bonds in the decisive cyclization step. (An exception is the Eglinton alkyne–alkyne coupling; see publications of Breslow[1] and Whitlock.[1]) The only macrobicyclic host to contain merely carbon atoms in its molecular framework, *25*, has been synthesized by sixfold malonic ester alkylation of the bismalonate *23* with the tribromide *24* (Figure 11.15).[25] Although this host does not form complexes with ionic guests such as 1,4-benzene bis(methylammonium) diiodide in aqueous solution, experiments with the hydrophobic aromatic compounds benzene, toluene, and mesitylene lead to drastic shifts of the guest signals in the ^{1}H-NMR spectra, indicating an inclusion inside the cavity of the host.

Figure 11.15. Synthesis of the bicyclic dodecacarboxylic acid *25* with an all-carbon framework.

11.7 Conclusions

The macrobicyclic ligands described in this review, with their frameworks opening some of the largest molecular cavities synthesized hitherto, demonstrate in principle that it is possible to create tailormade receptor molecules for organic and inorganic substrates. The synthetic expense, which is often higher than for open-chained and monocyclic systems, is more than compensated in many cases by new selectivities, higher binding constants, and better photochemical properties.

The chemistry of synthetic host compounds has its origin in the development of chelating ligands like crown ethers and cryptands for the complexation of metal cations. The state of research allows the interest to be focused more and more on the planned and selective molecular recognition of organic species. The forward view is fixed on the separation of chiral substances by corresponding receptor systems, the investigation of transport phenomena (especially with medical applications like tumor imaging or drug transport), the photolysis of water, and the development of enzyme models and synthetic catalysts.

To reach this aim, it will be necessary to investigate more precisely the interactions of host and guest on a molecular level, specifically in regard to hydrogen bonding; π-stacking (face-to-face), edge-to-face, and dipole–dipole interactions; charge transfer phenomena (electron donor-acceptor interactions); and the influence of the solvent. It is to be especially hoped that the application of efficient computational methods will help to find the optimally matching host/guest and catalyst/substrate systems.

References

1. (a) Ebmeyer, F.; Vögtle, F. In *Bioorganic Chemistry Frontiers*; Dugas, H., Ed.; Springer Verlag; Berlin, p 143, Vol. 1.
   (b) Smithrud, D. B.; Diederich, F. *J. Am. Chem. Soc.* **1986**, *108*, 2273 and references therein.
   (c) Diederich, F. *Angew. Chem.* **1988**, *100*, 372; *Angew. Chem. Int. Ed. Engl.* **1988**, *27*, 362.
   (d) Benson, D. R.; Valentekovich, R.; Diederich, F. *Angew. Chem.* **1990**, *102*, 213; *Angew. Chem. Int. Ed. Engl.* **1990**, *29*, 119.
   (e) O'Krongly, D.; Denmeade, S. R.; Chiang, M. Y.; Breslow, R.; *J. Am. Chem. Soc.* **1985**, *107*, 5544.
   (f) Whitlock, B. J.; Whitlock, H. W. *J. Am. Chem. Soc.* **1990**, *112*, 3910 and references therein.
   (g) Friederichsen, B. P.; Whitlock, H. W. *J. Am. Chem. Soc.* **1989**, *111*, 9132.
   (h) Canceill, J.; Collet, A. *J. Chem. Soc., Chem. Commun.* **1988**, 583 and references therein.
   (i) Saigo, K.; Kihara, N.; Hashimoto, Y.; Lin, R.-J.; Fujimura, H.; Suzuki, Y.; Hasegawa, M. *J. Am. Chem. Soc.* **1990**, *112*, 1144.
   (j) Sanderson, P. E. J.; Kilburn, J. D.; Still, W. C. *J. Am. Chem. Soc.* **1989**, *111*, 8314.
   (k) Jazwinsky, J.; Lehn. J.-M.; Lilienbaum, D.; Ziessel, R.; Guilhem, J.; Pascard, C. *J. Chem. Soc., Chem. Commun.* **1987**, 1691 and references therein.
   (l) Collman, J. P.; Brauman, J. I.; Fitzgerald, J. P.; Hampton, P. D.; Naruta, Y.; Sparapany, J. W.; Ibers, J. A. *J. Am. Chem. Soc.* **1990**, *112*, 5356.
   (m) For a review of "capped" porphyrines see: Morgan, B.; Dolphin, D. *Struct. Bonding* **1987**, *64*, 115.

2. (a) Pollack, J. R.; Neilands, J. B. *Biophys. Biochem. Res. Comm.* **1970**, *38*, 989.
   (b) O'Brien, I. G.; Gibson, F. *Biochim. Biophys. Acta* **1970**, *215*, 393.

3. (a) Harris, W. R.; Carrano, C. J.; Cooper, S. R.; Sofen, S. R.; Avdeef, A.; McArdle, J. V.; Raymond, K. N. *J. Am. Chem. Soc.* **1979**, *101*, 6097.

4. Review: Raymond, K. N.; Müller, G.; Matzanke, B. F. *Top. Curr. Chem.* **1984**, *123*, 49.

5. (a) Harris, W. R.; Weitl, F. L.; Raymond, K. N. *J. Chem. Soc., Chem. Commun.* **1979**, 177.
   (b) Venuti, M. C.; Rastetter, W. H.; Erickson, T. J.; Neilands, J. B. *J. Med. Chem.* **1979**, *22*, 123.

6. In concord with the systematization of the names of ligands (ending: "-and") and complexes (ending: "-plex") by Cram[10], iron binding ligands have to be called "siderands" and their Fe-complexes "sideroplexes." The natural siderophores then are a subclass of the siderands.

7. (a) Kiggen, W.; Vögtle, F. *Angew. Chem.* **1984**, *96*, 712; *Angew. Chem. Int. Ed. Engl.* **1984**, *23*, 714. We are grateful to Dr. R. C. Hider (Kings College London), A. D. Hall, and P. D. Taylor for the determination of the association constants.
   (b) Kiggen, W.; Vögtle, F.; Franken, S.; Puff, H. *Tetrahedron* **1986**, *42*, 1859.
   (c) For a comparable ligand see: McMurry, T. J.; Hosseini, M. W.; Garret, T. M.; Hahn, F. E.; Reyes, Z. E.; Raymond, K. N. *J. Am. Chem. Soc.* **1987**, *109*, 7196.

8. Stutte, P.; Kiggen, W.; Vögtle, F. *Tetrahedron* **1987**, *43*, 2065.

9. Peter-Katalinic, J.; Ebmeyer, F.; Seel, Ch.; Vögtle, F. *Chem. Ber.* **1989**, *122*, 2391.

10. Cram, D. J. *Angew. Chem.* **1986**, *25*, 1041; *Angew. Chem. Int. Ed. Engl.* **1986**, *98*, 1039.

11. Ebmeyer, F.; Vögtle, F. *Chem. Ber.* **1989**, *122*, 1725.

12. Belser, P.; De Cola, L.; v.Zelewsky, A. *J. Chem. Soc., Chem. Commun.* **1988**, 1057.

13. (a) Rodrigues-Ubis, J. C.; Alpha, B.; Plancherel, D.; Lehn, J.-M. *Helv. Chim. Acta.* **1984**, *67*, 2264.
    (b) Alpha, B.; Anklam, E.; Deschenaux, R.; Lehn, J.-M.; Pietraskiewicz, M. *Helv. Chim. Acta.* **1988**, *71*, 1042.

14. Dürr, H.; Zengerle, K.; Trierweiler, H.-P. *Z. Naturforsch.* **1988**, *43b*, 361.

15. Grammenudi, S.; Vögtle, F. *Angew. Chem.* **1986**, *98*, 1119; *Angew. Chem. Int. Ed. Engl.* **1986**, *25*, 1122.

16. Juris, A.; Balzani, V.; Barigelletti, F.; Campagna, S.; Belser, P.; v.Zelewsky, A. *Coord. Chem. Rev.* **1988**, *84*, 85.

17. (a) De Cola, L.; Barigelletti, F.; Balzani, V.; Belser, P.; v.Zelewsky, A.; Vögtle, F.; Ebmeyer, F.; Grammenudi, S. *J. Am. Chem. Soc.* **1988**, *110*, 7210.
    (b) De Cola, L.; Belser, P.; Ebmeyer, F.; Barigelletti, F.; Vögtle, F.; v.Zelewsky, A.; Balzani, V. *Inorg. Chem.,* **1990**, *29*, 495.

18. Ebmeyer, F.; Vögtle, F. *Angew. Chem.* **1989**, *101*, 95; *Angew. Chem. Int. Ed. Engl.* **1989**, *28*, 79.

19. Wester, N.; Vögtle, F. *Chem. Ber.* **1980**, *113*, 1487.

20. Wallon, A.; Werner, U.; Müller, W. M.; Nieger, M.; Vögtle, F. *Chem. Ber.* **1990**, *123*, 859.

21. Odashima, K.; Ita, A.; Iitaka, Y.; Koga, K. *J. Org. Chem.,* **1985**, *50*, 4478.

22. Wallon, A.; Peter-Katalinic, J.; Werner, U.; Müller, W. M.; Vögtle, F. *Chem. Ber.* **1990**, *123*, 375.

23. Vögtle, F.; Müller, W. M.; Werner, U.; Losensky, H.-W. *Angew. Chem.* **1987**, *99*, 930; *Angew. Chem. Int. Ed. Engl.* **1987**, *26*, 901.

24. (a) Franke, J.; Vögtle, F. *Angew. Chem.* **1985**, *97*; *Angew. Chem. Int. Ed. Engl.* **1985**, *24*, 219.
    (b) Schrage, H.; Franke, M.; Vögtle, F.; Steckhan, E. *Angew. Chem.* **1986**, *98*, 335; *Angew. Chem. Int. Ed. Engl.* **1986**, *25*, 336.

25. Merz, T.; Wirtz, H.; Vögtle, F. *Angew. Chem.* **1986**, *98*, 549; *Angew. Chem. Int. Ed. Engl.* **1986**, *25*, 567.

CHAPTER 12

Enantiomeric Recognition in Macrocycle-Primary Ammonium Cation Systems

Reed M. Izatt, Cheng Y. Zhu, Peter Huszthy, and Jerald S. Bradshaw

Department of Chemistry
Brigham Young University
Provo, Utah
U.S.A.

12.1 Introduction

Molecular recognition is ubiquitous in Nature. Examples include antibody–antigen interactions, biochemical catalysis reactions, the DNA double helix, and incorporation of single enantiomeric forms of amino acids and sugars in metabolic pathways. It was believed in the not-too-distant past that these biochemical phenomena were the result of the unique properties of biological macromolecules. However, recent successes in imitating such phenomena using small synthetic compounds has shown that biological behavior can be engineered into simple molecules. Crown ethers, for example, exhibit excellent ability to selectively bind cationic guests[1,2] and have gained much popularity as enzyme models.[3-5] Therefore, molecular recognition at the nonbiomolecular level has been an active, expanding field of research. To a great extent present activity in this field is driven by the inherent interest in elucidating the remarkable ability of molecules to recognize one another with subsequent interaction and formation of stable organized structures.[6] Although this interest was evident from many early studies, it received a significant impetus when Pedersen[7,8] published his work concerning synthesis of a large number of crown ethers, in which he identified their abilities to differentiate among similar metal cations. The rapid development of the field of molecular recognition as applied to macrocycles was recognized by the awarding of Nobel Prizes in 1987 to three of its pioneers, Pedersen,[9] Cram,[10] and Lehn.[11]

Macrocycles offer unusual opportunities for the study of molecular recognition. The work of Cram and Lehn provides a large number of examples.

Some idea of the explosion of interest in the field of macrocyclic chemistry is seen by comparing the number of thermodynamic values in two *Chemical Reviews* articles, one published in 1985[12] covering the literature through mid-1984 and another in 1991[13] covering the literature for the previous six years. The number of values reported in the latter had more than tripled.

Understanding molecular recognition requires that the interactions involved be quantitated. This quantification provides the basis for evaluating guest selectivity and binding strength. Correlation of the quantified properties of host–guest complexes with their molecular structures should provide the foundation for understanding host–guest recognition and predicting the ligands that should be synthesized in order to obtain desired selectivities. Although an extensive effort has been made to quantitate metal cation–macrocycle interactions, much less effort has been devoted to systems involving enantiomeric recognition of chiral organic cations by chiral macrocycles. The focus of our work has been to identify and understand on a molecular level the parameters governing enantiomeric recognition processes. We have found that the quantitation of these processes has been valuable in improving our capacity to predict the relative abilities of chiral hosts to recognize enantiomeric guests. This success has convinced us that continued effort in identifying and quantitating these parameters can lead to the effective future design of new organic molecules containing active sites for use in catalysis, sensing, and separation processes, and for the elucidation of biochemical pathways.

This chapter describes the work that has been carried out at BYU in identifying and quantitating several of the parameters involved in enantiomeric recognition of organic ammonium cation guests by chiral macrocyclic hosts. The underlying premise behind these studies was that quantitation could be useful in targeting those parameters that are most important in these processes. This knowledge, in turn, could be used as a guide in the synthesis of chiral receptors of improved selectivity for the enantiomers of interest. In the material that follows, the use of synthesis, calorimetry, NMR spectroscopy, X-ray crystallography, and molecular mechanics calculations to search for chiral hosts having superior enantiomeric recognition ability will be described. A section will be devoted to the presentation and description of directions this research on molecular recognition is expected to take. Finally, the potential uses for the design technique and for the resulting superior chiral receptors will be presented.

12.2 Background Studies

Enantiomeric recognition, as a special case of molecular recognition, involves the discrimination between enantiomers of the guest by a chiral receptor or a chiral matrix. The successful design, synthesis, and use of molecules capable of enantiomeric recognition of other species is of great interest to workers in asymmetric synthesis, enantiomeric separation, enzyme function, synthetic enzyme design, and other areas involving chiral recognition. The careful characterization of such synthetic systems that show chiral recognition could lead to a greatly improved understanding of natural systems. One area of recent interest is the enantiomeric recognition of chiral organic guests by chiral macrocyclic ligands. Several research groups have carried out work involving these host–guest systems. Cram and his co-workers first described the synthesis and characterization of a number of chiral crown ethers capable of enantiomeric recognition toward primary ammonium salts in 1973.[14,15] In those crown ethers, the hindered rotation of each binaphthyl moiety about its pivot bond played a crucial role both in creating the chirality of the ligand and in discriminating between the enantiomers of the primary ammonium cations. Later, further characterizations and applications of those chiral crown ethers were made using a solvent extraction technique,[16–18] transport of enantiomeric guests through liquid membranes,[19–21] and chromatographic resolution of enantiomers of various amino acids on a silica gel– or polystyrene-bound chiral host material.[22] Lehn and his co-workers have described the preparation of a number of chiral 18-crown-6 derivatives containing moieties from tartaric acid[23–25] and from α-amino acid[26] derivatives. They have also studied reactivity differences when certain p-nitrophenyl esters were thiolyzed while complexed with some of their chiral macrocyclic host molecules.[27] Still and his co-workers have reported significant chiral recognition toward certain enantiomeric amides by some macrobicyclic and macrotricyclic molecules of C_2 or C_3 symmetry.[28–30] Other research groups, including our own, have observed enantiomeric recognition of primary ammonium salts by chiral crown ethers derived from simple sugar molecules,[31,32] by chiral diazacrown ethers,[33] and by chiral crown ether containing pyridine and triazole subcyclic units.[34–36] We have shown[37] that many of the chiral pyridino-18-crown-6–type ligands formed complexes of appreciable stability with primary ammonium cations and displayed good chiral recognition toward enantiomers of these guests. For example, chiral ligand (S,S)-9 (Figure 12.1) exhibited chiral recognition toward the enantiomers of α-(1-naphthyl)ethylammonium (NapEt) perchlorate (Figure 12.2) by forming a complex with the (R) form of the guest, which was 2.6 times more stable than that formed with the (S) form in methanol. The stability constants for the complexes of (S,S)-9 with the (R) and (S) enantiomers of NapEt in methanol were found to be $10^{2.47}$ and $10^{2.06}$, respectively.[34] An excellent review of chiral crown ethers and their interactions with organic ammonium salts has been published.[38] In addition to the studies involving chiral crown ethers, other scientists have achieved enantiomeric recognition through various chromatographic techniques by using a variety of open chain, instead of macrocyclic, chiral ligands.[39–41] The successful separation of a large array of chiral guest molecules by Pirkle's chiral stationary phases (CSPs), which were derived from amino acid derivatives, is a notable example.[42,43]

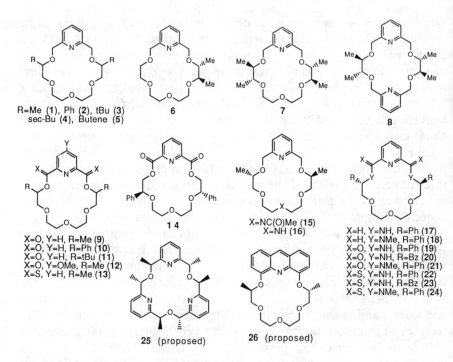

Figure 12.1. Chiral macrocyclic ligands.

In summary, research in the area of chiral recognition has been strong in synthesis, mediocre in application, and weak in characterization. Of the vast number of synthetic chiral hosts reported, most have not been well characterized from either a thermodynamic or a kinetic standpoint for their interactions with enantiomeric guests. The characterization of chiral host–guest interactions in terms of well-defined primary parameters such as equilibrium constants (K), heats of reaction (ΔH), and entropies of reaction (ΔS) is important for an accurate evaluation of the system's performance in chiral recognition. Such an accurate evaluation combined with knowledge about the structural features of the system should lead to improved understanding about chiral recognition in that system. This understanding, in turn, should lead to increased ability to design host–guest systems possessing improved chiral recognition.

12.3 In Search of Improved Chiral Recognition

This section describes the research conducted by our group during the past several years in search of improved chiral recognition. Our program features coordinated studies involving the design and synthesis of new chiral hosts, quantitation of chiral interactions using calorimetry and ¹H NMR spec-

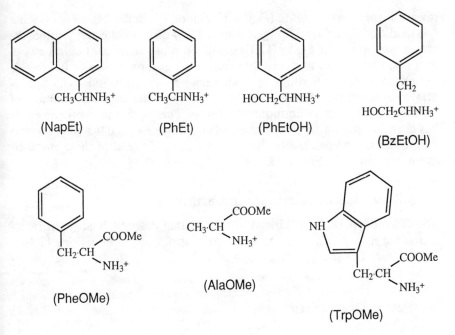

Figure 12.2. Chiral ammonium cations.

troscopy, and structural elucidation of new chiral host–guest complexes using X-ray crystallography, NMR spectroscopy, and molecular mechanics calculations.

12.3.1 Choice of Chiral Host-Guest Systems

For the study of factors governing enantiomeric recognition, the ideal candidates among chiral host–guest systems would be those that show relatively strong interaction and yet are sufficiently simple that synthetic modifications can be made easily. The choice of such systems should allow us to quantitate their interactions accurately using well-defined primary quantities such as K, ΔH, and ΔS. Proper choice of the host–guest systems should also allow us to examine in a systematic manner the effect on enantiomeric recognition of altering various structural features and environmental factors. It is well known that 18-crown-6–type ligands form stable complexes with ammonium and primary ammonium cations through a three-point (tripod) hydrogen bond interaction. Examples of chiral recognition involving chiral 18-crown-6–type ligands and chiral primary ammonium cations have been shown by us[37] and by others.[38] For instance, a chiral 18-crown-6–type ligand containing a pyridine subring, (S,S)-dimethylpyridino-18-crown-6 (ligand 1 in Figure 12.1), was found earlier by us[34] to display enantiomeric recognition toward NapEt and TrpOMe (Figure 12.2) ions. A similar ligand containing two additional carbonyl oxygen atoms (S,S)-dimethyldiesterpyridino-18-crown-6 (ligand 9 in Figure 12.1), exhibited more significant chiral recognition toward NapEt and AlaOMe (Figure 12.2) ions. It was of interest to extend these earlier findings by evaluating new chiral ligands with appropriate structural modifications for their ability to recognize enantiomers of selected chiral primary ammonium cations. Therefore, our synthetic effort has been focused on the preparation of chiral macrocyclic ligands in which various structural parameters have been altered. Several of these chiral ligands are shown in Figure 12.1.

12.3.2 Choice of Methods for Study

The thermodynamic and kinetic quantities that describe host–guest interaction are K, ΔH, ΔS, k_f, k_r, $\Delta G_f^{\ddagger}$, and $\Delta G_r^{\ddagger}$, as defined in Equations 12.1–5. Other quantities sometimes used to describe chiral host–chiral guest interactions such

$$\text{Host + Guest} \underset{k_r}{\overset{k_f}{\rightleftharpoons}} \text{[Host–Guest]} \tag{12.1}$$

$$K = \frac{[\text{Host·Guest}]}{[\text{Host}]\,[\text{Guest}]} \tag{12.2}$$

$$-2.303RT \log K = \Delta H - T\Delta S \tag{12.3}$$

$$\Delta G_f^{\ddagger} = -RT \ln \frac{k_f^{\hbar}}{\kappa T} \tag{12.4}$$

$$\Delta G_r^{\ddagger} = -RT \ln \frac{k_r^{\hbar}}{\kappa T} \tag{12.5}$$

as the extraction constant, liquid membrane transport, and the chromatographic separation factor describe more than one interaction or process. Therefore, unless the correlations between these quantities and their constituent interactions have been well established, conclusions concerning the causes of chiral recognition cannot be made from such data.

Titration calorimetry is a reliable tool for determining both log K and ΔH values accurately from a single experiment, provided that the log K value lies within a certain range[44–46] and the reaction produces or absorbs significant amounts of heat. A disadvantage of calorimetry is that a large amount of sample, usually tens or hundreds of milligrams, is required for a single experiment. Thus, calorimetry is not a practical means for performing a quick survey of host–guest interactions involving expensive chiral crown ethers. Many of the other methods for log K determination require much smaller sample amounts. Such methods include UV/VIS-spectroscopy,[47] potentiometry,[48] polarography,[49] and ^{1}H NMR spectroscopy.[50,51] The

^{1}H NMR method has proven to be effective in determining log K values for chiral crown ether–chiral primary ammonium cation interaction, especially for systems in which both host and guest molecules contain aromatic groups. The advantages of the ^{1}H NMR method are that the experiment can be carried out in a wide variety of solvents and that useful structural information can often be obtained. We have satisfactorily tested the ^{1}H NMR method against the calorimetric method for its accuracy in determining log K values[52] and have successfully employed this method to determine log K values for most of the chiral systems we have studied.

A knowledge of the mechanisms and the kinetic values k_f, k_r, $\Delta G_f^{\ddagger}$, and $\Delta G_r^{\ddagger}$ associated with chiral interactions should be important in leading to a better understanding of the origin of chiral recognition. A detailed kinetic analysis could reveal the mechanism of a chiral interaction and locate the step(s) that make(s) the major contribution to the chiral recognition. A large number of rate studies of complexation reactions involving macrocyclic ligands and metal ions has been published, and reviews of the subject are available.[12,13] However, no detailed kinetic study has been conducted for chiral crown ether–chiral primary ammonium cation interactions. A few kinetic studies concerning nonchiral crown ether–ammonium cation interactions have been reported.[53]

Measurements of $\Delta G_c^{\ddagger}$, the free energy of activation for complex dissociation at the ^{1}H NMR coalescence temperature, have been made using a variable temperature ^{1}H NMR procedure.[54,55] Since the dissociation rate for most of the crown ether–ammonium cation complexes is within the ^{1}H NMR time scale, the measurement of $\Delta G_c^{\ddagger}$ is convenient using the NMR technique. This method is advantageous because the required amount of sample is small (3–5 mg). However, $\Delta G_c^{\ddagger}$ cannot be used to quantitate crown ether–ammonium cation interactions. Since different crown ether–ammonium cation complexes have different coalescence temperatures, the corresponding $\Delta G_c^{\ddagger}$ values are not comparable with each other. In addition, without the knowledge of association rates of the complexes, it is inappropriate to correlate $\Delta G_c^{\ddagger}$ values with the thermodynamic ΔG values of the same systems. Also, the measurement of $\Delta G_c^{\ddagger}$ values provides no information about the mechanisms of host–guest interactions. Despite these limitations, $\Delta G_c^{\ddagger}$ values have been valuable in identifying the presence or absence of chiral recognition in host–guest pairs.

Determination of K values for chiral host–chiral guest interactions provides information about the capability of the chiral hosts to recognize enantiomers of the chiral guest under given sets of conditions. The correlation of the degree of recognition with the structural features of the host–guest complexes is essential in understanding the origin of the chiral recognition. Molecular structural features include both configurational and conformational aspects. Configurational structural features can be described accurately by a conventional two-dimensional (2D) representation of molecular structure, whereas the conformational structural features require a three-dimensional (3D) representation. The conventional 2D representation is of great value in depicting chemical bond connectivities and certain steric relationships within a molecule, but the use of a 2D representation alone is not sufficient to discuss all of the structural features of a molecule, and its use may lead to incorrect conclusions. Since a chiral host–guest interaction involves a complicated "steric matching" between the two molecules involved, an understanding of the chiral recognition requires knowledge of the three-dimensional structural features of the free and complexed host and guest molecules. Standard techniques for observing 3D structures in the crystalline state are X-ray crystallography and neutron diffraction crystallography. Although both techniques suffer some uncertainty in determining proton positions, they are generally satisfactory in providing a structure with fine detail. However, crystal structures do not necessarily represent those in solution. In fact, the crystal and solution structures may differ appreciably.[56] The development of modern NMR spectroscopic techniques, marked by the awarding of the 1991 Nobel Prize to the father of modern NMR, Dr. R. R. Ernst, makes it possible to observe important structural features of molecules in solution by employing appropriate (sometimes a combination of several) NMR experiment(s). For example, 1D and 2D NOE experiments are able to provide information about the spatial proximity between two atoms in a molecule as long as the motion of the molecule is not too fast for observation.[57] The structural information obtained from NMR spectroscopy, although neither complete nor in fine detail in terms of atomic coordinates, has still proven to be very helpful in understanding the molecular interactions in solution and in leading to better design and prediction of improved molecular recognition systems. In addition to NMR, other spectroscopic techniques such as IR/Raman and UV/VIS may provide useful structural information for these systems. The knowledge about the solution structures gained from the spectroscopic results may be improved with the aid of molecular mechanics calculations.[58,59] Successful molecular mechanics calculations should allow the identification of several of the most likely conformations of lowest total energies for the molecule under investigation. Usually, the actual conformations adopted in the solid state or in solution are among those found by the molecular mechanics calculations. The information obtained from spectroscopic results may then be used to locate and confirm the correct conformation from a pool of likely ones. Thus, by combining both spectroscopic and molecular mechanics calculation results, one may obtain a complete 3D picture of the solution structure of a molecule. Molecular mechanics calculations can also be used as a modeling technique for the design and preview of host–guest systems projected for synthesis. In contrast to conventional CPK models, these calculations can take into account parameters and functions for chemical interactions. Several reviews[60–62] describe in detail the procedures used in the application of molecular mechanics calculations and related concepts to macrocycle-guest interactions. Molecular mechanics calculations can sometimes even predict some experimentally measurable quantities for molecular interactions. One example is the calculation of the difference between the activation energies, $\Delta\Delta G^{\ddagger}$, for the dissociation of the complexes of a chiral host with the (R) and (S) enantiomers of a chiral guest.[63] The calculation involves the use of an empirical force field (EFF), which is an empirical function set developed by Lifson and his co-workers[64] for describing bond lengths, bond angles, torsional angles, and interatomic Coulombic and Lennard–Jones interactions. We have introduced the molecular mechanics calculations using the EFF approach into our investigation of chiral crown ether–chiral primary ammonium cation interactions and found good agreement between the calculated $\Delta\Delta G^{\ddagger}$ values and the $\Delta\Delta G_c^{\ddagger}$ values determined from variable temperature ^{1}H NMR experiments in many cases.[65,66] In the chiral systems studied, the EFF calculations have successfully predicted the chiral pair that showed greatest chiral recognition.

12.3.3 Present State of Research

Our research effort on enantiomeric recognition has focused on the quantitation of interactions involving the chiral pyridino- or diesterpyridino-18-crown-6–type ligands and various primary ammonium cations in a single solution phase. Systems showing good chiral recognition or those exhibiting appreciable interaction strength are subjected to further investigation of their structural features. The chiral ligands studied in our laboratory all have an 18-membered macrocyclic ring and contain a pyridine subunit. The choice of the 18-membered macrocyclic ring was based on the fact that 18-crown-6 generally forms more stable complexes than other crown ethers of different ring size with ammonium and primary ammonium cations in solvents other than water.[12,13] The incorporation of a pyridine subunit was considered to provide two favorable effects. First, the presence of pyridine nitrogen as one of the donor atoms of the ligand forces the primary ammonium cation to complex with the ligand through a three-point hydrogen bond interaction that involves the pyridine nitrogen. Interaction with the pyridine nitrogen is consistent with the observation that the N^+–H$\cdots$N hydrogen bond is generally stronger than the N^+–H$\cdots$O bond.[67] Second, the pyridine ring may make a secondary bond between the ligand and the cation possible through π–π interaction if the ammonium cation contains an aromatic group. In order to achieve good chiral recognition, the chiral host molecule should be designed in such a way that the interacting options available for the incoming chiral guest are limited. Preferably, the options would be limited to either "match" or "not match" interactions. The presence of the pyridine ring could be helpful in limiting the interaction options for the guest primary ammonium cations.

Table 12.1 lists the values of $\Delta G_c^{\ddagger}$ and T_c as determined by the variable temperature ^{1}H NMR method for the interactions of most of the chiral hosts shown in Figure 12.1 with enantiomers of several primary ammonium salts.

Table 12.1. $\Delta G_c^{\ddagger}$ $(T_c)^a$ values for the interactions of the chiral macrocyclic ligands with enantiomers of several ammonium cations[b] in CD_2Cl_2.

Ligand	Cation[b]	$\Delta G_c^{\ddagger}$	$\Delta\Delta G_c^{\ddagger c}$	Calculated $\Delta\Delta G^{\ddagger c}$	Reference
(S,S)-1	(R)-NapEt	43.1 (217.2)			34
	(S)-NapEt	36.4 (187.2)	6.7	7.1	34
	(R)-PheOMe	47.3 (233.2)			34
	(S)-PheOMe	41.8 (200.2)	5.5		34
(R,R)-2	(R)-NapEt	59.4 (279.2)			66
	(S)-NapEt	46.9 (239.2)	−12.5		66
(S,S)-3	(R)-NapEt	47.3 (235.2)			66
	(S)-NapEt	36.9 (178.2)	10.4	9.2	66
(R,R)-4	(R)-NapEt	52.3 (253.2)			65
	(S)-NapEt	55.6 (278.2)	3.3	7.1	65
	(R)-PhEt	45.2 (223.2)			65
	(S)-PhEt	50.6 (258.2)	4.4		65
	(R)-PheOMe	47.3 (231.2)			65
	(S)-PheOMe	45.6 (225.2)	−1.7		65
	(R)-PhEtOH	52.7 (243.2)			65
	(S)-PhEtOH	46.0 (234.2)	−6.7		65
(S,S)-5	(R)-NapEt	59.4 (302.2)			65
	(S)-NapEt	55.6 (283.2)	3.8		65
(R,R)-6	(R)-NapEt	62.3 (301.2)			66
	(S)-NapEt	61.1 (303.2)	−1.2		66
(R,R,R,R)-7	(R)-NapEt	56.0 (276.2)			65
	(S)-NapEt	59.8 (300.2)	3.8	3.8	65
	(R)-PhEt	50.6 (252.2)			65
	(S)-PhEt	59.8 (298.2)	9.2		65
(R,R,R,R)-8	(R)-NapEt	43.9 (221.2)			65
	(S)-NapEt	46.9 (230.2)	3.0		65

(continued)

Table 12.2 lists the values of log K and/or ΔH and ΔS as determined by either a ¹H NMR spectroscopic or a titration calorimetric method for various chiral macrocycle–chiral primary ammonium salt interactions. The data included in these tables provide information that makes possible the identification and evaluation of the relationship between the chiral ligand and chiral cation structures and the degree of chiral recognition. This information will now be examined.

12.3.3.1 Chiral Pyridino-18-Crown-6 Compounds

Chiral ligands *1–8* have in common identical macrocyclic backbones, which consist of one pyridine nitrogen (two in the case of *8*) and five (four in the case of *8*) ether oxygens as the donor atoms. Chiral ligands *1–5* (Figure 12.1) are structurally similar, with the R substituent being the only difference. The $\Delta\Delta G_c^{\ddagger}$ values in Table 12.1 indicate that these ligands have varied capabilities of recognizing enantiomers of the ammonium cations studied.

Table 12.1. (continued)

Ligand	Cation[b]	$\Delta G_c^{\ddagger}$	$\Delta\Delta G_c^{\ddagger c}$	Calculated $\Delta\Delta G^{\ddagger c}$	Reference
	(R)-PhEt	46.9 (235.2)			65
	(S)-PhEt	44.4 (227.2)	−2.5		65
(S,S)-9	(R)-NapEt	56.1 (285.2)			34
	(S)-NapEt	51.5 (254.2)	4.6	2.9	34
	(R)-PheOMe	50.6 (248.2)			34
	(S)-PheOMe	49.4 (237.2)	1.2		34
(R,R)-9	(R)-NapEt	52.3 (260.2)			34
	(S)-NapEt	56.1 (286.2)	3.8	2.9	34
	(R)-PheOMe	49.4 (237.2)			34
	(S)-PheOMe	50.6 (248.2)	1.2		34
(S,S)-10	(R)-NapEt	55.6 (284.2)			34
	(S)-NapEt	50.2 (238.2)	5.4	10.5	34
	(R)-PheOMe	49.8 (252.2)			34
	(S)-PheOMe	45.2 (228.2)	4.6		34
(S,S)-11	(R)-NapEt	43.1 (219.2)			66
	(S)-NapEt	———[d] (———)		10.5	66
(S,S)-13	(R)-NapEt	54.4 (273.2)			37
	(S)-NapEt	49.4 (243.2)	5.0		37
	(R)-PheOMe	47.7 (235.2)			37
	(S)-PheOMe	46.1 (212.2)	1.7		37
(S,S)-14	(R)-PheOMe	48.1 (240.2)			34
	(S)-PheOMe	48.5 (245.2)	−0.4		34

[a] $\Delta G_c^{\ddagger}$ and T_c values were determined by a variable temperature NMR technique. $\Delta G_c^{\ddagger}$ and T_c values are in units of kJ/mol and Kelvin, respectively.

[b] The perchlorate salts were used for all the ammonium cations listed in this table. The notations of the ammonium cations are defined in Figure 12.2.

[c] $\Delta\Delta G_c^{\ddagger}$ was defined as follows: $\Delta\Delta G_c^{\ddagger}(S,S\text{-}R) - \Delta G_c^{\ddagger}(S,S\text{-}S)$ or $\Delta G_c^{\ddagger}(R,R\text{-}S) - \Delta G_c^{\ddagger}(R,R\text{-}R)$. $\Delta\Delta G^{\ddagger}$ was defined similarly, except it was a predicted quantity by molecular mechanics calculations using empirical force field.[64]

[d] Measurement unsuccessful due to the very weak interaction.

The following are examples of significant recognition (given in terms of $\Delta\Delta G_c^{\ddagger}$, kJ/mol). NapEt: (S,S)-1-(R) over (S,S)-1-(S) by 6.7; (R,R)-2-(R) over (R,R)-2-(S) by 12.5; (S,S)-3-(R) over (S,S)-3-(S) by 10.4; PhEtOH: (R,R)-4-(R) over (R,R)-4-(S) by 6.7. The $\Delta\Delta G_c^{\ddagger}$ value of 10.4 for the (S,S)-3–NapEt system is in good agreement with the EFF-predicted $\Delta\Delta G^{\ddagger}$ value of 9.2.[37] Ligands 1–3 display much more chiral recognition toward enantiomers of NapEt than do ligands 4 and 5.

As mentioned earlier, $\Delta\Delta G_c^{\ddagger}$ is a kinetic quantity. It has no firm relationship to thermodynamic quantities like $\Delta\Delta G$ or $\Delta \log K$, although two studies found linear relationships between $\Delta G_c^{\ddagger}$ and ΔG.[53,68] Also, $\Delta\Delta G_c^{\ddagger}$ values have large uncertainties due to the problems discussed in section 12.3.2 and elsewhere.[68] Values of $\Delta\Delta G_c^{\ddagger}$ are most useful for preliminary surveys of chiral systems. If a significant $\Delta\Delta G_c^{\ddagger}$ value is found for a certain system, further studies including measurement of $\log K$ and/or ΔH are warranted.

Log K values for the interaction of (R,R)-2 with enantiomers of NapEt (Table 12.2) show that the ligand has little enantiomeric discrimination for

Table 12.2. Log K, ΔH, and ΔS values for the interactions of the macrocyclic ligands with enantiomers of several primary ammonium cations at 25°C.

Ligand	Cation[a]	Solvent[b]	Method[c]	Log K	ΔH^d	ΔS^d	Δ Log K^e	Reference
(S,S)-1	(R)-TrpOMe	M	Cal.	2.43	−14.4	−2.1		34
	(S)-TrpOMe	M	Cal.	2.29	−14.3	−4.5	0.14	34
(R,R)-2	(R)-NapEt	M	NMR	2.92				66
	(S)-NapEt	M	NMR	3.10			0.18	66
	(R)-PhEt	M	NMR	2.91				66
	(S)-PhEt	M	NMR	3.05			0.14	66
(S,S)-3	(R)-NapEt	$1M/9C$	NMR	1.33				66
	(S)-NapEt	$1M/9C$	NMR	0.62			0.71	66
(R,R)-6	(R)-NapEt	M	NMR	3.00				66
	(S)-NapEt	M	NMR	2.94			−0.06	66
(R,R,R,R)-7	(R)-PhEt	$1M/9C$	NMR	3.58				68
	(S)-PhEt	$1M/9C$	NMR	3.31			−0.27	68
(R,R,R,R)-8	(R)-NapEt	M	NMR	1.55				68
	(S)-NapEt	M	NMR	1.56			0.01	68
	(R)-PhEt	$1M/9C$	NMR	2.98				68
	(S)-PhEt	$1M/9C$	NMR	2.87			−0.11	68
(S,S)-9	(R)-NapEt	M	Cal.	2.47	−27.6	−10.8		34
	(S)-NapEt	M	Cal.	2.06	−26.4	−11.8	0.41	34
	(R)-AlaOMe	M	Cal.	2.02	−14.8	−2.6		34
	(S)-AlaOMe	M	Cal.	1.78	−14.6	−3.5	0.24	34
	(R)-TrpOMe	M	Cal.	1.73	−17.2	−24.8		34
	(S)-TrpOMe	M	Cal.	1.76	−19.2	−30.7	−0.03	34
(R,R)-9	(R)-NapEt	M	NMR	2.08				52
	(S)-NapEt	M	NMR	2.50			0.42	52
	(R)-NapEt	$1M/1C$	NMR	2.20				74
	(S)-NapEt	$1M/1C$	NMR	2.80			0.60	74
	(R)-NapEt	$1M/9C$	NMR	2.97				74
	(S)-NapEt	$1M/9C$	NMR	3.41			0.44	74
	(R)-NapEt	A	NMR	2.98				74
	(S)-NapEt	A	NMR	3.40			0.42	74
	(R)-NapEt	AN	NMR	3.80				74
	(S)-NapEt	AN	NMR	4.24			0.44	74
	(R)-NapEt	$DMSO$	NMR	NP[f]				74
	(S)-NapEt	$DMSO$	NMR	NR				74
	(R)-NapEt	NM	NMR	5.5				74
	(S)-NapEt	NM	NMR	> 6			> 0.5	74
	(R)-NapEt	$1M/1B$	NMR	2.55				74
	(S)-NapEt	$1M/1B$	NMR	2.99			0.44	74
	(R)-NapEt	$1E/1C$	NMR	2.08				74
	(S)-NapEt	$1E/1C$	NMR	2.78			0.70	74
	(R)-NapEt	$1iPr/1C$	NMR	2.17				74
	(S)-NapEt	$1iPr/1C$	NMR	2.77			0.60	74
	(R)-PhEt	M	NMR	1.88				74
	(S)-PhEt	M	NMR	2.33			0.45	74
(S,S)-10	(R)-NapEt	$7M/3C$	NMR	2.15				66
	(S)-NapEt	$7M/3C$	NMR	< 1.30			> 0.85	66
	(R)-PhEt	$1M/1C$	NMR	2.62				66

(continued)

Table 12.2. *(continued)*

Ligand	Cation[a]	Solvent[b]	Method[c]	Log K	ΔH[d]	ΔS[d]	Δ Log K[e]	Reference
	(S)-PhEt	1M/1C	NMR	2.06			0.56	66
	(R)-PhEtOH	1M/1C	NMR	2.24				66
	(S)-PhEtOH	1M/1C	NMR	2.95			−0.71	66
	(R)-BzEtOH	1M/1C	NMR	2.18				66
	(S)-BzEtOH	1M/1C	NMR	1.76			0.42	66
	(S)-PheOMe	1M/1C	NMR	1.60				66
	(S)-PheOMe	1M/1C	NMR	1.28			0.32	66
(S,S)-11	(R)-NapEt	1M/9C	NMR	NR				66
	(S)-NapEt	1M/9C	NMR	NR				66
(S,S)-12	(R)-NapEt	M	Cal.	2.71	−35.6	−67.7		74
	(S)-NapEt	M	Cal.	2.25	−38.5	−86.1	0.46	74
	(R)-NapEt	1M/1C	NMR	3.34				74
	(S)-NapEt	1M/1C	NMR	2.54			0.80	74
	(R)-PhEtOH	1M/1C	NMR	3.14				74
	(S)-PhEtOH	1M/1C	NMR	3.08			0.06	74
(S,S)-13	(R)-NapEt	M	Cal.	——[g]	——[g]			
	(S)-NapEt	M	Cal.	——[g]	——[g]			
(S,S)-14	(R)-AlaOMe	M	Cal.	1.85	−13.8	−2.6		34
	(S)-AlaOMe	M	Cal.	1.84	−14.0	−2.9	0.01	34
	(R)-TrpOMe	M	Cal.	1.96	−15.4	−14.2		34
	(S)-TrpOMe	M	Cal.	2.00	−16.7	−17.7	−0.04	34
(S,S)-15	(R)-NapEt	1M/9C	NMR	NR				68
	(S)-NapEt	1M/9C	NMR	NR				68
(S,S)-16	(R)-NapEt	M	NMR	1.51				68
	(S)-NapEt	M	NMR	1.49			0.02	68
(S,S)-17	(R)-NapEt	1M/1C	NMR	1.58				71
	(S)-NapEt	1M/1C	NMR	1.54			0.04	71
(S,S)-18	(R)-NapEt	1M/1C	NMR	3.17				71
	(S)-NapEt	1M/1C	NMR	3.30			−0.13	71
(S,S)-19	(R)-NapEt	1M/1C	NMR	< 1.0				71
	(S)-NapEt	1M/1C	NMR	NR				71
(S,S)-20	(R)-NapEt	1M/1C	NMR	NR				71
	(S)-NapEt	1M/1C	NMR	NR				71
(S,S)-21	(R)-NapEt	1M/1C	NMR	NR				71
	(S)-NapEt	1M/1C	NMR	NR				71
(S,S)-22	(R)-NapEt	1M/1C	NMR	1.39				71
	(S)-NapEt	1M/1C	NMR	1.02			0.37	71
(S,S)-23	(R)-NapEt	1M/9C	NMR	NR				71
	(S)-NapEt	1M/9C	NMR	NR				71
(S,S)-24	(R)-NapEt	1M/1C	NMR	< 1.0				71
	(S)-NapEt	1M/1C	NMR	< 1.0				71

[a]Perchlorate salts were used for all of the ammonium cations listed in this table. The notations for the ammonium cations are defined in Figure 12.2.

[b]M = CD_3OD (for NMR) or CH_3OH (for calorimetry); C = $CDCl_3$; 1M/1C = 50% CD_3OD–50% $CDCl_3$ (v/v); A = $CD_3C(O)CD_3$; AN = CD_3CN; $DMSO$ = $CD_3S(O)CD_3$; NM = CD_3NO_2; B = C_6D_6; E = C_2D_5OD; iPr = $(CD_3)_2CDOD$.

[c]NMR = ¹H NMR method; Cal. = Titration calorimetry.

[d]ΔH and ΔS values are in the units of kJ/mol and J/K·mol, respectively.

[e]Δ Log K = log K(S,S-R) − log K(S,S-S) or log K(R,R-S) − log K(R,R-R).

[f]NR = no observable interaction.

[g]The reaction heat was too small to allow accurate measurement.

NapEt in terms of $\Delta \log K$ (0.18) in M ($M = CD_3OD$). This is in sharp contrast to what was expected from the observed $\Delta\Delta G_c^{\ddagger}$ value (-12.5 kJ/mol). However, the log K value for (S,S)-3 with NapEt enantiomers indicate a good chiral recognition with a $\Delta \log K$ value of 0.71 in $1M/9C$ ($C = CDCl_3$). This is in good agreement with the expectation according to the $\Delta\Delta G_c^{\ddagger}$ value (10.4 kJ/mol). Despite the solvent difference, it is still evident that the bulkier the R substituent, the greater the chiral recognition by comparing the enantiomeric recognition of (R,R)-2 and (S,S)-3 for NapEt in terms of $\Delta \log K$. According to our study of the solvent effect on $\Delta \log K$ (discussed later in this section), the change of solvent from M to $1M/9C$ for (R,R)-2–NapEt system should not increase the $\Delta \log K$ value of 0.18 nearly enough to make it comparable to the value of 0.71 for the (S,S)-3–NapEt system in $1M/9C$. The enhanced chiral recognition in the $(S, S) = 3$–NapEt system compared to that in the (R,R)-2–NapEt system can be understood as follows. First, let us assume that both the (R) and (S) forms of NapEt form complexes with (S,S)-3 through a tripod hydrogen bond that must involve the pyridine nitrogen and two alternate oxygen atoms of the ligand (Figure 12.3, a), and the naphthyl group of the cation overlaps the pyridine ring of the ligand in both the (S,S)-(S) and (S,S)-(R) complexes (Figure 12.3, b and c). Under these conditions, the (S,S)-(S) and the (S,S)-(R) complexes will each have two possible conformations. In the two conformations of the (S,S)-(S) complex (Figure 12.3, b and d), either the naphthyl or the methyl group of the cation must be located very close to the tert-butyl group protruding above the same side of the macrocyclic plane where the ammonium cation is seated, resulting in van der Waal repulsion between the tert-butyl group of the ligand and either the naphthyl or the methyl group of the cation. Such steric repulsion is avoided in one of the two possible conformations for the (S,S)-(R) complex (Figure 12.3, c). This is consistent with the larger log K value for the formation of the (S,S)-(R) complex. This steric repulsion is much more pronounced in the (S,S)-3-(S)-NapEt complex (where R is the bulky tert-butyl group) than in the (S,S)-1-(S)–NapEt or (S,S)-2-(S)–NapEt complex, where the R substituent is the methyl or phenyl group. Hence, the (S,S)-3-(S)-NapEt complex is the least stable one among the three. Therefore, among the three ligands (S,S)-1, (S,S)-2, and (S,S)-3, (S,S)-3 rejects the (S)–NapEt the most although it accepts the (R)-NapEt reasonably well, leading to better enantiomeric recognition toward NapEt than is found for either (S,S)-1 or (S,S)-2.

It is valid to question whether or not the tripod hydrogen bond always involves the pyridine nitrogen and whether or not the π–π interaction between the pyridine ring of the ligand and the α-aromatic group of the ammonium cation is always present in any complexes formed between pyridino-18-crown-6–type ligands and primary ammonium cations containing an α-aromatic group. Regarding the first concern, it has been confirmed that the tripod hydrogen bond does involve the pyridine nitrogen in each of the known crystal structures of the complexes involving pyridino- or diester-

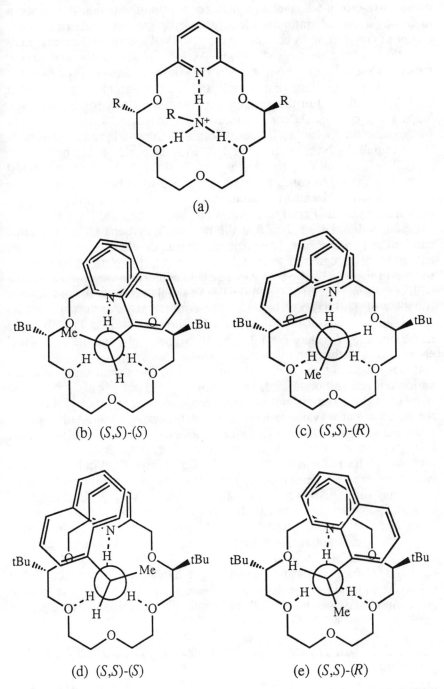

(a)

(b) (S,S)-(S)   (c) (S,S)-(R)

(d) (S,S)-(S)   (e) (S,S)-(R)

Figure 12.3. Illustrations of the possible conformations of the $(S,S)R_2P18C6$–(S)NapEt and $(S,S)R_2P18C6$–(R)NapEt complexes.

pyridino-18-crown-6 ligands and primary ammonium cations.[69,70] Work is underway in our laboratory to identify the hydrogen bond pattern for these complexes in solution by comparing the ^{15}N NMR spectra of the free ligands with those of the complexed ligands. As for the second concern, crystal structures and solution ^{1}H NMR spectra of a limited number of systems[56,69,70] indicate that the pyridine ring of the ligand and the α-aromatic group (either naphthyl or phenyl) of the ammonium cation are close, sometimes overlapping each other, in the complexes involving pyridino-18-crown-6 ligands. In the complexes involving diesterpyridino-18-crown-6 ligands, the α-naphthyl group of the NapEt cation overlaps with the pyridine ring both in the crystalline and solution (methanol and methanol/chloroform mixtures) states. On the other hand, the α-phenyl group of either the PhEt or the PhEtOH cation is located on the opposite side of, and far away from, the pyridine ring of the ligand in solution but not in the solid state.[56,70]

Ligand 6 differs from 1 in that the two methyl groups are on the same side of the pyridine ring. It is expected that 6 should display little or no enantiomeric recognition toward NapEt because the steric repulsion described in Figures 12.2b and d can be easily avoided in the complexes of 6 with enantiomers of NapEt, and the ligand should accept both enantiomers equally well. This has been confirmed by both the $\Delta G_c^{\ddagger}$ and the log K values for the interactions of 6 with (R) and (S) forms of NapEt. The observed $\Delta \Delta G_c^{\ddagger}$ and Δ log K values are 1.2 and 0.06, respectively, and are both small enough to be within their uncertainty range.

It was expected that (R,R,R,R)-7 and (R,R,R,R)-8 would show good enantiomeric recognition toward either NapEt or PhEt because of the increased number of chiral centers and the enhanced molecular symmetry. However, the data currently available have not met this expectation. Although the $\Delta \Delta G_c^{\ddagger}$ value of 9.2 kJ/mol (Table 12.1) indicates that (R,R,R,R)-7 should have great enantiomeric recognition toward PhEt in favor of the (S) rather than the (R) form, the Δ log K value of -0.27 (Table 12.2) indicates otherwise. This example demonstrates again that the information provided by $\Delta \Delta G_c^{\ddagger}$ may differ markedly from that provided by Δ log K. In the case of (R,R,R,R)-8 complexes with enantiomers of NapEt, neither $\Delta \Delta G_c^{\ddagger}$ nor Δ log K suggests a large chiral recognition. It is known that the actual conformation of (R,R,R,R)-8 is not as symmetrical as that indicated by the 2D drawing of the structure. The crystal structure of (R,R,R,R)-8[65] displayed a rather rigid and strained conformation resulting in a distorted C_2 symmetry. This distortion in ligand symmetry may decrease the ability of the host to form good tripod hydrogen bonds with the ammonium cation and hence reduce the effectiveness of chiral recognition.

12.3.3.2 Chiral Diester and Dithionoesterpyridino-18-Crown-6 Compounds

(S,S)-9 is one of the earliest chiral ligands established by calorimetry to show significant chiral recognition toward enantiomers of NapEt (Δ log K

value, 0.41 in favor of the (R) form of NapEt in methanol).[34] The crystal structures of the (S,S)-9-(R)–NapEt and (S,S)-9-S)–NapEt complexes have been determined.[69] Ligand (S,S)-9 differs from (S,S)-1 in that two carbonyl oxygens are present. The two carbonyl oxygens are expected to be in π conjugation with the pyridine ring of the ligand; therefore, the π area in the (S,S)-9 molecule should be larger than that in (S,S)-1, and the π–π interaction with the α-aromatic group of the ammonium guest could be enhanced. The crystal structures of the (S,S)-9 complexes with enantiomers of NapEt (Figure 12.4) show that the naphthyl group of the cation overlaps with the pyridine ring of the ligand in both the (S,S)-(R) and (S,S)-(S) complexes (Figure 12.4).[69] The pyridine nitrogen is also involved in the tripod hydrogen bonding in both cases. These two interactions are necessary to position the guest. However, the direct source of chiral recognition is believed to be the difference in the steric repulsion between the methyl group of the ligand and the naphthyl group of the cation in the two cases. The (R,R)-9 host should show an equal extent of enantiomeric recognition toward NapEt, but in favor of the (S) instead of the (R) form. This has been proven to be true by both log K and $\Delta G_c^{\ddagger}$ values (Tables 12.1 and 12.2).

(R,R)-9 was also found to exhibit good chiral recognition toward enantiomers of PhEt, with $\Delta \log K$ (CH$_3$OH) being 0.45 in favor of the (S) form of PhEt. However, the origin of the chiral recognition in this system is uncertain. Unlike the complexes with NapEt, the π–π overlap or π–π interaction between the pyridine ring of the ligand and the α-aromatic group of the cation is probably absent in the complexes with PhEt enantiomers,

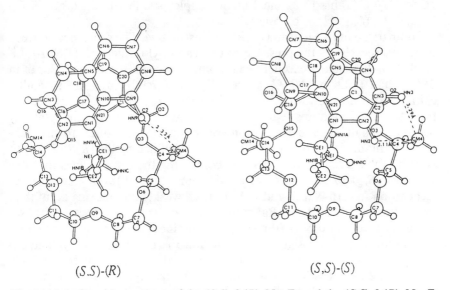

$(S.S)\text{-}(R)$ $\qquad\qquad\qquad$ $(S,S)\text{-}(S)$

Figure 12.4. Crystal structures of the (S,S)-9-(R)–NapEt and the (S,S)-9-(R)–NapEt complexes (reproduced by permission from reference 69).

because the phenyl group of the cation was found to be directed away from the pyridine ring of the ligand in CD_3OD and $CD_3OD/CDCl_3$ mixtures.[56] The ^{1}H NMR spectra of (R,R)-9 showed significant downfield shifts for the pyridine signals of the ligand upon formation of the complexes with PhEt enantiomers in the latter solvents, which strongly suggested that the phenyl group of the ammonium cation was not overlapping with, and even not close to, the pyridine ring of the ligand. Attempts to determine the crystal and solution structures of the complexes are underway. The directing of the phenyl group away from the pyridine ring is probably a result of the solvation of the carbonyl oxygen atoms.

Ligand (S,S)-10 was found to show excellent enantiomeric recognition toward NapEt by a $\Delta \log K$ value of >0.85 in $3M/7C$. Compared to the $\Delta \log K$ values of 0.60 in $1M/1C$ and 0.44 in $1M/9C$ for the (R,R)-9–NapEt system (Table 12.2), ligand (S,S)-10 appears to be better than (R,R)-9 in recognizing NapEt enantiomers. The greater size of the phenyl group compared to that of the methyl group provides a reasonable rationale for the enhanced chiral recognition by (S,S)-10. Ligand (S,S)-10 was also found to recognize the enantiomers of PhEt and PhEtOH (Figure 12.2) very well. Ligand (S,S)-11 is expected to be even better in chiral recognition since the tert-butyl group has greater bulk than the phenyl group. However, the negative effect of this size increase is that the overall complex stabilities are reduced significantly. A dramatic drop in complex stability can be seen by comparing the $\log K$ values for the interactions of NapEt with (R,R)-9, (S,S)-10, and (S,S)-11 (Table 12.2). Taking into account the steady $\log K$ increase for the (R,R)-9–NapEt interaction as the solvent is changed from M to $1M/1C$ to $1M/9C$ (Table 12.2), the drop of complex stability in the order (R,R)-9, (S,S)-10, (S,S)-11 is convincing.

Comparison of (S,S)-10 and (R,R)-2 is instructive. The difference between these macrocycles is the presence of two carbonyl oxygens in (S,S)-10. Ligand (S,S)-10 displays excellent enantiomeric recognition toward NapEt in $3M/7C$ and PhEt in $1M/1C$, whereas ligand (R,R)-2 shows much smaller chiral recognition toward either NapEt or PhEt in M. Change of solvent from M to M/C mixtures results in a significant increase in complex stability but only a small increase in enantiomeric recognition for chiral crown ether–chiral ammonium cation interactions, as exemplified by the (R,R)-9–NapEt system (Table 12.2). It is evident by comparing the $\log K$ values in Table 12.2 that (S,S)-10 is superior as a chiral host to (R,R)-2 in enantiomeric recognition toward NapEt and PhEt but is a weaker complexing agent than (R,R)-2 toward NapEt and PhEt. In other words, the presence of the two carbonyl oxygens in (S,S)-10 results in a significant enhancement in chiral recognition but also a significant reduction in complex stability. The enantiomeric recognition and complex stability differences between (S,S)-10 and (R,R)-2 can possibly be explained by the fact that (S,S)-10 forms less stable complexes with ammonium cations because the carbonyl oxygens reduce the basicity of the pyridine nitrogen through electron withdrawal, resulting in a reduction in the strength of the N^+–H$\cdots$N hydrogen bond. As well, the carbonyl oxygens make the ligand less flexible and therefore unable to adjust its conformation well enough to accommodate the guest ammonium cation for the best tripod hydrogen bonding. (S,S)-2 displays less enantiomeric recognition than (S,S)-10, probably because (S,S)-2 is flexible enough to adjust its conformation to relieve the strain energy caused by the steric repulsions described earlier.

Ligand (S,S)-12 differs from (S,S)-9 by having a 4-methoxyl group on the pyridine ring. One effect of the 4-methoxyl group is to increase the basicity of the pyridine nitrogen, thereby increasing the N^+–H$\cdots$N hydrogen bond strength and the stability of the complex. The result of this effect is seen by comparing the log K values for the interactions of the NapEt enantiomers with (S,S)-12 and (S,S)-9. (S,S)-12 does form a more stable complex (by ~0.2 log K units) than does (S,S)-9 with both the (R) and (S) forms of NapEt in methanol, but it shows little if any improvement in enantiomeric recognition toward NapEt (0.45 log K units) relative to that of (S,S)-9 (0.42 log K units) in the same solvent.

Ligand (S,S)-13 has been shown to recognize the (R) rather than the (S) form of NapEt in terms of $\Delta G_c^{\ddagger}$ values (54.4 to 49.4, $\Delta\Delta G_c^{\ddagger}$ 5.0 kJ/mol). The $\Delta G_c^{\ddagger}$ and $\Delta\Delta G_c^{\ddagger}$ values indicate that (S,S)-13 and (S,S)-9 exhibit little difference in recognizing enantiomers of NapEt. Unfortunately, the determination of the log K and ΔH values by calorimetry for the interactions of (S,S)-13 with NapEt enantiomers in methanol has been unsuccessful because the reaction heats were too small to allow accurate measurement. The attempt to determine the log K values by the ^{1}H NMR method is underway.

Ligand (S,S)-14 differs from (S,S)-10 in the position on the macroring of the phenyl groups. Data for a direct comparison between the two ligands in enantiomeric recognition toward a common cation are unavailable. However, comparison between (S,S)-14 and (S,S)-9 can be made from their interactions with enantiomers of AlaOMe (Figure 12.2). Both (S,S)-9 and (S,S)-14 were found to form reasonably stable complexes (1.7 < log K < 2.1) with the enantiomers of AlaOMe in methanol; however, the former exhibited a chiral recognition of 0.24 log K unit, but the latter exhibited none (Table 12.2).

12.3.3.3 Other Chiral Macrocyclic Compounds

The chiral crown ethers discussed earlier contain five oxygen and one pyridine nitrogen donor atoms. In ligands 15 through 24, at least one of the five oxygen atoms is replaced by a nitrogen atom. It is of interest to investigate the effect of this substitution on enantiomeric recognition by these ligands. Unfortunately, most of these ligands were found to be poor receptors for primary ammonium cations even in 90% $CDCl_3$–10% CD_3OD (v/v), with log K values < 1.0.[71] Ligands 16, 17, and 18 are the exceptions in that they form relatively stable complexes in CD_3OD and 50% CD_3OD–50%

CDCl$_3$ (v/v). However, these three ligands show little or no chiral recognition toward NapEt in CD$_3$OD and 50% CD$_3$OD–50% CDCl$_3$ (v/v). A possible explanation for the fact that most of ligands *15* to *24* do not form stable complexes with NapEt enantiomers is that the conformation of the macrocyclic ring is distorted by the replacement of oxygen atoms(s) by nitrogen(s); hence, the ligand is not suitable for the formation of a good tripod hydrogen bond with the ammonium cation. Molecular mechanics calculations using MMX protocol[72,73] for some of these ligands and their complexes with NapEt or PhEt provide support for this explanation. As seen in the MMX-calculated structure of the (*S,S*)-*15*–PhEt complex (Figure 12.5, *a*), two of the three ammonium hydrogens form nonlinear hydrogen

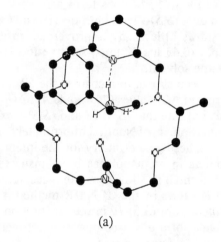

(a)

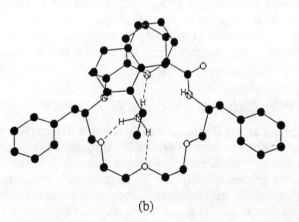

(b)

Figure 12.5. Computer drawings of the (*S,S*)-*16*-PhEt and the (*S,S*)-*20*–NapEt complexes as calculated by molecular mechanics calculations.

bonds with the donor atoms of the ligand, while the third ammonium hydrogen is too far away from any available ligand donor atoms to form a hydrogen bond. In the MMX structure of the (S,S)-20–NapEt complex (Figure 12.5, b), three hydrogen bonds are present, but two of them are apparently not linear, and the hydrogen bond pattern is not the tripod hydrogen bond we defined earlier. It is interesting that in 1M/1C, NapEt forms more stable complexes with (S,S)-18 than with any of the other ligands 15–24. A probable reason for the difference is that the two NMe groups in the (S,S)-18 molecule are involved in the hydrogen bonding with the ammonium cation. A ^{15}N NMR experiment may be crucial in judging which nitrogen atom(s) is/are involved in the hydrogen bonding in this case. Further investigation of the (S,S)-18 complexes using ^{15}N NMR spectroscopy is underway.

12.3.3.4 The Effect of Solvent on Chiral Recognition

As is seen in Table 12.2, various solvents were used in the determination of log K values. In most cases, the choice of solvent was based on our attempt to identify a reaction medium in which the log K values were between about 5 and 1. Outside of this range, log K values could not be determined accurately.[44–46] In order to compare data from different solvents, a knowledge about the variation of log K values for given reactions in a number of solvents is needed. Also, increased understanding of solvent effects on chiral host–guest interactions is needed.

The effect of solvent on the interactions of (R,R)-9 with NapEt enantiomers has been examined using a number of different solvents and solvent mixtures.[74] The log K values as determined by the ^{1}H NMR method for these interactions in different solvents and solvent mixtures are listed in Table 12.2. Two general facts can be visualized from these data. First, the extent of chiral recognition in terms of Δ log K is similar from solvent to solvent for methanol, acetone, and acetonitrile. Because the main factors that are expected to determine chiral recognition (i.e., steric repulsion, conformational strain, and π–π interaction) should have little solvent dependence, this similarity is understandable. However, the lack of a solvent effect on chiral recognition is by no means established. More results are needed before a final conclusion can be reached. Although the solvation energies for the two free guest enantiomers are expected to be identical, those for the complexes of the two enantiomers with a common chiral ligand may not be the same. Further investigation to estimate this solvation energy difference is needed. Such estimation of solvation energy differences may lead to improved understanding of the solvent effect on enantiomeric recognition. Second, the stabilities of complexes vary significantly as the solvent changes. This variation in log K values is illustrated in Table 12.3 for the interaction of (R,R)-9 with (S)- and (R)-NapEt in a number of solvents and solvent mixtures. The solvent parameters that may have an effect on

Table 12.3. Log K values for the interaction of (R,R)-9 with (S)-NapEt and (R)-NapEt in various deuterated solvents and solvent mixtures at 25°C.

Solvent[a]	log K[b]		Solvent Parameters[c]			
	(S)	(R)	ϵ_r	E_T	DN	AN
DMSO	NR	NR	46.45	0.444	29.8	19.3
M	2.50	2.08	32.66	0.762	19.0	41.5
1M/1C	2.80	2.20				
1M/9C	3.41	2.97				
C			4.81	0.259	4.0	23.1
A	3.40	2.98	20.56	0.355	17.0	12.5
AN	4.24	3.80	35.94	0.460	14.1	18.9
NM	> 6.0	5.5	35.94	0.481	2.7	20.5

[a]See footnote b of Table 2 for solvent notations.
[b]NR = no observable interaction.
[c]ϵ_r = dielectric constant; E_T = empirical polarity; DN = Gutmann donor number; AN = Gutmann acceptor number.

complex stabilities (i.e., dielectric constant,[75] empirical solvent polarity,[75] and Gutmann donor and acceptor numbers[76]) are also listed in Table 12.3. Examination of these various solvent parameters indicates that the stability of the complex is determined by the donor number, rather than by either the dielectric constant, empirical solvent polarity, or acceptor number of the solvents. The complex is more stable in solvents with smaller Gutmann donor numbers. This is evidenced by the gradual log K increase for either the (R,R)-9-(S)–NapEt or the (R,R)-9-(R)–NapEt interaction as the solvent changes in the order DMSO, M, A, AN, NM. The Gutmann donor numbers of these solvents decrease gradually in the same order (Table 12.3). The log K increase is also observed from M to $1M/1C$ to $1M/9C$. This increase in log K may also be attributed to the decrease in donor number. Since the data in Table 12.3 show chloroform to be a much weaker donor solvent than methanol, the donicities of methanol/chloroform mixtures should be smaller than that of pure methanol. Since the log K values for the crown ether–ammonium cation interaction in the solvents studied are not directly related to the dielectric constants, the ion–dipole interaction, which is electrostatic in nature, seems not to be the major contributor to the formation of the host–guest complex. Instead, hydrogen bonding is the main contributor. The solvents of higher donicity have stronger solvation power toward ammonium cations. Hence, they are able to destablize the complex formed between a crown ether ligand and an ammonium cation.

12.3.3.5 Enthalpy and Entropy of Chiral Interactions

A knowledge of enthalpic and entropic contributions to log K is helpful in understanding the driving forces of host–guest interactions. In the case of chiral interactions, this knowledge should be helpful in understanding the origins of chiral recognition. Therefore, determination of ΔH and ΔS for chiral crown ether–chiral ammonium salt interactions is desirable. Although it is possible to estimate ΔH and ΔS values from the variation of log K as a function of temperature, much greater accuracy can be obtained for these values using the calorimetric method.[52] However, as mentioned earlier, the calorimetric method requires a relatively large amount of sample. Values of ΔH and ΔS have been determined calorimetrically for several of the systems studied and are listed in Table 12.2. A general trend among these values is that both ΔH and ΔS favor the formation of the complex; if one enantiomer of the guest is favored over the other by the chiral host in terms of log K, it is also favored in terms of both ΔH (more negative in value) and ΔS (less negative in value). In some cases, the entropic contribution seems to be more important than the enthalpic contribution. With the limited number of ΔH and ΔS values available, it is difficult to tell whether the enthalpic and entropic contributions to the enantiomeric recognition are based more on the differentiation in solvation or on configurational and conformational differences in the complexes.

12.3.4 Conclusions

In summary, our knowledge at the present stage about the achievement of enantiomeric recognition in systems involving chiral crown ethers and primary ammonium cations is as follows. First, the ligand should be able to form stable complexes with the chiral guests. If the conformation of a chiral ligand is so twisted that a good three-point hydrogen bond cannot be formed, the ability of the ligand to recognize guest enantiomers will also be poor. Second, the ligand molecule should be rather rigid. If the ligand is too flexible, it will be able to match both enantiomers of a guest equally well by adjusting its conformation. The role of secondary and tertiary interactions (in addition to the three-point hydrogen bonding) between the ligand and the cation and the role of solvent in enantiomeric recognition remained inconclusive pending further experiments. For the systems discussed in this chapter, the pyridino-18-crown-6 ligands generally form more stable complexes with ammonium cations than do the diesterpyridino-18-crown-6 ligands. However, the latter display much improved enantiomeric recognition toward chiral primary ammonium cations. The origin of the improved chiral recognition is probably associated with the increased rigidity of the diesterpyridino-18-crown-6 ligands. Chiral crown ethers with one of more ether oxygen(s) replaced by nitrogen(s) have much decreased interactions with primary ammonium cations.

12.4 Future Prospects: Research and Applications

Our research on chiral recognition involving macrocyclic ligands has emphasized the quantitation of chiral interactions. As the data accumulate which quantify enantiomeric recognition for chiral host–guest systems, correlations between structural features and chiral recognition become apparent. We intend to explore in three-dimensional space the structural features of additional chiral systems in order to probe further the origins of chiral recognition and to develop the design techniques necessary to produce superior chiral ligands. A major thrust of our future research program will be the elucidation of structural features of chiral systems by application of high resolution NMR spectroscopy, X-ray crystallography, and molecular mechanics calculations. In the past several years, our research has featured coordinated studies in the design of macrocyclic hosts, followed by their synthesis and thermodynamic studies of their interaction with selected guests. Structural analysis of the hosts and host–guest complexes has provided insight into the causes of enantiomeric preferences. In turn, the results have led to the design and synthesis of new hosts. Our future research program will continue to follow this track in order to answer unsolved questions and to find superior chiral hosts. Unanswered questions include (1) how much more strongly an ammonium cation hydrogen bonds to pyridine than to an ether, (2) the degree of importance of π–π interaction in improving chiral recognition, and (3) how the solvent affects the extent of chiral recognition.

Chiral ligands *25* and *26* (Figure 12.1) appear to have promise in chiral recognition and are projected for synthesis and study. Ligand *25* has three pyridine subunits. If hydrogen bonding between the ligand and primary ammonium cation predominantly takes place on the nitrogen atoms, the conformation of the complex would be fixed because of the symmetry of the ligand. In this case, the ligand may turn out to be effective in recognizing one enantiomer of the cation over the other with no other choice. Ligand *26* offers the opportunity for a strong π–π interaction between its large π group and the aromatic group of the guest molecule. A strengthened π–π interaction may lead to much improved enantiomeric recognition.

As mentioned earlier, a knowledge of the mechanisms and kinetics of chiral interactions could be important in leading to a better understanding of the origin of enantiomeric differentiation. Hence, mechanistic and kinetic studies on chiral interactions are needed to determine which step of the interaction makes the largest contribution to chiral recognition.

The discovery of effective chiral receptors is one thing; the successful practical application of these ligands is another. An important goal of future research on chiral systems will be to direct the application of superior chiral receptors to the fields of enantiomeric resolution, purification, and sensing; the design of enzyme mimics and analogs; and hetergeneous catalysis in asymmetric reactions.

Acknowledgments

Appreciation for financial support of this research is expressed to the Office of Naval Research.

References

1. Pedersen, C. J. In *Synthetic Multidentate Macrocyclic Compounds*; Izatt, R. M., and J. J. Christensen, Eds.; Academic Press; New York, **1978**.

2. Lamb, J. D.; Izatt, R. M.; Christensen, J. J.; Eatough, D. J. In *Coordination Chemistry of Macrocyclic Compounds*; Melson, G. A., Ed.; Plenum Press; New York, **1979**.

3. Stoddart, J. F. In *The Chemistry of Enzyme Action*, Page, M. E., Ed.; Elsevier Science Publishers; Amsterdam, **1984**.

4. Lehn, J.-M. *Science* **1985**, *227*, 846.

5. Cram, D. J. *Science* **1983**, *219*, 1177.

6. Lehn, J.-M.; Mascal, M.; DeCian, A.; Fischer, J. *J. Chem. Soc., Chem. Commun.* **1990**, 479.

7. Pedersen, C. J. *J. Am. Chem. Soc.* **1967**, *89*, 2495.

8. Pedersen, C. J. *J. Am. Chem. Soc.* **1967**, *89*, 7017.

9. Pedersen, C. J. *J. Inclusion Phenom.* **1988**, *6*, 337.

10. Cram, D. J. *J. Inclusion Phenom.* **1988**, *6*, 397.

11. Lehn, J.-M. *J. Inclusion Phenom.* **1988**, *6*, 351.

12. Izatt, R. M.; Bradshaw, J. S.; Nielsen, S. A.; Lamb, J. D.; Christensen, J. J.; Sen, D. *Chem. Rev.* **1985**, *85*, 271.

13. Izatt, R. M.; Pawlak, K.; Bradshaw, J. S.; Bruening, R. L. *Chem. Rev.* **1991**, *91*, 1721.

14. Kyba, E. P.; Siegel, M. G.; Sousa, L. R.; Sogah, G. D. Y. Cram, D. J. *J. Am. Chem. Soc.* **1973**, *95*, 2691.

15. Kyba, E. P.; Koga, K.; Sousa, L. R.; Siegel, M. G.; Cram, D. J. *J. Am. Chem. Soc.* **1973**, *95*, 2692.

16. Gokel, G. W.; Timko, J. M.; Cram, D. J. *J. Chem. Soc., Chem. Commun.* **1975**, 394.

17. Gokel, G. W.; Timko, J. M.; Cram, D. J. *J. Chem. Soc., Chem. Commun.* **1975**, 444.

18. Peacock, S. C.; Cram, D. J. *J. Chem. Soc., Chem. Commun.* **1976**, 282.

19. Sogah, G. D. Y.; Cram, D. J. *J. Chem. Soc., Chem. Commun.* **1975**, 1259.

20. Sogah, G. D. Y.; Cram, D. J. *J. Chem. Soc., Chem. Commun.* **1976**, 3038.

21. Newcomb, M.; Helgeson, R. C.; Cram, D. J. *J. Chem. Soc., Chem. Commun.* **1974**, 7367.

22. Shibukawa, A.; Nakagawa, T. In *Chiral Separations by HPLC*, Krystulovič, A. M., Ed.; Ellis Horwood Limited; Chicester, **1989**.

23. Girodeau, J.-M.; Lehn, J.-M.; Sauvage, J.-P. *Angew. Chem. Int. Ed. Engl.* **1975**, *14*, 764.

24. Behr, J.-P.; Lehn, J.-M.; Vierling, P. *Helv. Chim. Acta* **1982**, *65*, 1853.

25. Behr, J.-P.; Lehn, J.-M.; Moras, D.; Thierry, J. C. *J. Am. Chem. Soc.* **1981**, *103*, 701.

26. Dietrich, B.; Lehn, J.-M.; Simon, J. *Angew. Chem. Int. Ed. Engl.* **1974**, *13*, 406.

27. Lehn, J.-M.; Sirlin, C. *J. Chem. Soc., Chem. Commun.* **1978**, 949.

28. Sanderson, P. E. J.; Kilburn, J. D.; Still, W. C. *J. Am. Chem. Soc.* **1989**, *111*, 8314.

29. Liu, R.; Sanderson, P. E. J.; Still, W. C. *J. Org. Chem.* **1990**, *55*, 5184.

30. Hong, J. I.; Namgoong, S. K.; Bernardi, A.; Still, W. C. *J. Am. Chem. Soc.* **1991**, *113*, 5111.

31. Curtis, W. D.; Laidler, D. A.; Stoddart, J. F.; Jones, G. H. *J. Chem. Soc., Perkin Trans. 1* **1977**, 1756.

32. Stoddart, J. F. In *Progress in Macrocyclic Chemistry*; Izatt, R. M., and J. J. Christensen, Eds.; Wiley-Interscience; New York, **1981**; Vol. 2.

33. Chadwick, D. J.; Cliffe, I. A.; Sutherland, I. O. *J. Chem. Soc., Chem. Commun.* **1981**, 992.

34. Davidson, R. B.; Bradshaw, J. S.; Jones, B. A.; Dalley, N. K.; Christensen, J. J.; Izatt, R. M.; Morin, F. G.; Grant, D. M. *J. Org. Chem.* **1984,** *49,* 353.

35. Bradshaw, J. S.; Thompson, P. K.; Izatt, R. M.; Morin, F G.; Grant, D. M. *J. Heterocyclic Chem.* **1984,** *21,* 897.

36. Bradshaw, J. S.; Chamberlin, D. A.; Harrison, P. E; Wilson, B. E.; Arena, G.; Dalley, N. K.; Lamb, J. D.; Izatt, R. M.; Morin, F. G.; Grant, D. M. *J. Org. Chem.* **1985,** *50,* 3065.

37. Bradshaw, J. S.; Huszthy, P.; McDaniel, C. W.; Oue, M.; Zhu, C. Y.; Izatt, R. M. *J. Coordination Chem. Section B,* in press.

38. Stoddart, J. F. *Topics in Stereochemistry*; Eliel, E. L., and S. H. Wilen, Eds.; Wiley-Interscience; New York, 1988; Vol. 17.

39. Okamoto, Y.; Hatada, K. In *Chiral Separations by HPLC*; Krystulovič, A. M., Ed.; Ellis Horwood Limited; Chichester, 1989.

40. Allenmark, A. In *Chiral Separations by HPLC*; Krystulovič, A. M., Ed.; Ellis Horwood Limited; Chichester, 1989.

41. Shibata, T.; Mori, K. In *Chiral Separations by HPLC*; Krystulovič, A. M., Ed.; Ellis Horwood Limited; Chichester, 1989.

42. Pirkle, W. H.; Mahle, G. S.; Pochapsky, T. C.; Hyun, M. H. *J. Chromatogr.* **1987,** *388,* 307.

43. Pirkle, W. H.; Pochapsky, T. C.; Mahler, G. S.; Corey, D. E.; Reno, D. S.; Alessi, D. M. *J. Org. Chem.* **1986,** *51,* 4991.

44. Christensen, J. J.; Ruckman, J.; Eatough, D. J.; Izatt, R. M. *Thermochim. Acta* **1972,** *3,* 203.

45. Eatough, D. J.; Christensen, J. J.; Izatt, R. M. *Thermochim. Acta* **1972,** *3,* 219.

46. Christensen, J. J.; Wrathall, D. P.; Oscarson, J. L.; Izatt, R. M. *Anal. Chem.* **1968,** *40,* 1713.

47. Yamabe, T.; Hori, K.; Akag, K.; Fukui, K. *Tetrahedron* **1979,** *35,* 1065.

48. Frensdorf, H. K. *J. Am. Chem. Soc.* **1971,** *93,* 606.

49. Kodama, M.; Kimura, E. *J. Chem. Soc., Dalton Trans.* **1978,** 1081.

50. de Jong, F.; Reinhoudt, D. N.; Smit, C. J.; Huis, R. *Tetrahedron Lett.* **1976,** 4783.

51. de Boer, J. A. A.; Reinhoudt, D. N. *J. Am. Chem. Soc.* **1985,** *107,* 5347.

52. Zhu, C. Y.; Bradshaw, J. S.; Oscarson, J. J.; Izatt, R. M. *J. Incl. Phenom.,* **1992,** *12,* 275.

53. Reinhoudt, D. N.; de Jong, F. In *Progress in Macrocyclic Chemistry*; Izatt, R. M., and J. J. Christensen, Eds.; Wiley-Interscience; New York, **1979**; Vol. 1.

54. Sutherland, I. O. *Annu. Rep. NMR Spectrosc.* **1971,** *4,* 71.

55. Baxter,S. L.; Bradshaw, J. S. *J. Heterocycl. Chem.* **1981,** *18,* 233.

56. Izatt, R. M.; Zhu, C. Y.; Dalley, N. K.; Curtis, J. C.; Bradshaw, J. S., *J. Phys. Org. Chem.,* in press.

57. Sanders, J. K. M.; Hunter, B. K. *Modern NMR Spectroscopy*; Oxford University; Oxford, **1987.**

58. Kollman, P. A. *Annu. Rep. Phys. Chem.* **1987,** *38,* 303.

59. Uiterwijk, Jos W. H. M.; Harkema, S.; Feil, D. *J. Chem. Soc., Perkin Trans. 2* **1987,** 721.

60. Hancock, R. D. *Acc. Chem. Res.* **1991,** *23,* 253.

61. Cremer, D.; Kraka, E. *Mol. Struct. Energ.* **1988,** *7,* 65.

62. Alder, R. W. *Chem. Rev.* **1989,** *89,* 1215.

63. Lifson, S.; Felder, C. E.; Shanzer, A. *J. Am. Chem. Soc.* **1983,** *105,* 3866.

64. Lifson, S.; Felder, C. E.; Shanzer, A.; Libman, J. In *Progress in Macrocyclic Chemistry*; Izatt, R. M., and J. J. Christensen, Eds.; Wiley-Interscience; New York, **1987**; Vol. 3.

65. Bradshaw, J. S.; Huszthy, P.; McDaniel, C. W.; Zhu, C. Y.; Dalley, N. K.; Izatt, R. M.; Lifson, S. *J. Org. Chem.* **1990,** *55,* 3129.

66. Huszthy, P.; Bradshaw, J. S.; Zhu, C. Y.; Izatt, R. M.; Lifson, S. *J. Org. Chem.* **1991,** *56,* 3330.

67. Vinogradov, S. N.; Linnell, K. H. *Hydrogen Bonding*; Van Nostrand Reinhold; New York, **1971**; Chapter 5.

68. Zhu, C. Y. Ph.D. Dissertation, Brigham Young University, **1990,** pp 42–44.

69. Davidson, R. B.; Dalley, N. K.; Izatt, R. M.; Christensen, J. J.; Bradshaw, J. S.; Campana, C. F. *Isr. J. Chem.* **1985,** *25,* 27.

70. Zhu, C. Y.; Izatt, R. M.; Bradshaw, J. S.; Dalley, N. K. *J. Incl. Phenom.,* **1992,** *13,* 17.

71. Huszthy, P.; Oue, M.; Bradshaw, J. S.; Zhu, C. Y.; Wang, T.; Dalley, N. K.; Curtis, J. C.; Izatt, R. M., *J. Org. Chem.,* in press.

72. Allinger, N. L.; Chang, S. H.-M.; Glaser, D. H.; Hoing, H. *Isr. J. Chem.* **1980,** *20,* 51.

73. Allinger, N. L. *J. Am. Chem. Soc.* **1977,** *9,* 8127.

74. Zhu, C. Y.; Izatt, R. M.; Hathaway, J.; Zhang, X.; Wang, T.; Huszthy, P.; Bradshaw, J. S., in preparation.

75. Reichardt, C. *Solvents and Solvent Effects in Organic Chemistry*; VCH Verlagsgesellschaft mbH; Weinheim (FRG), **1988**; pp 339–406.

76. Gutman, V.; Wychera, E. *Inorg. Nuclear Chem. Lett* **1966,** *2,* 257.

CHAPTER 13

Molecule and Cation Recognition by Synthetic Host Molecules

Ian O. Sutherland

Department of Chemistry
The University of Liverpool
PO Box 147
Liverpool L69 3BX UK

13.1 Introduction

The formation of molecular complexes due to attractive noncovalent interactions between two or more molecules is a central feature of biological processes[1,2] but was for a long time a relatively neglected area of chemical research, which was focused to a much greater extent on the formation of covalent bonds. During the 1950s the attention of organic chemists was drawn to the cyclodextrins,[3] which were shown to form complexes in aqueous solutions with a wide variety of organic compounds, particularly derivatives of benzene. At the same time, organized arrangements of two different organic molecules in the crystalline state were recognized, in which one molecular type occupied spaces in the crystal lattice of the second species; these solid-state complexes were called *clathrates*.[4] Both of these topics continue to attract the attention of chemists, but they have been joined by many other studies of complexation involving synthetic host and guest molecules. This high level of current activity can be traced back to the discovery of the crown ethers in 1967 by the late Charles Pedersen.[5]

Pedersen's now famous contribution showed that macrocyclic derivatives of polyethylene glycol with metal cations and alkylammonium cations form 1:1 complexes that can be obtained in the crystalline state. Subsequently the list of guest species was extended to other cationic organic species, such as diazonium cations and a few neutral molecules that act as hydrogen bond donors. The enhanced solubility of metal salts and alkylammonium salts in organic solvents in the presence of crown ethers also provided evidence for complexation in solution. In all cases the binding interaction between the host crown ether and the guest species was recognized to be essentially electrostatic in character and to show the expected dependence on solvent polarity. These results provided inspiration for organic chemists who have since developed[6-14] host molecules that function in organic solvents and water and that form complexes with an ever-increasing range of smaller guest molecules. In addition, many of these host molecules show a high degree of selection in their choice of guests, which can range from simple metal cations to complex organic molecules such as polycyclic aromatics, nucleoside bases, and even carbohydrates. This chapter will concentrate upon the design and synthesis of selective host molecules, their application in analytical science, and other applications that lie in the future.

13.2 Crown Ether Complexes and Host Design

Crown ethers, such as 18-crown-6 (*1*) (Fig. 13.1), were a product of serendipity and the observations and deductive powers of their discoverer.[10] Many of their derivatives that have since been synthesized may well have arisen by a similar route, but the ambition of synthetic chemists working in this area must be the design and synthesis of host molecules that will be selective for just a single guest species or for a chosen range of guests. The most successful designs have been based upon space-filling molecular models and information obtained from crystal structures, combined with imagination and intuition.[11,12] However, the rationalization of the results obtained from this approach is now being derived from a more systematic approach based upon molecular graphics, molecular mechanics,[15] and molecular dynamics. Even if such studies do not provide inspiration, they have the merit of revealing the complexity of even the simplest examples of complexation, such as the formation of the cation complex (*2*), and providing a test bed for ideas that complements the more traditional space-filling models.

Very many papers have appeared describing the application of molecular mechanics and molecular dynamics to the study of crown ethers and their

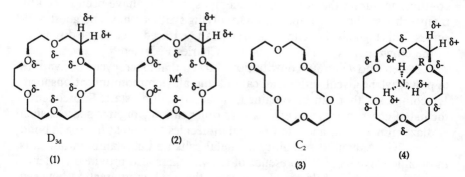

Figure 13.1. 18-crown-6 in its coordinating form (of D_{3d} symmetry) and as the free ligand (of C_1 symmetry).

complexes as well as to other host molecules. Only the principal features of these studies will be discussed here, as they indicate the computational requirements for accurate calculations of host–guest binding energies.

There are three major obstacles to good computer-based models of crown ethers and their complexes, assuming that good procedures are available for the minimization of steric energy. The first is the multiplicity of conformations with rather similar steric energies that are found for most macrocyclic compounds. The second is the need for accurate evaluation of the energy of intermolecular and intramolecular electrostatic interactions between polarized host and guest molecules. The final obstacle is the evaluation of solvation and entropic effects. These difficulties and methods for their solution have been the subject of recent computational studies of 18-crown-6 and its complexes.

The best-known conformations[16,17] of 18-crown-6 are the C_i conformation (3), found in the crystal structure of the free crown ether, and the C_{3D} conformation (1), found in the crystalline state of complexes of metal cations (2) and primary alkylammonium cations (4).[18] Other conformations of 18-crown-6 are found in the crystal structures of other complexes and these, together with (1) and (3), were considered in early molecular mechanics investigations[18–20] of the steric energies of crown ethers, but recent, more complete examinations[21] of the low-energy conformations of 18-crown-6 reveal nearly 200 conformations that lie in a narrow range of steric energies as calculated using the MM2 or AMBER force fields. (For details of the MM3 force field relevant for crown ethers and aza crown ethers, see references 22 and 23.) Nine of these conformations have steric energies close to those of (1) and (3). Because calculated steric energies depend upon force field parameters and polar compounds such as crown ethers particularly depend upon the 1,4-electrostatic interactions between the partly charged oxygen atoms (see 1), it is clear that a large number of these 200 conformations may contribute to the conformational situation for 18-crown-6 in solution. Although the complexing conformation (1) is included in the possible solution conformations of the free crown ether, it is evident that flexible host molecules such as the crown ethers are far from preorganized for complexation. These results are comparable with a recent examination by a variety of procedures[24,25] of the conformational space of cycloheptadecane, which revealed 262 conformations of this cycloalkane with MM2 energies within 3 kcal mol^{-1} of the global minimum.

It appears that even the most symmetrical macrocyclic host molecules may be conformationally very complex unless their structures are based upon rigid structural units. Host molecules with lower symmetry than the simple crown ethers, such as the aza (5) and diaza (6) crown ethers (Fig. 13.2), present an even more complex problem, and this is further complicated by the possibility of conformational inversion at any saturated nitrogen centers in the macrocycle. Thus for a diaza crown ether (6), a full conformational search must include both the *cis* (6a) and *trans* (6b) species.

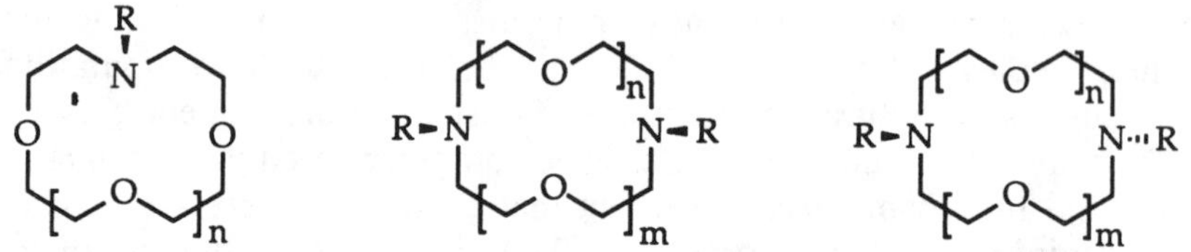

Figure 13.2. Structures of mono- and diaza crown ethers.

Relative steric energies obtained from molecular mechanics calculations may be equated with relative conformational potential energies in the gas phase. Recent approaches to conformational energies have also included solvation effects,[26] and potential energies have been extended to free energies. Such complete studies are expensive in computer time because, for example, relatively large numbers of solvent molecules must be included in the calculations, and they are not yet widely performed.

The prediction of selectivity for the complexation of guest cations by a host macrocycle of the crown ether type is most simply based upon the match between cavity diameter and cation diameter. The host macrocycle can best solvate a guest cation that fits into the molecular cavity of a low-energy conformation with optimum distances between the metal cation and the surrounding solvating hetero-atoms. More detailed analysis of cation selectivity requires a full molecular mechanics or molecular dynamics study of complexation that includes an adequate number of solvent molecules to simulate solvation,[27,28] although it has been noted that relatively few solvent molecules[29] may serve as a computationally inexpensive solvation model. These theoretical studies have been extended[30] to complexes between 18-crown-6 and alkylammonium cations based upon a simple electrostatic model [see charge distribution in (1) and (4)] with some explicit consideration of the nonbonded interaction potential for atoms linked by hydrogen bonds. Although this simple model did not include solvation effects or entropic effects, it was able to account for a number of trends identified for alkylammonium cation recognition in solution, generally in organic solvents such as chloroform.

It is to be expected that refinements in the force field and improvements in the calculation of the electrostatic terms will result in improved prediction, or rationalization, of molecular recognition in crown ether–alkylammonium cation complexes and that these predictions will become quantitative when solvation and entropic effects are included. In addition to studies of crown ethers, there are many published discussions of the conformational properties and binding behavior of other host molecules, and most workers in host–guest chemistry[31] now use computational models (in addition to the more traditional molecular models) to examine the potential of new host molecules. However, analogy with the crystal structures of known host molecules in their free and complexed states also provides important guidelines[11,32] for the design of new hosts. This article will present some experimental studies of selective complexation by designed host molecules in which host design has been based upon analogy with known hosts combined with, in two cases, the advantages of combining two simple receptor sites in a single host molecule.

13.3 Molecular Recognition by Ditopic Receptors

Our initial target in crown ether chemistry was to simulate the high degree of guest recognition that is shown by some biological host molecules.[33] It is true that many biological hosts actually show a rather low degree of guest recognition and that recognition is only required for the relatively limited range of possible guest molecules that may be present in a biological system. Even antibodies, which might be expected to be highly discriminating in their selection of guest molecules, are not required to discriminate against unimportant unnatural substrates. Thus the level of recognition achievable by synthetic host molecules could in principle exceed that shown by many biological hosts.

There are two possible strategies for enhancing the rather low discrimination shown[34] by simple crown ethers, such as 18-crown-6, for guest alkylammonium cations RNH_3^+ that have different groups R. The first approach was elegantly explored by D. J. Cram and his co-workers in their pioneering studies[35-37] of enantioselective crown ether hosts. In this work, selectivity was enhanced through the introduction of steric barriers within the crown ether structure that interacted with the group R of the guest cation as shown diagrammatically in (7) (Fig. 13.3). The alternative approach was adopted by J.-M. Lehn's group in their seminal work[38] on cation selection by cryptands; in this approach, the guest cation is enclosed in the three-dimensional cavity within the host structure. However, although metal cations with a spherical charge distribution are relatively easy to enclose in a three-dimensional cavity lined with electron rich hetero-atoms, as in (8), it is relatively difficult to attract organic molecules into a molecular cavity by the weak attractive forces between electrically neutral molecular fragments.

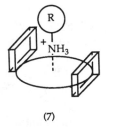

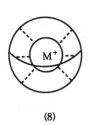

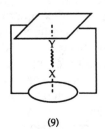

(7)   (8)   (9)

Figure 13.3. Approaches to enhancing binding selectivity (see text).

This problem can be overcome by solvent effects. For example, hydrophobic and related effects[13,14,39-41] can be used for ordered solvents such as water and alcohols; bulky solvents that do not effectively solvate the interior of small cavities can also be used.[42-45] For crown ether systems, which function best in organic solvents such as chloroform, the binding forces for the NH_3^+ group of alkylammonium cations are relatively strong, but the group R is not attracted to the host by similarly strong forces. Molecular enclosure in cases such as this is best achieved by two or more functional groups in the guest molecule [see X and Y in (9)], which are strongly bound by a complementary pair of receptor sites [the ellipse and the rectangle in (9)] in the host cavity. It is obvious that this strategy has a number of advantages and, not surprisingly, it is almost certainly used by many biological host molecules that select polyfunctional substrates. In a reverse sense, it is also inherent in the design of many biologically active synthetic compounds[33] that function as agonists or antagonists for the analogous natural polyfunctional substrate of a biological receptor. Two examples of the use of this strategy for the design of selective host molecules will be described.

13.3.1 Bis-crown Ethers

The formation of complexes of alkylammonium cations RNH_3^+ by all oxygen crown ethers was shown in our initial studies[46] to be extendable to the monoaza crown ethers (5) and diaza crown ethers (6). It was also shown[47,48] that, in the case of the 12- and 15-membered crown ether macrocycles, these complexes were formed stereoselectively as either the *syn*-stereoisomer (10) (Fig. 13.4) or the *syn,syn*-stereoisomer (11) (Fig. 13.4). The tertiary nitrogen atoms of the diaza crown ethers (6) provide stereochemically flexible points for the attachment of the links required for the synthesis of the ditopic systems (12) (Fig. 13.4) and a series of such compounds (13, a and b) (Fig. 13.4) were synthesized[49,50] by a simple one-step synthesis from the parent crown ethers (6, R=H). A further series of compounds (13c) was described by J.-M. Lehn and co-workers.[51] Subsequently, asymmetrical hosts of this type (13, d and e) were synthesized by a more lengthy procedure.[52] The compounds (13) were designed as selective hosts for guest bis-alkylammonium cations $H_3N^+(CH_2)_xNH_3^+$, and the rigid $-CH_2-Ar-CH_2-$ bridges were selected to provide a relatively well-defined spacing between the two crown ether macrocycles.

The synthetic hosts (13) proved[49-53] to be gratifyingly selective hosts for the bis-cations $H_3N^+(CH_2)_xNH_3^+$, and in many cases the addition or subtraction of just a single CH_2 group in the polymethylene chain of the guest results in a change in association constant for the formation of the complex (14) in chloroform or methylene chloride by a factor of 10 or more. This high selectivity is a consequence of both serendipity and design. Thus, simple comparisons of the fixed separation *l* in the host and *d*, the separation of the two N^+ centers in the fully extended all-anti conformation of the preferred guest, are consistent with a structure of type (*14*) for the complex (Fig 13.5).

The fully extended all-anti conformation of the $-(CH_2)_x-$ chain of the guest dication in (*14*) is required by the molecular cavity, which is relatively narrow as defined by the Ar⋯Ar separation and is unable to accept a guest conformation with one or more gauche bonds. In addition, the extended conformation of the guest dication minimizes the repulsion between the two N^+ centers. This very high level of chain length recognition contrasts with the bis-porphyrin systems described in the next section, which have a

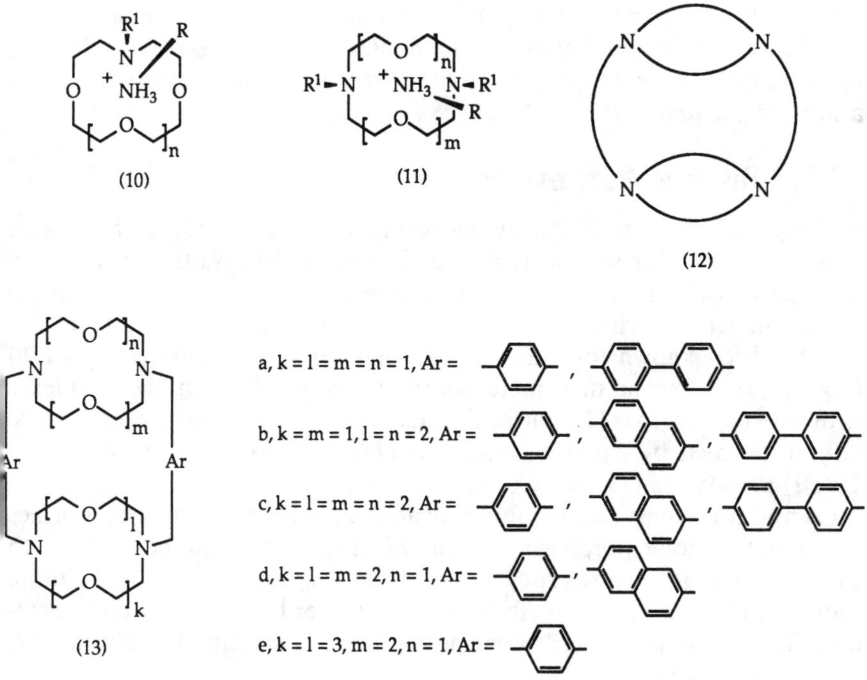

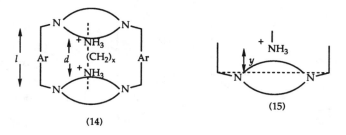

(10)

(11)

(12)

(13)

a, k = l = m = n = 1, Ar =

b, k = m = 1, l = n = 2, Ar =

c, k = l = m = n = 2, Ar =

d, k = l = m = 2, n = 1, Ar =

e, k = l = 3, m = 2, n = 1, Ar =

Figure 13.4. Crowns used to complex ammonium cations (see text).

(14)

(15)

Figure 13.5. Metrical features of bis(ammonium) cation complexes.

wide cavity and form complexes with the neutral diamines $H_2N(CH_2)_nNH_2$ rather than with the dications. Guest selection is not only a function of the parameter l, as expected; it is also a function of y [see (*15*)], the extent to which the $-NH_3^+$ group penetrates the crown ether macrocycles for optimum complexation. The value of y depends upon the size of the crown ether macrocycle and provides "fine tuning" of the selectivity, with the distance l serving as a "coarse tuning" device. Inclusion complexation, as in (*14*), is enforced by the relationship between the two functional groups in the guest and the two crown ether receptor sites in the host. It is also encouraged for hosts (*13a*) and (*13b*) with 12- and 15-membered crown ether macrocycles by preferential formation of a *syn,syn*-complex [see (*11*)], and these two series of hosts also form 1:1 (*16*) (Fig. 13.6) and 1:2 (*17*) complexes with the methylammonium cation. Such complexes are not necessarily formed by hosts based upon the larger crown ether macrocycles because these can also form the *anti,anti*-complexes (*18*).

13.3.2 Bis-zinc Porphyrins

Although this account is largely concerned with crown ethers and related host molecules, this section is included because the results contrast with those described in the previous section and also because they illustrate important general principles for the design of selective host molecules.

It has been known for many years[54] that some metalloporphyrins (*19*) (Fig. 13.7) can expand their metal coordination shell by forming complexes with one (*20*) or two (*21*) guest ligands. In particular, complexes of type (*20*) are formed by zinc(II) porphyrins (*19*, M=Zn) and of type (*21*) by Co(III) porphyrins (*19*, M=Co–L).

Complexes of type (*20*) are formally analogous to crown ether complexes (*4*) so that a ditopic porphyrin system (*22*) (Fig. 13.8) might be expected to show complexation properties for bidentate ligands analogous to those shown by the ditopic crown ether systems (*13*) for bis-alkylammonium cations. The bis-porphrins (*23*) were prepared from the diarylporphyrin (*24*) to test this idea.[55,56]

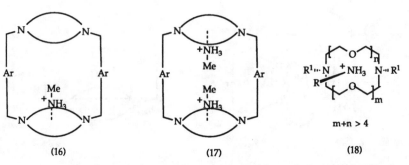

Figure 13.6. Formation of 1:1 and 1:2 complexes with methylammonium cations.

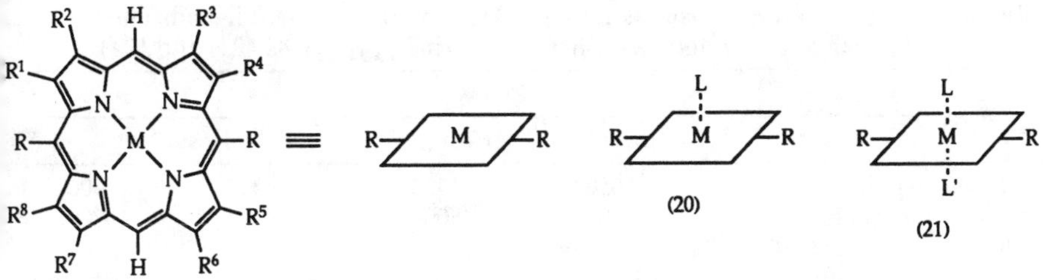

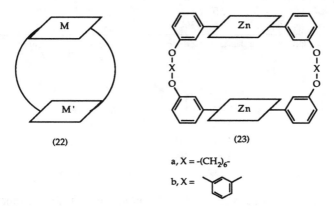

(24) $R^1 = R^4 = R^5 = R^8 = Me$, $R^2 = R^3 = R^6 = R^7 = Et$

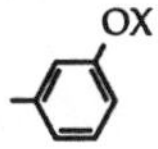

Figure 13.7. Association of metalloporphyrins with one or more guest molecules.

a, $X = -(CH_2)_6-$

b, $X =$

Figure 13.8. Ditopic porphyrins for complexation with bidentate ligands.

The hosts (*23*) were found to form complexes with a variety of diamines, and values of association constants K_a in methylene chloride are reported in Table 13.1. The results for the series of guest diamines $H_2N(CH_2)_nNH_2$ (entries *5–13*) contrast with those observed for the complexes of the bis-crown ethers (*14*) with the guest bis-alkylammonium cations $H_3N^+(CH_2)_nNH_3^+$. Both of the bis-porphyrin hosts (*23*) show relatively little recognition, through a change in K_a, of the chain length of the diamine, although the host (*23b*) with the more rigid bridging group does show rather greater recognition than the host (*23a*), which has a flexible bridging group. Even diamines such as octane-1,8-diamine, which are too long in their fully extended conformation to fit into the gap between the Zn atoms, are strongly complexed. In this case, the broad cavity allows the guest diamine to adopt conformations containing one or more gauche C–C bonds, thus reducing

Table 13.1. Association constants for complexation of bidentate ligands and analogous monodentate ligands by zinc porphyrins (*23*) and (*24*).

No.	Guest	K_a in CH_2Cl_2 $(M^{-1})^a$			
		Host (*23a*)	Host (*23b*)	Host (*24a*)	
1	pyridine	1050	1400	1900	(1900)
2	butylamine	4300	5000	5800	(5800)
3	imidazole (*25*)^b	1.0×10^4		8400	(8400)
4	quinuclidine (*26*)^c	4.1×10^4		8.5×10^4	(8.5×10^4)
5	$NH_2(CH_2)_2NH_2$	1.2×10^6	6×10^5	1.1×10^4	(5500)
6	$NH_2(CH_2)_3NH_2$	1.7×10^6		1.4×10^4	(7000)
7	$NH_2(CH_2)_4NH_2$	5.0×10^6	6×10^5		
8	$NH_2(CH_2)_5NH_2$	5.5×10^6	1×10^6	1.4×10^4	(7000)
9	$NH_2(CH_2)_6NH_2$	4.5×10^6	3×10^6		
10	$NH_2(CH_2)_7NH_2$	2.5×10^6	2×10^6	1.6×10^4	(8000)
11	$NH_2(CH_2)_8NH_2$	1.9×10^6	3×10^6		
12	$NH_2(CH_2)_9NH_2$	2.6×10^6		1.6×10^4	(8000)
13	$NH_2(CH_2)_{10}NH_2$	1.4×10^6			
14	4,4'-dipyridyl (*27*)^d	2.4×10^7	3×10^7	3500	(1750)
15	DABCO (*28*)^e	2.0×10^8		1.5×10^5	(7.5×10^4)
16	Histamine (*29*)^f	7.5×10^6		2.5×10^4	

^{a}In CH_2Cl_2 at 25°C, based upon change in absorbance at 407 and 418 nm on addition of guest to a solution of host (10^{-6} M).

^{b}(*25*) R = H
^{f}(*29*) R = $CH_2CH_2NH_2$

^{c}(*26*) X = CH
^{e}(*28*) X = N

^{d}(*27*)

the N···N separation as compared with the all-anti conformation. In fact, the diamines would not be expected to exist exclusively in the all-anti conformation, particularly the longer members of the series in which the statistical weighting of conformations containing a single gauche C–C bond would outweigh the enthalpic factor which favors the all-anti conformation.

The data in Table 13.1 also illustrate another important principle that can be used to enhance molecular recognition of difunctional or polyfunctional substrates. The simple zinc porphyrin (*24*) shows the sequence of recognition DABCO > quinuclidine > histamine > diamines > imidazole > butylamine > dipyridyl > pyridine. Since some of the selectivity may be accounted for by the two binding sites of bidentate guests, the numbers in parentheses in the last column refer to binding of a single nitrogen atom, assuming that the K_as for both the first and second binding sites are identical. This correction gives a binding sequence quinuclidine, DABCO > monoamines, diamines $\approx$ imidazole > pyridine, dipyridyl. The ditopic hosts (*23*) show a binding sequence for guests that for the most part could be expected: DABCO > dipyridyl > histamine > diamines ≫ monodentate guests; but it is evident that dipyridyl (entry *14*) is bound more strongly than expected as compared with the other guests. This is not just a consequence of a more appropriate $N \cdots N$ separation in dipyridyl, because this must be matched by at least one member of the diamine series. The anomalous position of dipyridyl in the selectivity sequence appears to be a consequence of the chelation effect.

It is known that the total free energy of binding of two ligands A and B (ΔG_A and ΔG_B) to a protein or other receptor containing two binding sites may be significantly less than the free energy of binding (ΔG_{AB}) of a ligand $A \cdots B$ in which A and B are covalently connected,[57,58] provided that the connection is consistent with the geometrical relationship between the two binding sites. Expressed in terms of association constants, this means the possibility that $K_{AB} > K_A \cdot K_B$, although there are rather few results for synthetic host molecules that unambiguously demonstrates the correctness of this proposal.

The chelation effect strictly applies to noncovalent binding at two sites. It is related to the high effective molarity[59] shown in many intramolecular reactions, but in these cases a transition state is involved with at least partly formed covalent bonds. Two-site noncovalent binding is perhaps better compared with intramolecular reactions with loosely bonded transition states, such as those involving general acid and general base catalysis, which show relatively low effective molarities.

A recent examination[60] of antifluorescyl antibodies shows that even for a conformationally well-defined substrate (and probably for rigid and selective proteins) the chelation effect ($K_{AB}/K_A \cdot K_B$) is only of the order 10–100 M. The chelation effect results from the smaller loss of rotational and translational entropy that accompanies binding of a guest $A \cdots B$ as compared with the two analogous guests A and B, and it could in theory be as high as 10^8 M.[57,59] It would be of interest to determine the maximum value that can be observed for a well-defined synthetic ditopic receptor and a bidentate guest ligand.

Estimation of the chelation effects for complexation of diamines by the bis-porphyrins (*23*) from the data in Table 13.1 generally gives values that are < 1 M for all guests except dipyridyl (*14*) for which values of 10–100 M are obtained depending upon the choice of standard for binding a single pyridine unit. These values are appropriate for a rigid guest, a fairly flexible host that can adapt to give an appropriate $Zn \cdots Zn$ separation (≈ 11.6 Å), the characteristics of $Zn \cdots N$ binding in which the Zn,N linkage is relatively long (≈ 2.2 Å) and probably flexible, and solvation by methylene chloride.

Recent publications[61-71] describe results showing that comparable ditopic systems give similar values for the chelation effect. The requirement for values > 1 M seems to be that at least one component, host or guest, should have a rather rigid structure and that there should be a good geometrical match between the disposition of the functional groups and the binding sites in the receptor.

Thus molecular recognition in ditopic systems is not only a function of guest–host fit but is also a function of guest and host rigidity. The development of rigid synthetic host frameworks, such as those found in spherands,[72,73] calixarenes,[7,74,75] rigidly linked metalloporphyrins,[65,67,76] rigid molecular tweezers,[61–63] and the natural cyclodextrins[3,64] should clarify the potential of the chelation effect and further illustrate its application for enhanced molecular recognition. The development of such rigid and well-defined host molecules will also solve the difficulties that are associated with conformationally complex flexible receptors such as the simple crown ethers.

13.4 Chromogenic Reagents and Chromoionophores

The continued success and development of host–guest chemistry will depend upon demonstrations of its ability to contribute to the social and economic needs of the 21st century, in addition to its intrinsic scientific interest. It is therefore appropriate to consider applications of molecular recognition by synthetic host molecules in this article.

Complex host molecules that show a high level of guest recognition, such as those described in the preceding sections of this article, often require long and expensive synthetic procedures and are more suitable for small scale use (μg to mg scale) than for large scale applications (g or kg scale). Their application as analytical reagents,[77–81] particularly for transduction or transport in molecular sensors, matches this requirement, and both natural and synthetic ionophores have been exploited in analytical devices such as ion-selective electrodes, ion-selective membranes, and chromogenic reagents for cations.[78–86] Work at Liverpool has been directed towards their use in optical fiber–based ion sensors.[87] An optical fiber–based sensor, shown diagramatically in Figure 13.9, is based upon the change in absorption (or reflection) spectrum at the tip of an optical fiber that is coated with ion-sensitive material. In work at Liverpool[88,89] carried out in collaboration with Professor J. F. Alder and Dr. R. Narayanaswamy of the Department of Analytical and Instrumental Science at UMIST, the ion-sensitive material consists of a selective chromogenic reagent absorbed on the surface of cross-linked XAD2 polystyrene beads (200 mesh). The performance requirements for the chromogenic reagent are dictated by the measurement for which the sensor is designed.

Target cations for sensors of this type include the metal cations that are found[1,2] at mM concentration in intra- and extracellular fluid (Table 13.2). One of the major problems in these measurements is the estimation of K^+ in extracellular fluids (such as blood) in the presence of much higher concentrations of Na^+ and comparable concentrations of Ca^{2+} and Mg^{2+}. This requires a chromogenic reagent that is sufficiently sensitive to respond to aqueous K^+ at mM concentrations and sufficiently selective to show no detectable response to Na^+, Mg^{2+}, and Ca^{2+}.

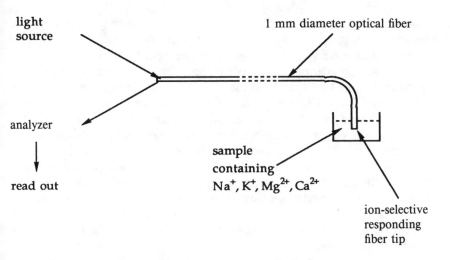

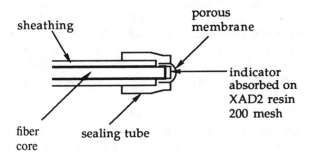

Figure 13.9. An optical fiber–based cation sensor. The top diagram shows the fiber connections, and the lower diagram shows details of the optical fiber tip.

Table 13.2. Ionic concentrations in intra- and extracellular fluid.[a]

	Mammalian tissue[b]		Squid axon[c]	
Ion	Cell	Blood	Cell	Blood
K+	139	4	400	20
Na+	12	145	50	440
Mg²⁺	0.8	1.5		
Ca²⁺	< 1 μM	1.8	0.3 μM	10

[a]Data from ref. 2; concentrations mM unless otherwise stated.
[b]Typical values for vertebrates.
[c]Typical values for invertebrates.

The requirements for a chromogenic reagent that can be used in an optical fiber–based sensor are somewhat different from those that are required for reagents that are designed to recognize cation type. It is important that the color change upon cation capture occurs only for the chosen cation and that the other cations that may be present in the target samples give minimal response (see Table 13.2 for extracellular fluid as the target). The extent of this color change should be proportional to the concentration of the selected cation, and the reagent should not become saturated within the concentration range that is likely to be measured. Furthermore, response of the sensor to changes in concentration of the cation should be adequate and rapid; otherwise the advantage of "real time" sensing, as compared with the more traditional "batch mode" sensing, is lost. Finally, it is important that the shelf life and operating lifetime of the sensor should be as long as possible; the chromogenic reagent must therefore be stable under the conditions of storage and measurement.

A number of chromogenic reagents for cations have been described in the literature. In particular, the chromophore-modified aza crown ethers such as (30) (Fig. 13.10), described by F. Vögtle and his co-workers,[90] show color changes in organic solvents that are characteristic of the cation type. Unfortunately, these compounds do not show adequate sensitivity for K^+ when used as the chromogenic reagent at an optical fiber tip, and the color changes on complexation are limited as compared with the ionizable systems described later.

The analogous chromogenic reagent (31) is potentially more sensitive than (30) because the chromophore provides a cation binding site rather than a site of cation repulsion [see partial charges in (30) and (31)], but it also fails

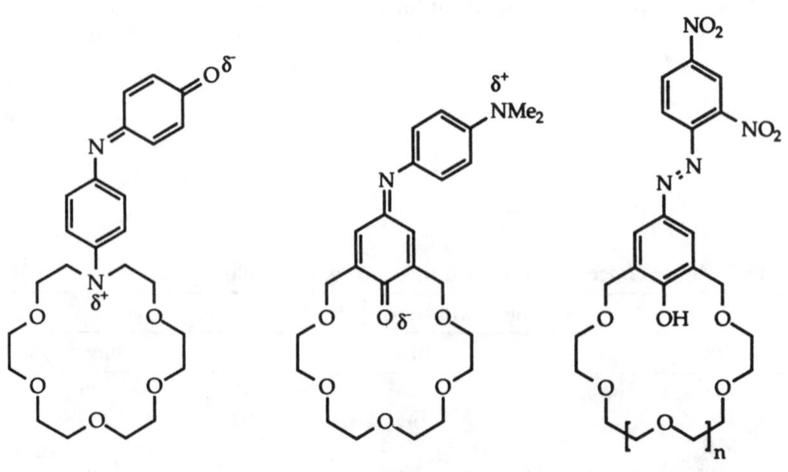

(30)   (31)   (34) a, n = 0
                         b, n = 1

Figure 13.10. Chromogenic crown ethers for detection and assay of cations.

to show adequate sensitivity. Ionizable chromogenic reagents, which have been extensively investigated by Takagi and Ueno and their co-workers,[79] proved to be more suitable. A typical reagent of this type, the 18-crown-6 analogue (*32*), was immobilized at the tip of an optical fiber under the conditions indicated by Figure 13.10. It responded[88] to mM K$^+$ in aqueous solution in the pH range 6–9 with maximum sensitivity at pH 8. Typical response curves are shown in Figure 13.11. The response of this reagent depends upon the formation of the salt (*33*) by the two phase process summarized in Figure 13.12. the "organic phase" may be either a solvent phase

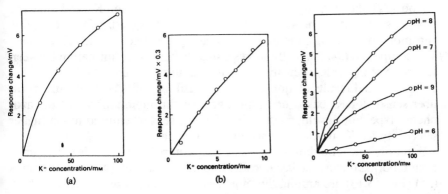

Figure 13.11. Optical fiber sensor response to aqueous K$^+$ at pH 8 in concentration range (*a*) 0–100 mM and (*b*) 0–10 mM and (*c*) to aqueous K$^+$ over the pH range 6–9.

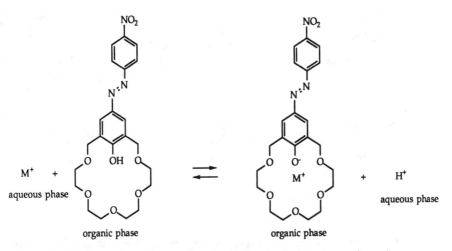

Figure 13.12. Extraction of a metal cation by an ionizable chromoionophore.

(for example, methylene chloride) or the surface of polystyrene beads as shown in Figure 13.10.

It is known that the related compounds (*34*), first studied by Kaneda and Misumi and their co-workers,[91,92] respond to a range of cations in chloroform solution in the presence of a considerable excess of pyridine. Although highly sensitive under these single-phase conditions, compounds (*34*) were reported to give rather poor results for the extraction of alkali metal cations from an aqueous phase into an organic phase. Nevertheless, compound (*32*) shows a good linear response to mM K$^+$ ions in aqueous solution as indicated in Figure 13.12*c* and, in terms of sensitivity, meets the requirements for sensing K$^+$ in extracellular fluid. However, further work demonstrated that compound (*32*) shows a higher sensitivity for Ca^{2+} than for K$^+$ and that the relative sensitivities for K$^+$ and Na$^+$ are too close [K_e(K$^+$)/K_e(Na$^+$) 6.4]; this chromoionophore (*32*), therefore, has inadequate selectivity for K$^+$ sensing. The compound (*34b*) actually shows a slightly better discrimination between K$^+$ and Na$^+$ but is less sensitive and is therefore also unsuitable.

These initial results demonstrated the feasibility of developing an optical fiber sensor for K$^+$ based upon an ionizable chromoionophore of the azophenol type, but development of an effective sensor required much higher selectivity for K$^+$ in the presence of Na$^+$ and Ca^{2+}.

A number of ionophores that have the required level of selectivity have been reported during the last decade. In particular the cryptahemispherands (*35*) (Fig. 13.13) are among the most selective ionophores for K$^+$ and Na$^+$. They have been developed[84] as highly selective chromogenic reagents (*36*) for these two cations, but the reported color changes are less suitable for sensing mM K$^+$ using an optical fiber method than are the color changes

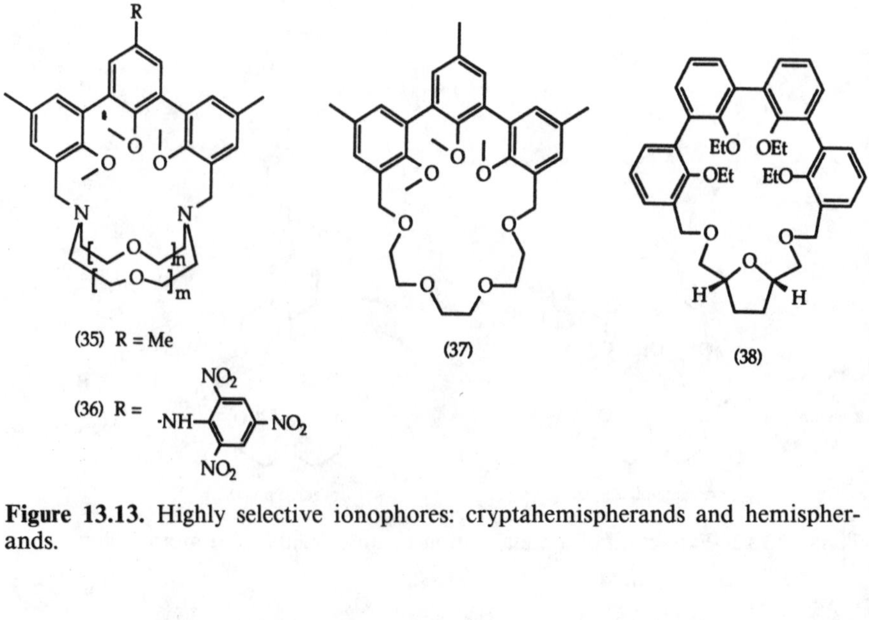

(35) R = Me

(36) R = ·NH—(NO₂, NO₂, NO₂ phenyl)

(37)

(38)

Figure 13.13. Highly selective ionophores: cryptahemispherands and hemispherands.

associated with the azophenol system (*32*), although excellent K^+/Na^+ selectivity has been reported. The hemispherand systems (*37*)[93] and (*38*),[94] although rather less selective than the cryptahemispherands (*35*), have binding constants in the same range as 18-crown-6 and were expected to provide analogues of the azophenol dye (*32*) with similar sensitivity but much improved selectivity for K^+. On this basis, the synthetically more accessible hemispherand (*39a*) (Fig. 13.14) was selected as the next synthetic target. The synthesis of the hemispherand quinone (*40a*) was carried out by a conventional route, and it was converted into the required chromoionophore (*39a*) by reaction with 2,4-dinitrophenylhydrazine. The product showed some of the required properties for K^+ sensing; for example, it

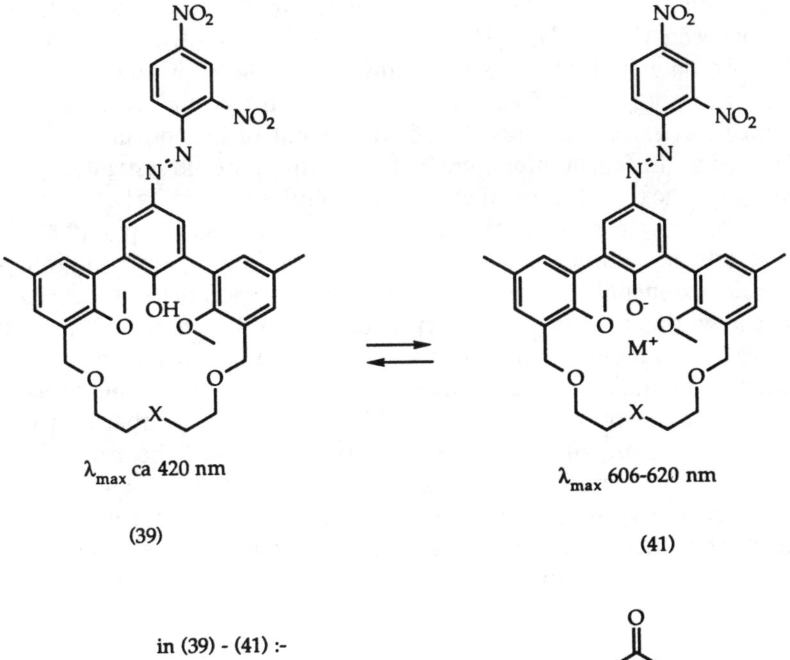

λ_{max} ca 420 nm

(39)

λ_{max} 606-620 nm

(41)

in (39) - (41) :-

a, X = $-OCH_2CH_2O-$

b, X = 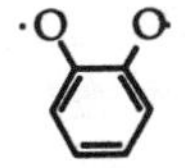

c, X = -O-

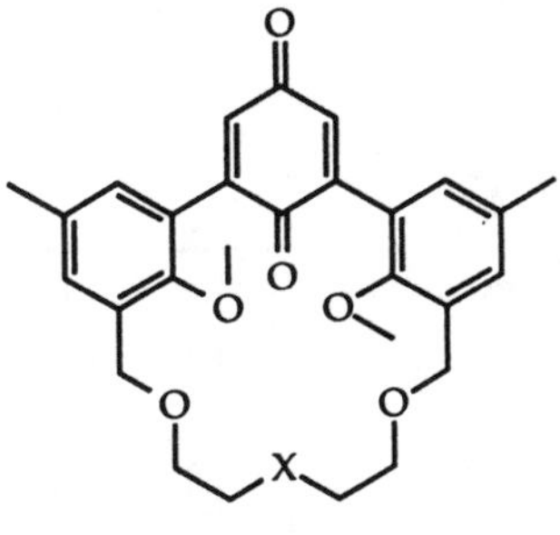

(40)

Figure 13.14. Complexation by hemispherands bearing chromophores and redox-active groups.

showed virtually no response to Mg^{2+} and Ca^{2+} in solvent extraction experiments and showed sensitivity for K^+ similar to that shown by (32). Unfortunately the selectivity for K^+ as compared to that for Na^+ was inadequate for a K^+ sensor [$K_e(K^+)/K_e(Na^+) = 26$] in spite of the high ratio of $K_a(K^+)/K_a(Na^+)$ reported for the parent hemispherand system (36). The related, more rigid structure of the closely related benzo-fused chromoionophore (39b), prepared from the corresponding quinone (40b), proved to be rather more satisfactory, and the ratio of extraction coefficients for K^+ and Na^+ shows that it is potentially suitable for sensing K^+ in extracellular fluid [no detectable extraction of Mg^{2+} and Ca^{2+}, and $K_e(K^+)/K_e(Na^+) = 170$]. The extraction coefficients for alkali metal cations by the chromoionophores (39a) and (39b) are summarized in Table 13.3; this table also includes data for the related chromoionophore (39c) which, as expected from the published results for the related hemispherand,[95] shows no ability to discriminate between K^+ and Na^+. The spectra which are shown in Figure 13.15 reflect the data in Table 13.3 and show clearly the high selectivity of the chromoionophore (39b) for K^+ as compared with the other cations that are present in extracellular fluid. The development of an optical fiber sensor for K^+ using this chromoionophore is currently under investigation.

Although the reported use of chromoionophores has largely been restricted to response towards metal cations, there have also been reports[96–98] of response towards amines, particularly for chromoionophores containing an ionizable azophenol chromophore. Preliminary experiments at Liverpool[99] have shown that a methylene chloride solution of the azophenol derivative (34b) extracts amines from aqueous solution at pH 7 with a spectroscopic response that indicates the formation of a salt (saltex).[96] This process is selective for the primary amines RNH_2, and selectivity appears to be a function of the pK_a of the amine, its lipophilicity, and the group R. The association constants for complexation of amines in single-phase systems (organic solvents) have been measured; they are dependent upon solvent polarity and amine structure and are related to the selectivity shown in the two-phase system. On the basis of these results it is clear that optical fiber

Table 13.3. Extraction coefficients[a] for alkali metal cations and azophenols (39).

Metal ion	$\log_{10}K_e$ (± 0.2) (λ_{max})					
	(39c)[a]	(λ_{max})[b]	(39a)[a]	(λ_{max})[b]	(39b)[a]	(λ_{max})[b]
Li^+	−9.34	(606)	−9.34	(594)	−9.91	(586)
Na^+	−7.06	(610)	−8.40	(606)	−9.57	(610)
K^+	−7.14	(612)	−6.99	(616)	−7.34	(612)
Rb^+	−8.38	(614)	−6.67	(620)	−6.63	(620)
Cs^+	−9.14	(614)	−6.53	(620)	−6.64	(620)

[a]Calculated with the expression $K_e = [H^+]_{aq} \cdot [ML]_{org} \cdot [HL]_{org}$ for a solution of (39) at $\approx 6 \times 10^{-5}$ M in CH_2Cl_2 and M^+Cl^- in H_2O at pH 8.
[b]For extraction of aqueous MOH (1 M) by a solution of (39) at $\approx 6 \times 10^{-5}$ M.

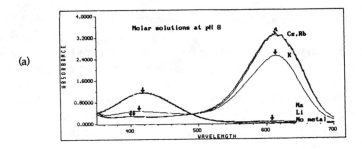

(a)

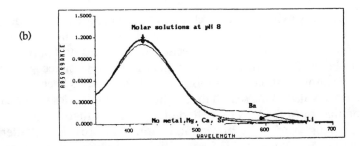

(b)

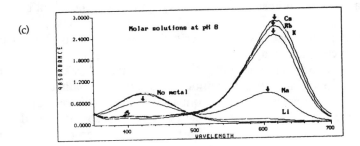

(c)

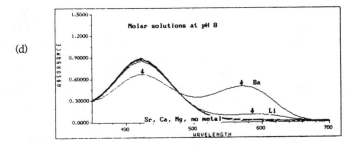

(d)

Figure 13.15. Spectra associated with the equilibrium (*39*) ⇌ (*41*) for aqueous solutions of metal cations (pH 8, 1 M) and solutions of chromoionophores (*39b*) [spectra (*a*) and (*b*)] and (*39a*) [spectra (*c*) and (*d*)] in CH_2Cl_2 ($\approx 10^{-5}$ M). The absorbance at ≈ 600 nm is due to the salt (*41*).

sensors based upon amine-selective ionizable chromoionophores could be developed for selective sensing of primary amines and their derivatives. Because optical fiber sensors of the type shown in Figure 13.9 use the sensing agent under conditions that approximate an organic phase, it should also be possible to develop selective sensors for di- and polyfunctional compounds using di- and polytopic receptors of the type that have been discussed in the previous sections. The outlook for optical fiber sensors based upon synthetic host molecules is therefore extremely promising.

13.5 Applications of Supramolecular Chemistry

Most of the very extensive research that has been carried out in the general area of host–guest chemistry has been motivated by a desire to discover more about the nature of intermolecular interactions and ways in which they can be exploited for achieving a high level of guest recognition by a relatively simple synthetic host molecule. There has also been considerable effort[100-106] devoted to the search for synthetic host molecules with a catalytic function that show the selectivity and high efficiency that is associated with enzymes. This search will, no doubt, continue to provide a challenge for the chemist and has also led to imaginative research into topics such as self replication.[106] There are many other areas that may be relevant to chemists with an interest in host–guest chemistry, and two of these will be discussed here.

The preceding account has described synthetic host molecules that are selective complexing agents for amines and cations. Incorporation of a responding chromoionophore into these hosts gives substances that respond optically to the presence of amines and cations in aqueous solution; in addition, they may be switched from a neutral to an ionized state. Synthetic host molecules are also used in electrochemical sensors to provide selective transport of the sensed species across a lipophilic membrane or to provide an ionophore[107] that shows a change in redox properties upon cation complexation.

Thus, the response of host molecules to cations may involve their electrochemical, optical, or chemical properties. This response, if cation selective, can be used in sensors and also provides a lead to other potential applications of synthetic hosts in electrical or optical devices and, ultimately, as the components of a molecule-based computer. Such an application requires molecules that can transport electrons, ions, or chemical compounds and that can be "switched" from one state to another. The additional requirement for designed emplacement of such molecules in a two- or three-dimensional assembly constitutes an aspect of supramolecular chemistry that is currently being addressed in a number of ways[108-112] but is beyond the scope of this contribution. The problem of switching and reading the states of host molecules is closer to the topics that have been described, and a number of methods can be perceived that have been discussed in the literatures.[107,113] This is a topic that is highly appropriate for responsive host molecules, particularly those of the crown ether type, and is within the reach of synthetic chemistry.

The chromoionophores demonstrate optical switching by cation capture. They can potentially "read" the state of cation transport at the molecular level. Alternatively, the switching can be electrochemical[107,114-116] if the responsive ionophore is a quinone such as (40), which will undergo a change in redox potential (enhanced stability of the reduced, radical anion or dianion states) when a cation is captured. Molecules such as the spiroindoline (42) (Fig. 13.16), described recently,[116] also undergo a significant change in molecular geometry as well as optical properties upon cation complexation. Thus, response to cation transport at the molecular level is achievable in a number of ways.

If switching is cation induced, then a further component of a molecular computer would be a switchable cation-transport system that can be isolated from molecular components other than the detecting system. Such a transport system would most simply be based upon a switchable ion channel. Ion channels are a feature of ion transport across cell membranes in biology and although they have not yet received much attention from synthetic chemists, some promising results have been reported.[117-119]

The use of synthetic host molecules in the construction of molecular sensors for medical, environmental, and industrial applications promises to be a major incentive for developing hosts that show high selectivity and a variety of responses. The high sensitivity of a fluorescent response[80,81,85,86,120-128] is particularly attractive and has been used very successfully for the detection of ions in solution. The rather longer-term research that is required for the design and synthesis of supramolecular systems that may be used as the components of a molecular computer will

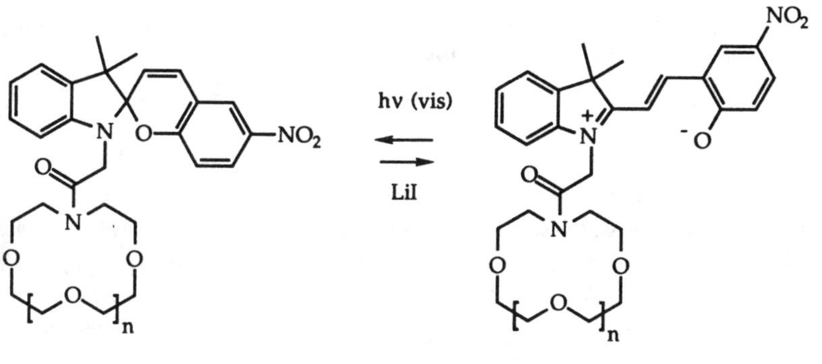

(42) n = 2,3

Figure 13.16. Reversible structural changes in a spiroindoline upon cation complexation.

provide further incentives for the expansion of the research area that has been the subject of this article, which can be traced back to the important discoveries of the late 1960s.

References

1. Stryer, L. *Biochemistry*; 2nd ed.; W. H. Freeman; New York, 1981.

2. Darnall, J.; Lodish, H.; Baltimore, D. *Molecular Cell Biology*; Scientific American Books; New York, 1986.

3. Bender, M. L.; Komiyama, M. *Cyclodextrin Chemistry*; Springer Verlag; New York, 1978.

4. Weber, E. *Topics Curr. Chem.*, **1987**, *140*, 1.

5. Pedersen, C. J. *J. Am. Chem. Soc.* **1967**, *89*, 2495, 7017.

6. Rebek, J., Jr. *Topics Curr. Chem.* **1988**, *149*, 189.

7. Gutsche, C. D. *Calixarenes*; Royal Society of Chemistry; Cambridge, 1989.

8. Hamilton, A. D. In *Advances in Supramolecular Chemistry*; Gokel, G. W., Ed.; JAI; Greenwich, Connecticut, 1990; p 1.

9. Schmidtchen, F. P. *Topics Curr. Chem.* **1986**, *132*, 101.

10. Pedersen, C. J. *Angew. Chem. Int. Ed. Engl.* **1988**, *27*, 1021.

11. Cram, D. J. *Angew. Chem. Int. Ed. Engl.* **1988**, *27*, 1009.

12. Lehn, J.-M. *Angew. Chem. Int. Ed. Engl.* **1988**, *27*, 89.

13. Diederich, F. *Angew. Chem. Int. Ed. Engl.* **1988**, *27*, 362.

14. Murakami, Y.; Kikuchi, J.; Ohno, T. In *Advances in Supramolecular Chemistry*; Gokel, G. W., Ed.; JAI; Greenwich, Connecticut, 1990; Vol. 1, p 109.

15. Clark, T. *A Handbook of Computational Chemistry*; Wiley; New York, 1985.

16. Dunitz, J. D.; Dobler, M.; Seiler, P.; Phizackerley, R. P. *Acta Cryst. B* **1974**, *30*, 2739.

17. Dunitz, J. D.; Seiler, P. *Acta Cryst. B* **1974**, *30*, 2739.

18. Bovill, M. J.; Chadwick, D. J.; Sutherland, I. O.; Watkins, D. J. *J. Chem. Soc., Perkin Trans. 2* **1980**, 1529.

19. Uiterwijk, J. W. H. M.; Harkema, S.; van de Waal, B. W.; Gobel, F.; Nibbeling, H. T. M. *J. Chem. Soc., Perkin Trans. 2* **1983**, 1843.

20. Weiner, S. J.; Kollman, P. A.; Case, D. A.; Singh, U. C.; Ghio, C.; Alagona, G.; Profeta, S.; Weiner, P. J. *J. Am. Chem. Soc.* **1984**, *106*, 765.

21. Billeter, M.; Howard, A. E.; Kuntz, I. D.; Killman, P. A. *J. Am. Chem. Soc.* **1988**, *110*, 8385.

22. Allinger, N. L.; Rahman, M.; Lii, J.-H. *J. Am. Chem. Soc.* **1990**, *112*, 8293.

23. Schmitz, L. R.; Allinger, N. L. *J. Am. Chem. Soc.* **1990**, *112*, 8307.

24. Chang, G.; Guida, W. C.; Still, W. C. *J. Am. Chem. Soc.* **1989**, *111*, 4379.

25. Saunders, M.; Houk, K. N.; Wu, Y.-D.; Still, W. C.; Lipton, M.; Chang, G.; Guida, W. C. *J. Am. Chem. Soc.* **1990**, *112*, 1419.

26. Ranghino, G.; Romano, S.; Lehn, J.-M.; Wipff, G. *J. Am. Chem. Soc.* **1985**, *107*, 7873.

27. Wipff, G.; Weiner, P.; Kollman, P. A. *J. Am. Chem. Soc.* **1982**, *104*, 3249.

28. Dang, L. X.; Kollman, P. A. *J. Am. Chem. Soc.* **1990**, *112*, 5716.

29. Grootenhuis, P. D. J.; Kollman, P. A. *J. Am. Chem. Soc.* **1989**, *111*, 2152.

30. Gehin, D.; Kollman, P. A.; Wipff, G. *J. Am. Chem. Soc.* **1989**, *111*, 3011.

31. Grootenhuis, P. D. J.; Kollman, P. A.; Groenen, L. C.; Reinhoudt, D. N.; van Hummel, G. J.; Ugozzoli, F.; Andreetti, G. D. *J. Am. Chem. Soc.* **1990**, *112*, 4165.

32. Cram, D. J.; Trueblood, K. N. *Topics Curr. Chem.* **1981**, *98*, 43.

33. Roberts, S. M. *Molecular Recognition: Chemical and Biochemical Problems*; Royal Society of Chemistry; Cambridge, 1989.

34. de Jong, F.; Reinhoudt, D. N. *Stability and Reactivity of Crown Ether Complexes*; Academic; London, 1981.

35. Kyba, E. P.; Timko, J. M.; Kaplan, L. J.; de Jong, F.; Gokel, G. W.; Cram, D. J. *J. Am. Chem. Soc.* **1978**, *100*, 4555.

36. Sousa, L. R.; Sogah, G. D. Y.; Hoffman, D. M.; Cram, D. J. *J. Am. Chem. Soc.* **1978**, *100*, 4569.

37. Knobler, C. B.; Gaeta, F. C. A.; Cram, D. J. *J. Am. Chem. Soc.* **1988**, *110*, 330.

38. Dietrich, B.; Lehn, J.-M.; Sauvage, J. P. *Tetrahedron Lett.* **1969**, 2885, 2889.

39. Odashima, K.; Koga, K. In *Cyclophanes*; Keehn, P. M., and S. Rosenfeld, Eds.; Academic; New York, 1983; p 627.

40. Smithrud, D. B.; Diederich, F. *J. Am. Chem. Soc.* **1990**, *112*, 339.

41. Schneider, H. J.; Guttes, D.; Schneider, V. *Angew. Chem. Int. Ed. Engl.* **1986**, *25*, 647.

42. Canceill, J.; Lacombe, L.; Collet, A. *J. Chem. Soc., Chem. Commun.* **1987**, 219.

43. Canceill, J.; Collet, A. *J. Chem. Soc., Chem. Commun.* **1988**, 582.

44. Cram, D. J.; Karbach, S.; Kim, H.-E.; Knobler, C. B.; Maverick, E. F.; Ericson, J. L.; Helgeson, R. C. *J. Am. Chem. Soc.* **1988**, *110*, 2229.

45. Tucker, J. A.; Knobler, C. B.; Trueblood, K. N.; Cram, D. J. *J. Am. Chem. Soc.* **1989**, *111*, 3688.

46. Leigh, S. J.; Sutherland, I. O. *J. Chem. Soc., Chem. Commun.* **1975**, 414.

47. Johnson, M. R.; Sutherland, I. O.; Newton, R. F. *J. Chem. Soc., Perkin Trans. 1* **1980**, 586.

48. Johnson, M. R.; Sutherland, I. O.; Newton, R. F. *J. Chem. Soc., Perkin Trans. 1* **1979**, 357.

49. Johnson, M. R.; Sutherland, I. O.; Newton, R. F. *J. Chem. Soc., Chem. Commun.* **1979**, 309.

50. Mageswaran, R.; Mageswaran, S. M.; Sutherland, I. O. *J. Chem. Soc., Chem. Commun.* **1979**, 722.

51. Kotzyba-Hibert, F.; Lehn, J.-M.; Vierling, P. *Tetrahedron Lett.* **1980**, 941.

52. Jones, N. F.; Kumar, A.; Sutherland, I. O. *J. Chem. Soc., Chem. Commun.* **1981**, 990.

53. Kotzyba-Hibert, F.; Lehn, J.-M.; Saigo, K. *J. Am. Chem. Soc.* **1981**, *103*, 4226.

54. Smith, K. M. *Porphyrins and Metalloporphyrins*; Elsevier; Amsterdam, 1975.

55. Danks, I. P.; Sutherland, I. O.; Yap, C. H. *J. Chem. Soc., Perkin Trans. 1* **1990**, 421.

56. Lane, T. Ph.D. Thesis. Liverpool, 1990.

57. Page, M. I. In *Chemistry of Enzyme Action*; Page, M. I., Ed.; Elsevier; Amsterdam, 1984; Chapter 1.

58. Page, M. I. In *Enzyme Mechanisms*; Page, M. I., and A. Williams, Eds.; Royal Society of Chemistry; London, 1987; Chapter 1.

59. Kirby, A. J. In *Enzyme Mechanisms*; Page, M. I., and A. Williams, Eds.; Royal Society of Chemistry; London, 1987; Chapter 5.

60. Janjic, N.; Schloeder, D.; Tramontano, A. *J. Am. Chem. Soc.* **1989**, *111*, 6374.

61. Zimmerman, S. C.; Mrksich, M.; Baloga, M. *J. Am. Chem. Soc.* **1989**, *111*, 8528.

62. Zimmerman, S. C.; van Zyl, C. M. *J. Am. Chem. Soc.* **1987**, *109*, 7894.

63. Zimmerman, S. C.; van Zyl, C. M.; Hamilton, G. S. *J. Am. Chem. Soc.* **1989**, *111*, 1373.

64. Breslow, R.; Greenspoon, N.; Guo, T.; Zorzycki, R. *J. Am. Chem. Soc.* **1989**, *111*, 8296.

65. Kugimiya, S. I. *J. Chem. Soc., Chem. Commun.* **1990**, 432.

66. Garcia-Tellado, F.; Goswami, S.; Chang, S.-K.; Geib, S. J.; Hamilton, A. D. *J. Am. Chem. Soc.* **1990**, *112*, 7393.

67. Anderson, H. L.; Sanders, J. K. M. *J. Chem. Soc., Chem. Commun.*, **1989**, 1714.

68. Adrian, J. C., Jr.; Wilcox, C. S. *J. Am. Chem. Soc.* **1989**, *111*, 8055.

69. Aoyama, Y.; Tanaka, Y.; Sugihara, S. *J. Am. Chem. Soc.* **1989**, *111*, 5397.

70. Tanaka, Y.; Kato, Y.; Aoyama, Y. *J. Am. Chem. Soc.* **1990**, *112*, 2807.

71. Aoyama, Y.; Asakawa, M.; Yamagishi, A.; Toi, H.; Ogoshi, H. *J. Am. Chem. Soc.* **1990**, *112*, 3145.

72. Cram, D. J.; Lein, G. M. *J. Am. Chem. Soc.* **1985**, *107*, 3657.

73. Cram, D. J.; Kaneda, T.; Helgeson, R. C.; Brown, S. B.; Knobler, C. B.; Maverick, E.; Trueblood, K. N. *J. Am. Chem. Soc.* **1985**, *107*, 3645.

74. Shinkai, S. In *Bioorganic Chemistry Frontiers*; Dugas, H., Ed.; Springer Verlag; Berlin, 1990; Vol. 1, pp 178–193.

75. Ghidini, E.; Ugozzoli, F.; Ungaro, R.; Harkema, S.; El-Fadl, A. A.; Reinhoudt, D. N. *J. Am. Chem. Soc.* **1985**, *107*, 4192.

76. Tabushi, I.; Kugimiya, S.; Kinnaird, M. G.; Sasaki, T. *J. Am. Chem. Soc.* **1985**, *107*, 4192.

77. *Medical and Biological Applications of Analytical Devices*; Koryta, J., Ed.; Wiley: Chichester, 1980.

78. Takeda, Y. *Topics Curr. Chem.* **1984**, *121*, 1.

79. Takagi, M.; Ueno, K. *Topics Curr. Chem.* **1984**, *121*, 1.

80. Tsien, R. Y. *Methods in Cell Biology* **1989**, *30B*, 127.

81. Tsien, R. Y. *Ann. Rev. Neurosci.* **1989**, *12*, 227.

82. van Gent, J.; Sudholter, E. J. R.; Lambeck, P. V.; Popma, T. J. A.; Gerritsma, G. J.; Reinhoudt, D. N. *J. Chem. Soc., Chem. Commun.* **1988**, 893.

83. Cram, D. J.; Carmack, R. A.; Helgeson, R. C. *J. Am. Chem. Soc.* **1988**, *100*, 571.

84. Helgeson, R. C.; Czech, B. P.; Chapoteau, E.; Gebauer, C. R.; Kumar, A.; Cram, D. J. *J. Am. Chem. Soc.* **1989**, *111*, 6339.

85. Fages, F.; Desvergne, J. P.; Bouas-Laurent, H.; Lehn, J.-M.; Konopelski, J. P.; Marsau, P.; Barrons, Y. *J. Chem. Soc., Chem. Commun.* **1990**, 655.

86. de Silva, A. P.; Gunaratne, H. Q. N. *J. Chem. Soc., Chem. Commun.* **1990**, 186.

87. Kirkbright, G. F.; Narayanaswamy, R.; Welti, N. A. *Analyst.* **1984**, *109*, 1025.

88. Alder, J. F.; Ashworth, D. C.; Narayanaswamy, R.; Moss, R. E.; Sutherland, I. O. *Analyst.* **1987**, *112*, 1191.

89. Moss, R. E.; Sutherland, I. O. *Anal. Proc.* **1989**, *25*, 272.

90. Löhr, H. G.; Vögtle, F. *Acc. Chem. Res.* **1985**, *18*, 65.

91. Kaneda, T.; Sugihara, Y.; Kamiya, H.; Misumi, S. *Tetrahedron Lett.* **1981**, *22*, 4407.

92. Sugihara, K.; Kaneda, T.; Misumi, S. *Heterocycles.* **1982**, *18*, 57.

93. Dijkstra, P. J.; den Hertog, H. J.; van Steen, B. J.; Zijlstra, S.; Skawronska-Prasinska, M.; Reinhoudt, D. N.; Van Erden, J.; Harkema, S. *J. Org. Chem.* **1987**, *106*, 2160.

94. Artz, S. P.; Cram, D. J. *J. Am. Chem. Soc.* **1984,** *106,* 2160.

95. Koenig, K. E.; Lein, P. M.; Stuckler, P.; Kaneda, T.; Cram, D. J. *J. Am. Chem. Soc.* **1979,** *101,* 3553.

96. Kaneda, T.; Ishizaki, Y.; Misumi, S.; Kai, Y.; Hirao, G.; Kasai, N. *J. Am. Chem. Soc.* **1988,** *110,* 2970.

97. Kaneda, T.; Hirose, K.; Misumi, S. *J. Am. Chem. Soc.* **1989,** *111,* 743.

98. Kaneda, T.; Umeda, S.; Ishizaki, Y.; Kuo, H. S.; Misumi, S. *J. Am. Chem. Soc.* **1989,** *111,* 1881.

99. King, A. Ph.D. Thesis, Liverpool, 1991.

100. Stoddart, J. F. In *Enzyme Mechanisms*; Page, M. I., and A. Williams, Eds.; Royal Society of Chemistry; London, 1987; Chapter 3.

101. Dugas, H. *Biorganic Chemistry—A Chemical Approach to Enzyme Action*; 2nd ed.; Springer-Verlag; New York, 1989; Chapter 5.

102. Kelly, T. R.; Zhao, C.; Bridger, C. J. *J. Am. Chem. Soc.* **1989,** *111,* 3744.

103. Hosseini, M. W.; Blacker, A. J.; Lehn, J.-M. *J. Am. Chem. Soc.* **1990,** *112,* 3896.

104. Diederich, F. *Angew. Chem. Int. Ed. Engl.* **1988,** *27,* 362.

105. Cram, D. J.; Lam. P. Y. S.; Ho, S. P. *J. Am. Chem. Soc.* **1986,** *108,* 839.

106. Tjivikva, A.; Ballester, P.; Rebek, J., Jr. *J. Am. Chem. Soc.* **1990,** *112,* 1249.

107. Gokel, G. W.; Echegoyen, L. In *Bioorganic Chemistry Frontiers*; Dugas, H., Ed.; Springer-Verlag; Berlin, 1990; Vol. 1, p 115.

108. Whitesides, G. M. In *Molecular Recognition*; Roberts, S. M., Ed.; Royal Society of Chemistry; London, 1990; p 270.

109. Sagiv, J. *J. Am. Chem. Soc.* **1980,** *102,* 92.

110. Dubois, L. H.; Zegarski, B. R.; Nuzzo, R. G. *J. Am. Chem. Soc.* **1990,** *112,* 570.

111. Bain, C. D.; Whitesides, G. M. *J. Am. Chem. Soc.* **1989,** *111,* 7164.

112. Bain, C. D.; Evall, J.; Whitesides, G. M. *J. Am. Chem. Soc.* **1989,** *111,* 7155.

113. Shinkai, S. In *Biorganic Chemistry Frontiers*; Dugas, H., Ed.; Springer-Verlag; Berlin, 1990; Vol. 1, p 161.

114. Ozeki, E.; Kimura, S.; Imanishi, Y. *J. Chem. Soc., Chem. Commun.* **1988,** 1353.

115. Echegoyen, L. E.; Yoo, H. K.; Gatto, V. J.; Gokel, G. W.; Echegoyen, L. *J. Am. Chem. Soc.* **1989,** *111,* 2440.

116. Inouye, M.; Ueno, M.; Kitao, T.; Tsuchiya, K. *J. Am. Chem. Soc.* **1990,** *112,* 8977.

117. Fyles, T. M. In *Bioorganic Chemistry Frontiers*; Dugas, H., Ed.; Springer-Verlag; Berlin, 1990; Vol. 1, p 71.

118. Carmichael, V. E.; Dutton, P. J.; Fyles, T. M.; James, T. D.; Swan, J. A.; Zojaji, M. *J. Am. Chem. Soc.* **1989,** *111,* 767.

119. Nakano, A.; Xie, Q.; Mallen, J. V.; Echegoyen, L.; Gokel, G. W. *J. Am. Chem. Soc.* **1990,** *112,* 1287.

120. Huston, M. E.; Akkaya, E. U.; Czarnik, A. W. *J. Am. Chem. Soc.* **1989,** *111,* 8735.

121. Fages, F.; Desvergne, J.-P.; Bouas-Laurent, H.; Marsau, P.; Lehn, J.-M.; Kotzyba-Hibert, F.; Albrecht-Gary, A. M.; Al Joubbah, M. *J. Am. Chem. Soc.* **1989,** *111,* 8672.

122. Fery-Forgues, S.; LeBris, M. T.; Guette, J.-P.; Valeur, B. *J. Chem. Soc., Chem. Commun.* **1988,** 384.

123. Adams, S. R.; Kao, J. P. Y.; Tsien, R. Y. *J. Am. Chem. Soc.* **1989,** *111,* 7957.

124. Akkaya, E. U., Huston, M. E.; Czarnik, A. W. *J. Am. Chem. Soc.* **1990,** *112,* 3590.

125. Huston, M. E.; Engleman, C.; Czarnik, A. W. *J. Am. Chem. Soc.* **1990,** *112,* 7054.

126. de Silva, A. P.; Sandanayake, K. R. A. S. *Angew. Chem. Int. Ed. Engl.* **1990,** *29,* 1173.

127. de Silva, A. P.; Sandanayake, K. R. A. S. *J. Chem. Soc., Chem. Commun.* **1989,** 1183.

128. de Silva, A. P.; de Silva, S. A.; Dissanayake, A. S.; Sandanayake, K. R. A. S. *J. Chem. Soc., Chem. Commun.* **1989,** 1054.

CHAPTER 14

Synthesis and Chemistry of Macrocyclic Sulfides (Thiacrown Ethers)

Richard M. Kellogg
Department of Organic Chemistry
University of Groningen
Nijenborgh 4
Groningen 9747 AG
The Netherlands

14.1 Introduction

Dramatic occurrences in chemistry, as in the greater arena of history, are commonly seen only in retrospect. At the time only a presentient few recognize the implications. Although the language is devoid of any drama, the observation by Charles Pedersen[1] with regard to the conversion depicted in Equation 14.1, namely that "1 mole (360 g) of product can be synthesized in a volume of 5 L," marked the true opening of crown ether chemistry. Compounds like dibenzo-18-crown-6 (*3*) would not be laboratory curiosities but rather materials available via a remarkably simple synthesis in sufficient quantity to allow extensive study. The success that followed is history and needs no retelling. However, the fact that synthetic methodology was an irreplaceable key to that success does deserve emphasis here. Templating by the cation associated with the nucleophile (the bis-phenolate of *1*) seems to be the important extra factor that drives the macrocyclizations.[2]

Equation 14.1

Innumerable oxa- and azacrown ethers have now been prepared and studied. Templated syntheses have been of major importance in solving the problems of macrocyclization. The step to sulfur (as sulfide) as a component of macrocycles was obvious. However, one may anticipate immediately that templated syntheses, if dependent on complexation of sulfide to a hard cation, probably will be less efficacious than in the oxa- and aza-cases because of the lack of affinity of sulfides for the alkali metal cations usually employed.

Although macrocyclic sulfides had long been known, serious synthetic activity really began together with the other developments in crown ether chemistry. This chemistry has been well reviewed by Bradshaw.[3] Templated approaches indeed enjoyed little success. Syntheses by Lehn[4] and Vögtle,[5] for example, involved high-dilution techniques and/or the formation of nonsulfide bonds in the ring-closure step, such as amide bonds from the reaction under high-dilution conditions of amines with acid chlorides. In many cases[3] moderate to excellent yields could be obtained although the use of high dilution may be experimentally daunting, especially with regard to syntheses on a large scale. A representative example is shown in Equation 14.2[6]

Although doubtless an oversimplification, one is left with the impression that the synthetic effort required, the general unpleasantness of thiols as intermediates, and the absence of dramatic observations of truly unusual properties discouraged for some time any further systematic investigation of such compounds. Attention was focused on oxa- and azacrowns because they could be prepared without untoward difficulties and because the chemistry was so rewarding.

14.2 Synthetic Methodology

In this chapter I will try to sketch what I believe to be significant developments in and possibilities for macrocyclic sulfides. I will place our own work of recent years (and currently in progress) in the larger context of the work of others. The literature cited is meant only to be indicative and in

Equation 14.2

no fashion complete. Very significant work is not mentioned for reasons of economy of space.

We became interested in macrocyclic sulfides in general and thiacrown ethers (this term is used here very loosely to describe compounds with multiple, usually three or more, sulfide linkages) in particular as a test case for a synthetic approach that we had developed wherein cesium salts play a central role.[7] The general approach is either two (Equation 14.3) or one component (Equation 14.4); the two-component approach of Equation 14.3 involves, as an intermediate state, a single component as in Equation 14.4. Extension to thiolates was entirely logical. The cesium salts are formed directly in dimethylformamide (DMF) solution by deprotonation with Cs_2CO_3. Reasonably acidic reactants are required. Carboxylic acids and phenols can be used but not amines (these can be activated as tosylamides,

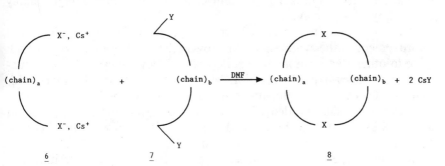

Equation 14.3

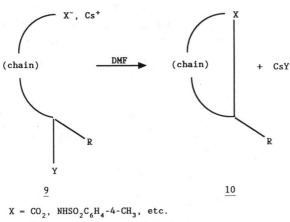

$$X = CO_2, \ NHSO_2C_6H_4\text{-}4\text{-}CH_3, \ etc.$$
$$Y = Br, \ Cl, \ OSO_2CH_3, \ OSO_2C_6H_4\text{-}4\text{-}CH_3$$
$$R = H, \ CH_3, \ aryl$$

Equation 14.4

which are acidic enough to be deprontonated by Cs_2CO_3 but not by Cs-$HCO_3)^8$ or alcohols. Both aryl and alkylthiols are sufficiently acidic to be deprononated readily in DMF by Cs_2CO_3 (two equivalents of thiol per Cs_2CO_3).

The situation with regard to the synthesis of macrocyclic sulfides including the (lack of) impact of templated syntheses has been sketched in the Introduction. The following gives an idea of the magnitude of the synthetic problems. The sulfur analog of 18-crown-6, hexathiaether *11,* had been isolated in 1934 by Meadow and Reid[9] in 1.7% yield when prepared by the tetracomponent route of Equation 14.5. Ochrymowycz[10] adopted a two-component approach (see Equation 14.3) using 3-thiapentane-1,5-dithiol and 1,11-dichloro-3,6,9-tetrathiaundecane in basic n-butanol and was able to raise the isolated yield to 33%. High-dilution conditions were not used. Despite this significant improvement, extensive purification was still required. The vesicant properties of some of the intermediates were justifiably stressed by Ochrymowycz.

We believe that cesium salts have offered in many cases a significant improvement in the synthesis of thiacrown ethers.[11] The improvement relative to other alkali cations was quite dramatic for the case of simple macrocyclic sulfides like *12* (Equation 14.6).[11] The synthesis of "mixed" 18-crown-6 derivative (*13*) illustrates the power of the method. A previous synthesis of *13* had afforded *13* in 29% yield.[12] Using the two-component approach exemplified in Equation 14.7, hexathia-18-crown-6 itself can be obtained on a 50 g scale in 84% isolated yield.[13]

We[7] and Cooper[14] have emphasized that it is unlikely that cesium ions are involved in any templating of macrocyclization. Most likely the cesium thiolates are present in DMF as solvent-separated, highly reactive ion pairs. The yields obtained are roughly equal to or somewhat higher than those one would anticipate purely on the basis of the concentrations used and the estimated effective molarities[15] for the rings involved. For 18-membered lactones, the effective molarity for formation is about 10^{-2} mol.[15] The rates of oligomerization and cyclization are by definition equal at this concentration, and one anticipates as a first order approximation a 50% yield of

$$\frac{C_2H_5OH}{C_2H_5O^-, \ Na^+}$$

11

Equation 14.5

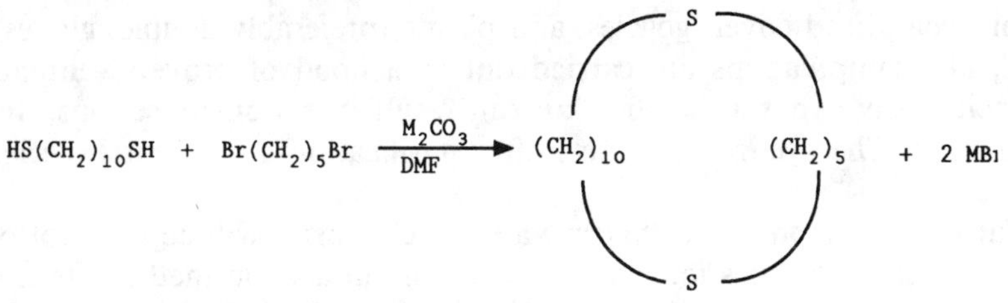

$$HS(CH_2)_{10}SH \ + \ Br(CH_2)_5Br \ \xrightarrow[\text{DMF}]{M_2CO_3} \ $$

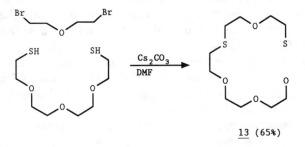

12

M_2CO_3	yield (%)
Cs_2CO_3	90% (isolated)
Rb_2CO_3	84% (glpc)
K_2CO_3	48% (glpc)
Na_2CO_3	33% (glpc)
Li_2CO_3	0% (glpc)

Equation 14.6

Equation 14.7

macrocycle.[15] Such an argument ignores cation effects, of course. We generally hold the concentration under 10^{-2} mol by adding the reactants by means of a dropping funnel to hot DMF. Several macrocyclizations are done one after the other in the same batch of solvent to make optimal use of DMF.[16] Difficult cyclizations seem invariably to proceed better with Cs_2CO_3; less demanding cyclizations can often be carried out satisfactorily with cheaper K_2CO_3 in DMF.

The various approaches available to thiacrown ethers/macrocyclic sulfides have been carefully summarized by Cooper.[14] A special word of caution is necessary at this point, however. The procedures developed by Ochrymowycz[13] and by us often involve the use of structural units with the segment $-SCH_2CH_2X$ where X is halide or another leaving group. Such compounds generally are vesicants and/or mustards and must be handled with care. We require as standard laboratory precaution a full-body laboratory coat, head cover, goggles, and plastic (preferably double) gloves; as well, all manipulations are carried out in a hood of proven ventilation capacity. Any exposure to the skin can result in extremely serious blister formation. The cyclized products after purification are in our experience innocuous.

Further development of thiacrown ether chemistry will depend not only on successful strategies for macrocyclization but also on methods to introduce functionality into the macrocyclic periphery. To this end, commercially available isobutenyl dichloride (*14*) has been examined as a possible building block for the macrocyclic skeletons. The unique reactivity of disubstituted isobutenyl systems, in particular the (possibly aromatic) dianion, has been emphasized.[17] Moreover, the carbon skeleton is that of trimethylenemethane, which has been applied with remarkable success in some transition metal catalyzed processes.[18]

Isobutenyldichloride (*14*) reacts cleanly with, for example, *15,* to provide in 80% yield macrocycle *16* (Equation 14.8). Compounds *17–20,* obtained in 52–100% yields,[19] illustrate the structural variation possible. An alternative approach to these types of macrocyclic systems is through *23,* readily prepared as shown in Equation 14.9. Although the use of cesium thioacetate (*21*) as nucleophile is not mandatory, the material does have the advantage of being a readily handled, stable solid that is quite soluble in DMF (concentrations of about 1 mol are used, which leads to a high rate of nucleophilic substitution). It is prepared from Cs_2CO_3 and thiolacetic acid in CH_3OH, followed by removal of solvent. The dithiol (*23*) forms in DMF a fairly soluble dicesium salt, which may be used in macrocyclizations with appropriate long chain dihalides, dimesylates, or ditosylates.

To our pleasant surprise, 1,3-dichloroacetone (*24*) is also very suitable as a component in macrocyclizations. A representative example is *26,* prepared as illustrated in Equation 14.10.[20] Compounds *27–30,* obtained in 70–80% yields, give an indication of some of the structural variation that can be achieved using the branched structures *14* and *24*. Compound *31* represents

Equation 14.8

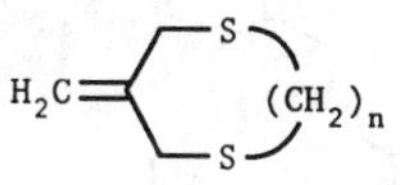

17
a) n = 5
b) n = 10
c) n = 12
d) n = 16

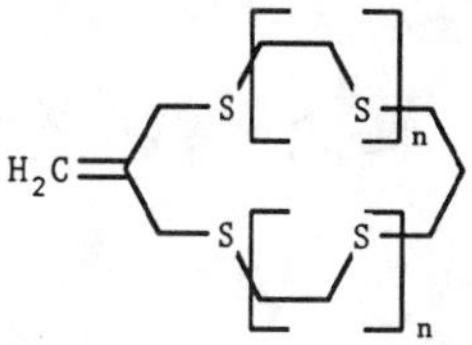

18
a) n = 1
b) n = 2

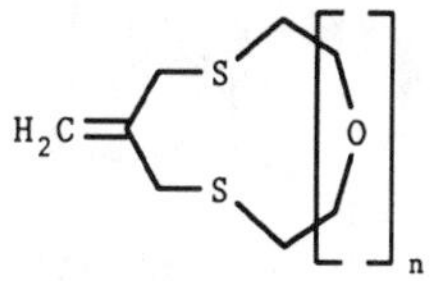

19
a) n = 1
b) n = 2

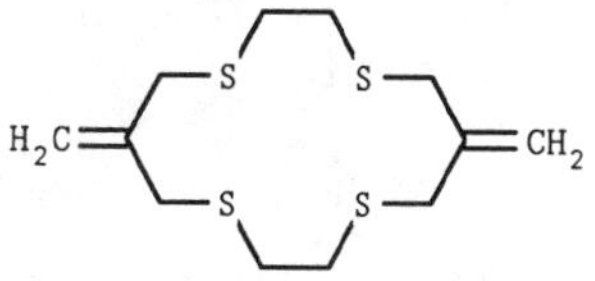

20

Compounds 17–20

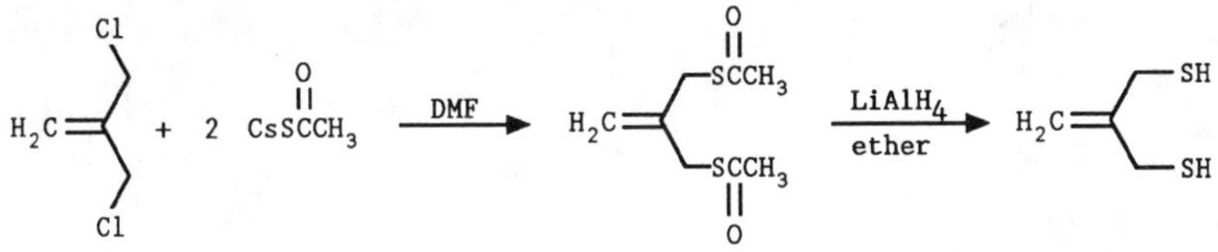

14   **21**   **22**   **23**

Equation 14.9

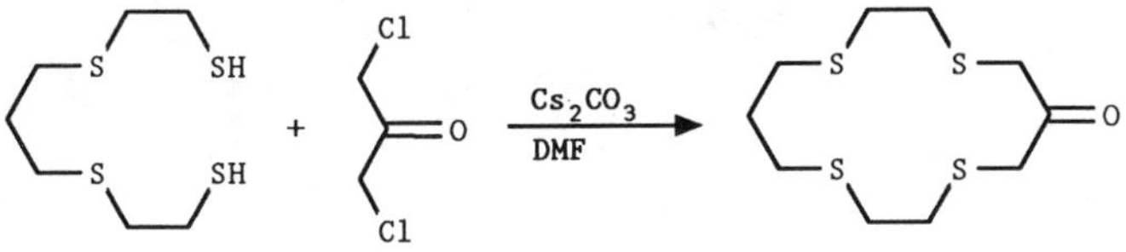

25   **24**   **26**

Equation 14.10

a combination of *14* and *24* in a single macrocycle. Apparently the reason that *24* can be used with such success is that cyclization occurs under basic, aprotic conditions under which thioketal formation cannot take place. However, the conditions are not basic enough to induce Favorskii chemistry (Equation 14.11).

Care must be taken, however. Diketone (*37*) is prepared as shown in Equation 14.12; in this case the keto functionality must be protected until

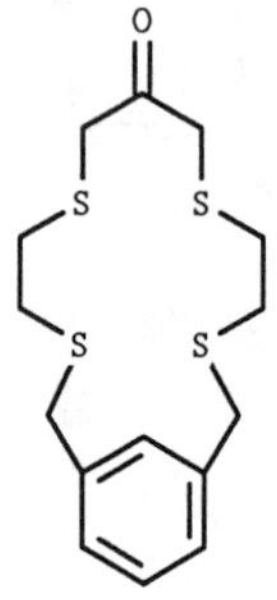

27 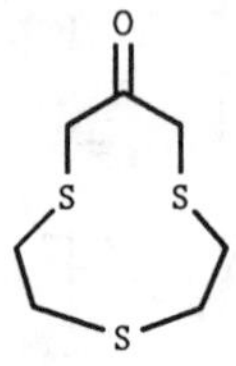   28 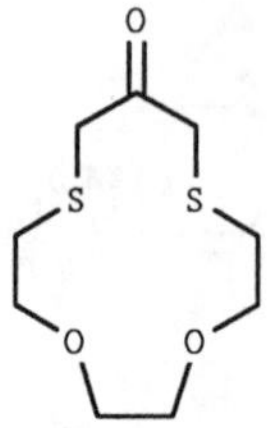   29

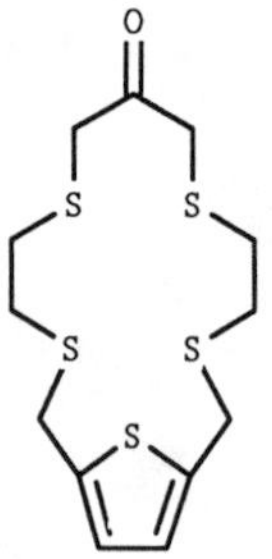

30 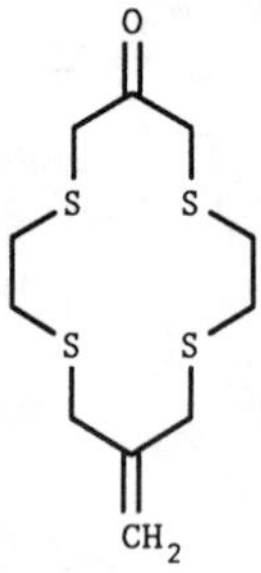   31

Compounds 27–30

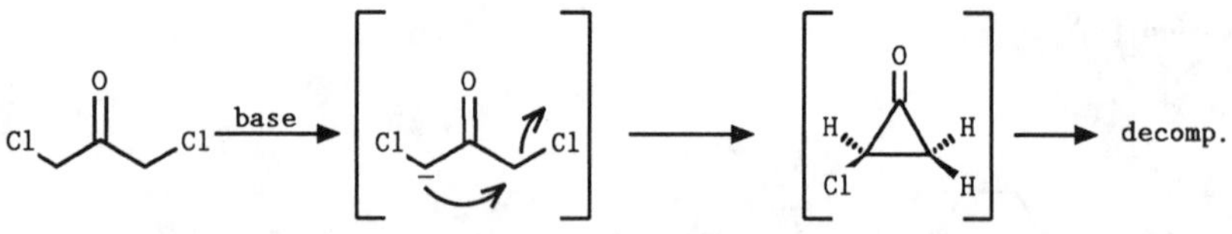

24

Equation 14.11

the final step to prevent thioketal formation. A key intermediate is *32*, which is readily formed but is rather difficult to substitute because of its neopentyl character. Unprotected *38* was prepared, but this was totally unsuitable for macrocylizations because of instability probably caused for the reasons illustrated in Equation 14.13.

14.3 Chiral Nonracemic Macrocyclic Sulfides

Another synthetic problem that has to be met is the introduction of chiral, nonracemic units into the macrocyclic framework. Tartaric acid (*39*) is read-

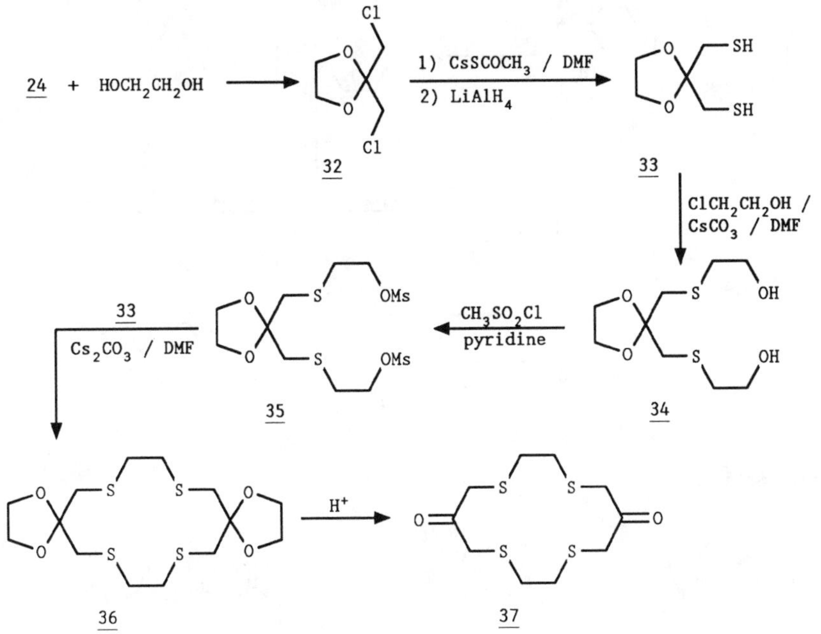

Equation 14.12

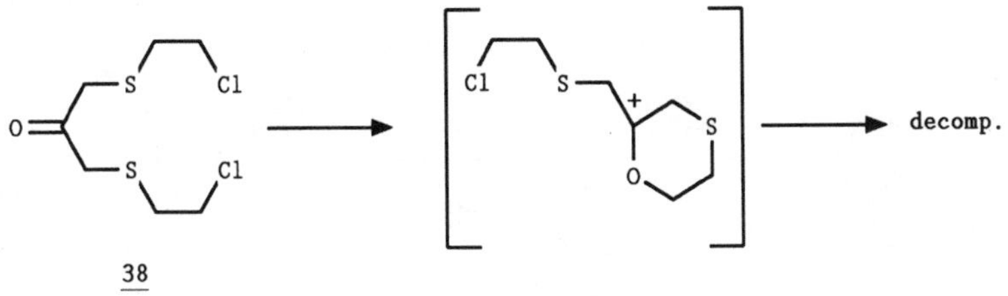

Equation 14.13

ily converted to the dimesylate (40), which can be used to form, for example, 41 and 42 (Equation 14.14).[21]

Amino acids may also be incorporated as chiral components. An example, namely 46, wherein phenylalanine (43) is used is shown in Equation 14.15.[21] In a similar fashion (Equation 14.16) L-cysteine (47) was transformed into 49. This synthesis was particularly arduous;[21] the intermediate 48 is very reactive, difficult to handle, prone to racemization, and is an especially strong vesicant.

The final type of chiral, nonracemic unit that we have examined is the well-known bis-naphthol (50). Following the general procedures given in

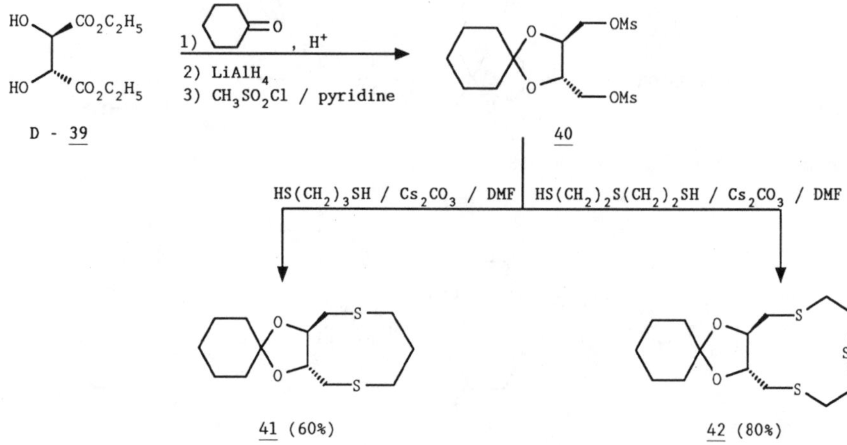

Equation 14.14

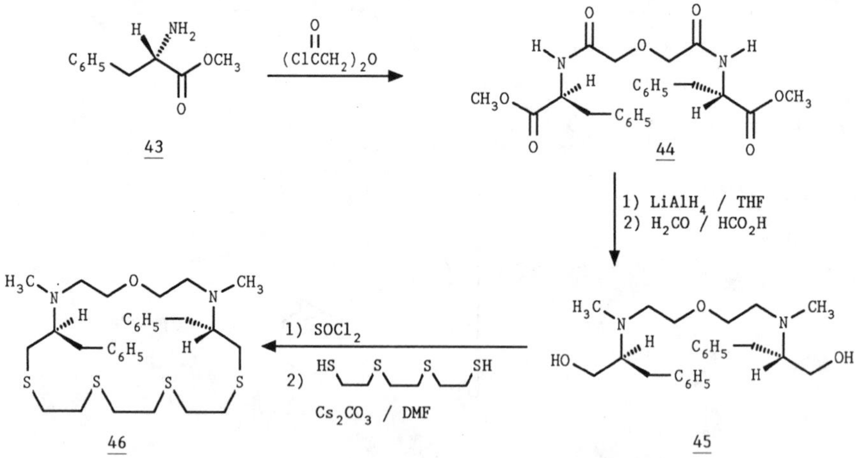

Equation 14.15

the foregoing equation, for example, *51* has been prepared in optically pure form (Equation 14.17).[22]

14.4 Demonstrated Chemistry and Potential Chemistry

What sort of chemistry may one expect from these thiacrown ethers? Cooper and Rawle[23] have expertly discussed the structural aspects of the coordination chemistry of known thiacrown ethers. Cr, Mn, Fe, In, Cu, Au, and Ag ions, just to name some examples, can complex with soft (relative to ether oxygen) sulfide linkages, and the effect of multiple sulfide sites can be

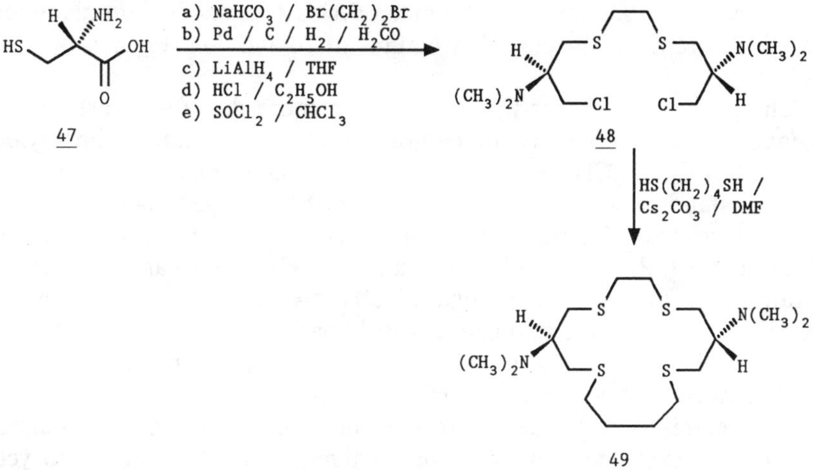

Equation 14.16

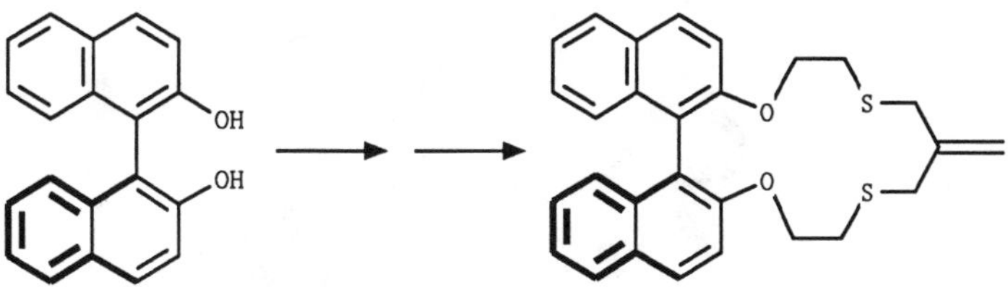

Equation 14.17

cooperative. The macrocyclic effects tend to be of lesser magnitude than those observed with oxa- and azacrown ethers, probably for conformational reasons. Cooper,[23] on the basis of accumulated crystallographic data, has noted the pronounced tendency of $-SCH_2CH_2S-$ units to adopt anti conformations instead of the gauche conformations favored in crown ethers. This factor causes many thiacrown ethers that contain this segment to adopt conformations with the sulfurs placed "outside" (exodentate)[24] the macrocyclic ring. Extensive conformational reorganization is required to place a metal ion above the cavity of the macrocycle. This exodentate placement is well illustrated in the crystal structures of *26* and *18a* (Figures 14.1 and 14.2).

Conformations of trimethylene units are more difficult to predict. An essential question for us is the conformation of substituted trimethylene units $-SCH_2C(=X)CH_2S-$, where X is CH_2 and O. In Figures 14.1 and 14.2 this unit stands essentially perpendicular to the averaged plane formed by the sulfur atoms. This places the sulfurs *syn* with respect to the exocyclic double bond (*52*); from CPK molecular models it is clear that *anti,anti* conformation (*53*) is virtually unattainable owing to steric repulsions between the sulfur atoms, as illustrated in Equation 14.18. Conformation *54* (*syn, anti*) should in principle be accessible as well as several conformations not illustrated here wherein the thiasubstituents are rotated out of the plane of the page. It is difficult to ascertain whether the occurrence of conformation *52* is dictated exclusively by conformational requirements of the macrocyclic ring or whether this reflects an intrinsic stereochemical property of the grouping $-SCH_2C(=X)CH_2S-$.

That some conformational reorganization occurs on the binding of a metal ion is seen in the Ag⁺ complexes of *18a*; in one case (Figure 14.3) complexation of carbonyl oxygen is observed, but when triflate is the anion (Figure 14.4) an infinite chain of units where in each Ag⁺ is coordinated by four different sulfides from four individual crown ethers is obtained.[20] "Holes" are created between the molecules, and these are filled with triflate anions. Whether oxygen is complexed or not ($AgClO_4$ compared to Ag-

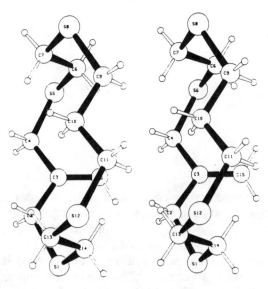

Figure 14.1. Crystal structures of compound *26*. Hydrogen atoms have been omitted for clarity.

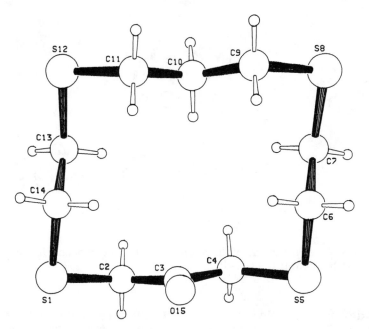

Figure 14.2. Crystal structure of compound *18a*.

52 syn, syn

53 anti, anti

54 syn, anti

(X = CH$_2$, O)

Equation 14.18

OSO$_2$CF$_3$) the conformation *52* about –CH$_2$(C=O)CH$_2$– is maintained, and this unit remains roughly perpendicular to the plane of the sulfides. However, in neither structure is the metal ion held in the ring; complexation has occurred exterior to the macrocycle.

The functionalized thiacrown ethers that we have prepared clearly offer possibilities in the field of metal ion coordination. The conformational and structural aspects will be extensions of the discussion already adequately given by Cooper.[23] There are, however, other chemical aspects that complement the aspect of ion complexation and which give a unique dimension to thiacrown ethers.

We begin with the exomethylene function obtained on use of *14* as a building block. Further development of the structure of the methylene unit

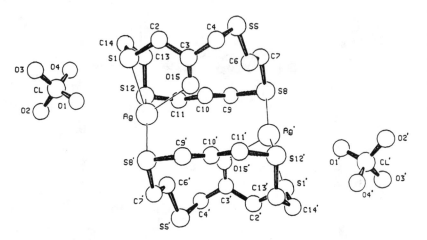

Figure 14.3. Crystal structure of the AgClO$_4$ complex of compound *18a*. Hydrogen atoms have been omitted for clarity.

via hydroboration is obvious (Equation 14.19). This would result, as illustrated, in the classical anti-Markovnikov addition of H$_2$O. The conversion of the trialkylborane to alcohol with H$_2$O$_2$ is, however, a reason for concern owing to the sensitivity of sulfides to oxidation. On the other hand, sulfides complex with BH$_3$ to form dipolar bonds, the stable complex (*55*) of BH$_3$ with (CH$_3$)$_2$S being an example.[25-27] There appears to be little known about the behavior of such complexes toward oxidizing agents. Our experience[19] has been that complexation does protect the sulfide linkages from oxidation. For example, *26* is cleanly converted to *56* (Equation 14.20) with no significant formation of sulfoxides or sulfones. The intermediate borane is stable to prolonged exposure to air; the stoichiometry is unfortunately variable, and the material has resisted investigation. Thiacrowns like *57* form also a very stable but nonstoichiometric complex with BH$_3$. This reverts to *57* on exposure to borane acceptors like alkenes and carbonyl compounds. Although this and other complexes that we have made are generally extremely stable and can be handled without any precautions (note for example that the simple borane sulfide complex *55* is pyrophoric), the insolubility has precluded structural characterization. A macrocyclic effect definitely seems to be involved; open-chain analogs of, for example, *57,* lose BH$_3$ on evaporation of THF solvent.

Numerous applications of these materials can be imagined. Enantioselective reductions with chiral thiacrown ethers, as BH$_3$ complexes, are particularly attractive. However, the obtainment of stoichiometric, structurally defined complexes looks to be a prerequisite for further study. A more shielded environment wherein the borane units are prevented from intermolecular association seems a logical solution to this problem. Bis-naphthol units have proven capabilities in cavity formation. Compound *58,* for ex-

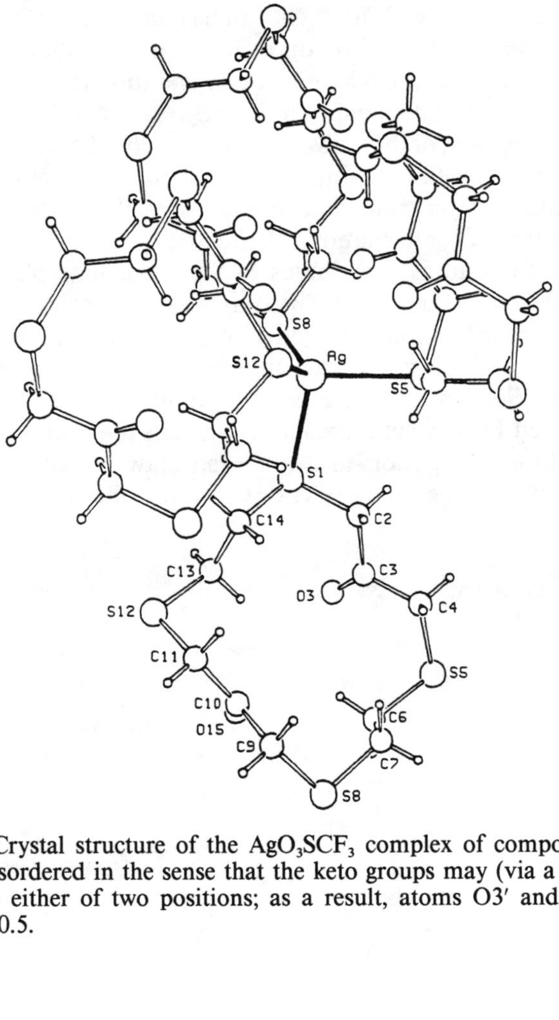

Figure 14.4. Crystal structure of the AgO$_3$SCF$_3$ complex of compound *18a*. The structure is disordered in the sense that the keto groups may (via a pseudo-C$_2$ operation) be in either of two positions; as a result, atoms O3' and O15 have an occupancy of 0.5.

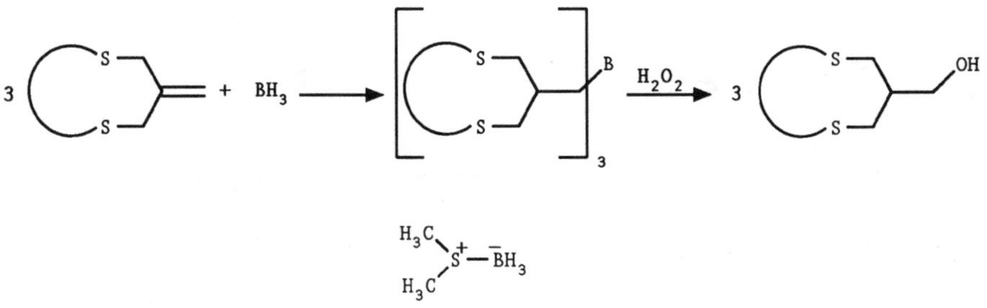

Equation 14.19

ample, forms a complex with BH_3 that can be handled in solution (Equation 14.21).[22] The complex loses BH_3 on evaporation but study in solution is possible; examination of the possibilities for enantioselective reactions with carbonyl compounds and alkenes are in progress.

Further development of this and other forms of the chemistry of thiacrown ethers will depend on the obtainment of information about how multiple sulfide sites can work in concert. There are several aspects to such understanding. The first is conformation. The tendency of $-SCH_2CH_2S-$ units to adopt anti conformations contributes to the exodentate placement of the sulfides in a thiacrown ether.[23,28] This is a deleterious conformational effect with regard to coordination. Energetically expensive conformational reorganization must take place, or coordination will occur exterior to the macrocyclic rings. Solution of these conformational problems can probably best be achieved by synthetic strategies wherein the anti conformation is destabilized. More straightforwardly, one can draw the conclusion that it is better not to include the unit $-SCH_2CH_2S-$ but rather to use trimethylene

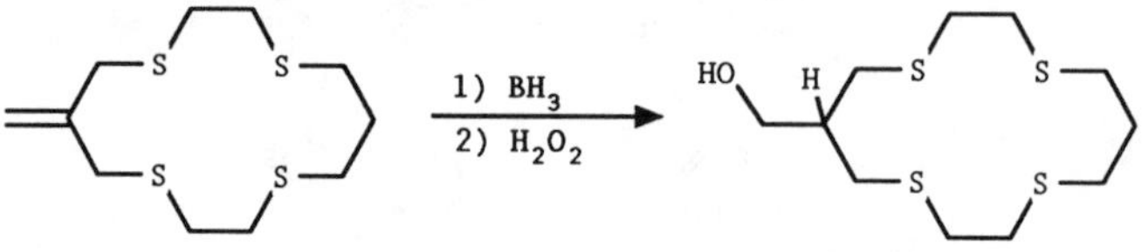

Equation 14.20

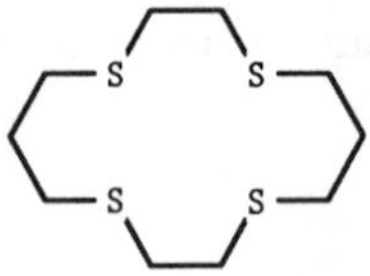

Structure 57

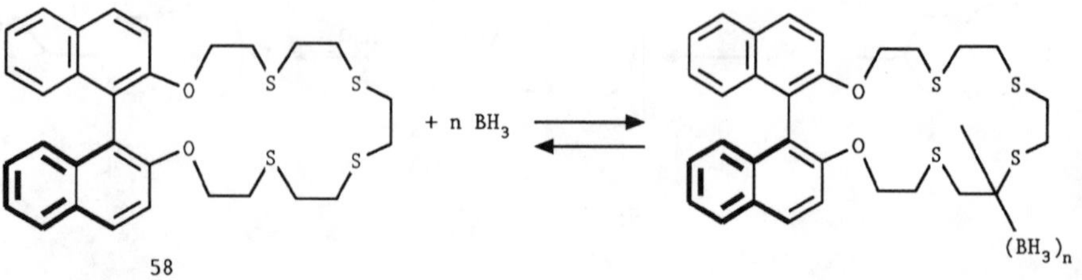

Equation 14.21

spacers. The resulting 1,5-placement of the sulfides is certainly generally better for ligation to metal ions. However, conformational predictions for these trimethylene units are not straightforward, chiefly because of lack of good structural information. Introduction of gem-dimethyl groups has been shown to force the sulfide units in *59* "inside" rather than "outside."[29] Other synthetic approaches to accomplish the same end are conceivable.

Direct bonding is another form of sulfur–sulfur interactions. In this regard, 1,5-dithiocane (*60*) and other 8-membered rings have provided a wealth of information about the possibilities for chemical interactions between sulfur sites. The key to understanding the chemistry is the realization that sulfide is a good electron donor toward electrophiles and oxidizing agents. Leonard[30] demonstrated, for example, that *61* delivers not the carbonium ion but rather a sulfonium ion (*62*) via 1,5-transannular participation (Equation 14.22). What was less anticipated was the readiness with which this type of transannular participation could be extended to the two sulfur atoms of *63* (Equation 14.23).[31] The conformation *60* is such that the dialkylated disulfide (*65*) can be generated, and this can participate in further redox chemistry.[32] One-electron oxidizing agents lead to the cation radical of *65*, which can also participate in various types of redox processes.[33] This

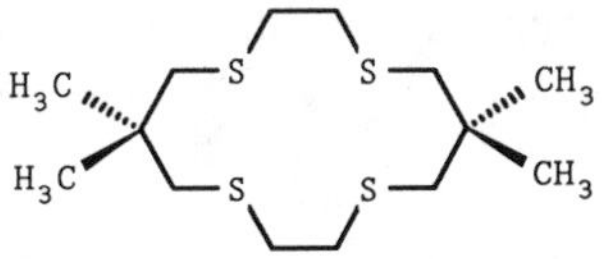

59

Structure 59

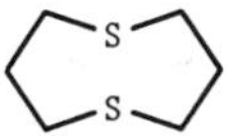

60

Structure 60

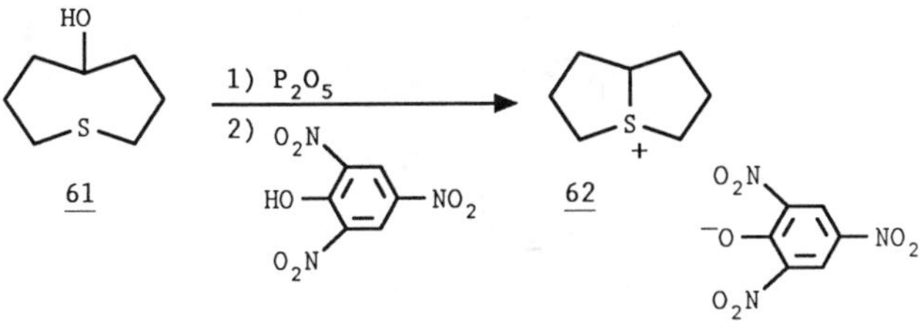

Equation 14.22

type of chemistry has received extensive attention; the references cited are purely indicative.[34]

From thiacrown ethers/macrocyclic sulfides we may anticipate the accessibility of a multitude of dialkylated disulfides and radical cations, the exact structures of which depend on conformations of the macrocycle. Systems with more than one functionality are imaginable as schematically indicated in *67*. With such intermediates an opportunity is offered to span a wide range of redox potentials, thereby increasing the scope for electron transfer in organic solution. Such systems may also be relevant to certain types of biological electron-transfer reactions. The redox chemistry of metal complexes should be equally rich.

Sulfide is also the precursor of sulfonium salts, sulfoxides, and sulfones. Little indeed is known about, for example, macrocyclic sulfonium salts. Is it possible to design multisulfonium receptors for tetracoordinate ions like PO_4^{3-} using the sulfonium as oxygen binder (*68*)? Thus far, ammonium ions have chiefly been used for this purpose.[35] Interesting challenges in the control of conformation and stereochemistry of the sulfonium sites must first be surmounted, however.

Similar arguments apply to sulfoxides. Problems of control of stereochemistry and conformation must again be dealt with successfully. The semipolar bonds may very well be able to act in a cooperative fashion to encapsulate cations as schematically illustrated for a tris-sulfoxide combi-

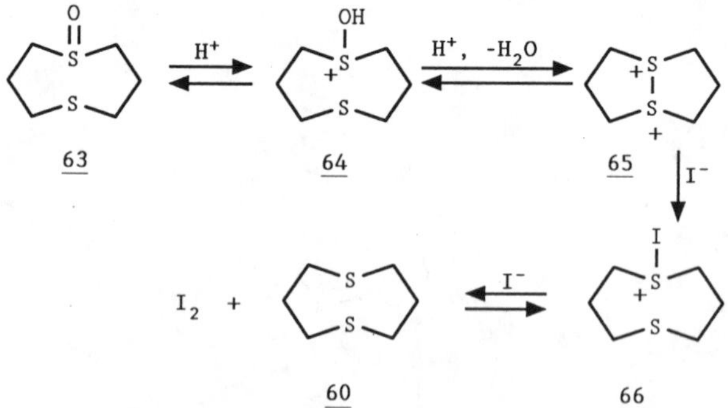

Equation 14.23

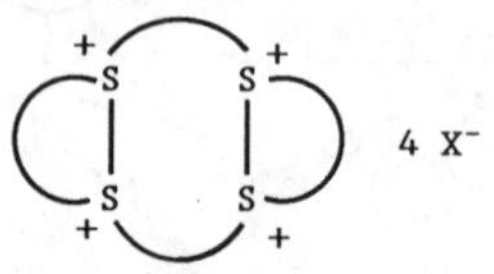

67 (or cation radical)

Structure 67

nation (*69*). Virtually nothing is known whether a number of sulfide sites in close proximity can all be oxidized to the sulfoxide oxidation state without intramolecular participation or complications arising from overoxidation to the sulfone oxidation level. The sulfones are, of course, also potentially interesting materials. For example, cyclic polyvinylsulfones (*70*) are in principle available by oxidation of the appropriate thiacrowns. The dearth of the information on medium ring and macrocyclic sulfones deserves careful note.[36]

A highly speculative application of macrocyclic sulfones is in Ramberg-Backlund[37] (Equation 14.24) and Stevens[38] (Equation 14.25) rearrangement chemistry. The Stevens rearrangement has received particularly extensive attention in cyclophane chemistry whereby (chiefly) benzylic positions are coupled. Will it be possible to carry out transformations like that schematically presented in Equation 14.26?

Three dimensionality should also be obtainable in thiacrown ethers. Compound *20* serves as an example to illustrate the potentialities (Equation 14.27). The exomethylene groups can be converted to, for example, hy-

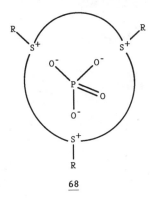

68

Structure 68

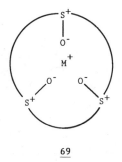

69

Structure 69

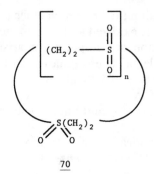

$\underline{70}$

Structure 70

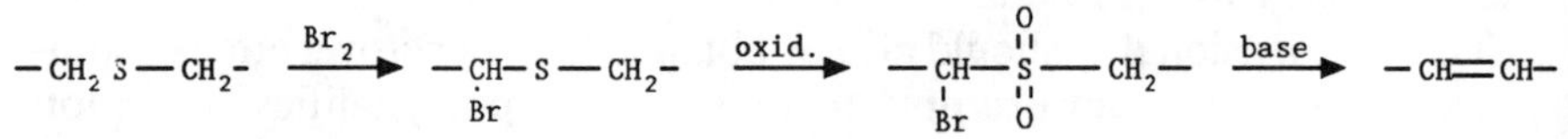

$$-CH_2-S-CH_2-\ \xrightarrow{Br_2}\ \underset{Br}{-CH-}S-CH_2-\ \xrightarrow{oxid.}\ \underset{Br}{-CH-}\overset{O}{\underset{O}{\overset{\|}{S}}}-CH_2-\ \xrightarrow{base}\ -CH\!=\!CH-$$

Equation 14.24

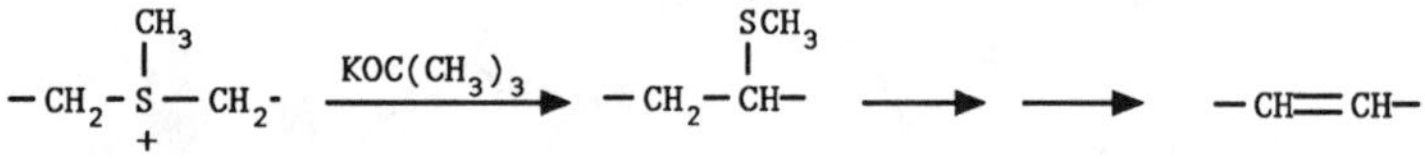

$$-CH_2-\overset{CH_3}{\underset{+}{S}}-CH_2-\ \xrightarrow{KOC(CH_3)_3}\ -CH_2-\overset{SCH_3}{CH-}\ \longrightarrow\ \longrightarrow\ -CH\!=\!CH-$$

Equation 14.25

$$\begin{bmatrix} S\!\begin{smallmatrix}CH_2-(CH_2)_nCH_2\\ \\ CH_2-(CH_2)_nCH_2\end{smallmatrix}\!S \end{bmatrix}_m \longrightarrow \longrightarrow \begin{bmatrix} (CH_2)_n \\ \\ (CH_2)_n \end{bmatrix}_m$$

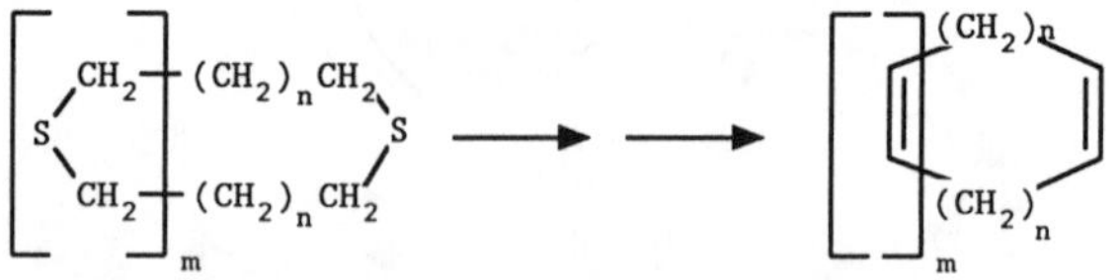

m = 1, 2, 3...
n = 0, 1, 2...

Equation 14.26

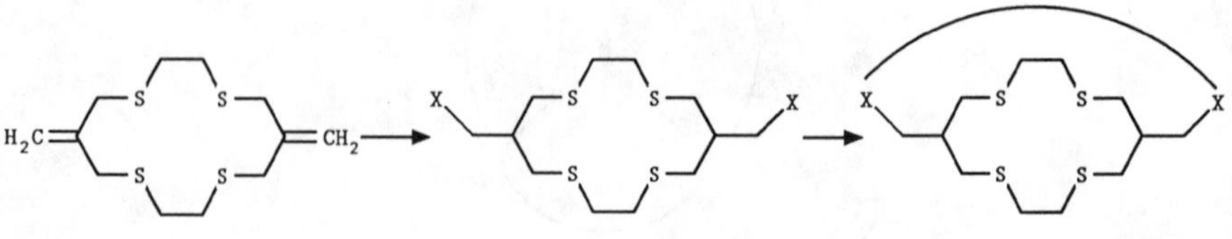

X = OH, SH, CO$_2$H, etc.

$\underline{20}$ $\underline{71}$ $\underline{72}$

Equation 14.27

droxymethylene (–CH$_2$OH) groupings (*71*). Bridging via ester formation, for example, provides a cryptate system (*72*). Again, control of stereochemistry, in this case at the tertiary carbon atoms, is essential. A logical way to achieve this is via intramolecular addition of borane so that *cis* stereochemistry is attained (Equation 14.28).

Preliminary investigations make clear that the –SCH$_2$COCH$_2$S– unit in, for example, *26* undergoes Mannich chemistry (Equation 14.29). The Mannich adducts should in general provide entry into a multitude of uniquely substituted thiamacrocycles.

14.5 Conclusions

I have emphasized the present-day maturity of synthetic techniques for the preparation of macrocyclic sulfides. Designed systems can be obtained in reasonable amounts without unreasonable investments in time for synthesis. Of course, there are restrictions having to do with the stability of inter-

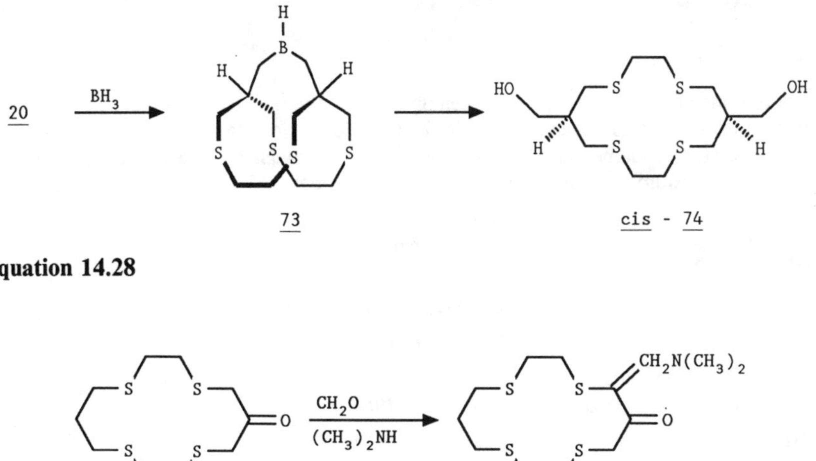

Equation 14.28

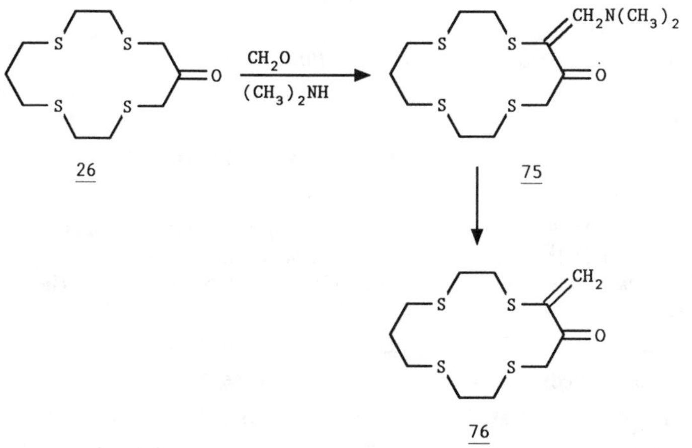

Equation 14.29

mediates required and the opportunity for unimpeded S_N2 substitution in the macrocylization step if the cesium salt approach is used.

Macrocyclic sulfides, in particular thiacrown ethers, will prove to be far removed in chemistry from their oxa- and aza-cousins. Analogies in complexation behavior are obvious, although the differences deserve perhaps more emphasis. Hard cations like those of the alkali metals and ammonium ions bind on the basis of electrostatic considerations. Metal ions like those of Ag, Hg, Pb, and so on have a far richer chemistry and stereochemistry. This is reflected in the structures found.

One may anticipate extensive use *catalytically* of metal ions bound in thiacrown ether cavities. The scope of application is in principle greater than that of oxa- and azacrown ethers. Whether this scope will be realized is a question for the future.

Macrocyclic sulfide and thiacrown ether chemistry will develop also on the basis of the extraordinary chemical potential of sulfur. Variable oxidation states, stabilization of radicals and radical cations, intramolecular reaction between sulfur sites, and possible applications of this sulfur chemistry in the development of new materials will dominate future developments.

Acknowledgments

Coworkers cited in the references have been involved in the chemistry described here. Seminal contributions were made by Marc Lemaire and Bindert Vriesema. It is a special pleasure to acknowledge the fundamental synthetic contributions through the years of Jan Buter, who developed the cesium salt approach to macrocyclic sulfides.

References

1. Pedersen, C. J. *J. Am. Chem. Soc.* **1967**, *89*, 2495, 7017.

2. Illuminati, G.; Mandolini, L.; Masai, B. *J. Am. Chem. Soc.* **1983**, *105*, 555.

3. Reviews: (a) Bradshaw, J. S.; Hui, J. Y. K. *J. Heterocycl. Chem.* **1974**, *11*, 679.
   (b) Bradshaw, J. S.; Stott, P. E. *Tetrahedron* **1989**, *36*, 461.

4. (a) Dietrich, B.; Lehn, J.-M.; Sauvage, J. P. *J. Chem. Soc., Chem. Commun.* **1970**, 1055.
   (b) Alberts, A. H.; Annunziata, R.; Lehn, J.-M. *J. Am. Chem. Soc.* **1977**, *99*, 8502.

5. For example: Vögtle, F.; Neumann, P.; Zuber, M. *Chem. Ber.* **1972**, *105*, 2955.

6. Vögtle, F.; Nätscher, R. *Chem. Ber.* **1976**, *109*, 994.

7. (a) Kruizinga, W. H.; Kellogg, R. M. *J. Am. Chem. Soc.* **1981**, *103*, 5183.
   (b Dijkstra, G. D. H.; Kellogg, R. M. *J. Org. Chem.* **1987**, *52*, 4230.
   (c) For an extremely useful variation, see Reinhoudt, D. N.; de Jong, F.; Thomassen, H. P. *Tetrahedron Lett.* **1979**, 2067.

8. Vriesema, B. K.; Buter, J.; Kellogg, R. M. *J. Org. Chem.* **1984**, *49*, 110.

9. Meadow, J. R.; Redi, E. E. *J. Am. Chem. Soc.* **1934**, *56*, 2177.

10. Ochrymowycz, L. A.; Mak, C.-P.; Michna, J. D. *J. Org. Chem.* **1974**, *39*, 2079.

11. (a) Buter, J.; Kellogg, R. M. *J. Chem. Soc., Chem. Commun.* **1980**, 466.
    (b) Buter, J.; Kellogg, R. M. *J. Org. Chem.* **1981**, *46*, 4481.

12. Bradshaw, J. S.; Hui, J. Y; Hayore, R. L.; Christensen, J. J.; Izatt, R. M. *J. Heterocycl. Chem.* **1973**, *10*, 1.

13. Wolf, Jr., R. E.; Hartman, J. R.; Ochrymowycz, L. A.; Cooper, S. R. *Inorg. Synth.* **1989**, *25*, 123.

14. Cooper, S. R. *Acct. Chem. Res.* **1988**, *21*, 141.

15. Galli, C.; Illuminati, G.; Mandolini, L.; Tamborra, P. *J. Am. Chem. Soc.* **1977**, *99*, 2591.

16. Buter, J.; Kellogg, R. M. *Org. Synth.* **1987**, *65*, 150.

17. Klein, J. *Tetrahedron* **1983**, *39*, 2733.

18. For example: (a) Trost, B. M.; Chan, D. M. T. *J. Am. Chem. Soc.* **1983**, *105*, 2315.
    (b) Trost, B. M.; MacPherson, D. T. *J. Am. Chem. Soc.* **1987**, *109*, 3483.
    (c) Albright, T. A. *Acc. Chem. Res.* **1982**, *15*, 149.

19. Buter, J.; Kellogg, R. M.; van Bolhuis, F. *J. Chem. Soc., Chem. Commun.* **1990**, 282.

20. Buter, J.; Kellogg, R. M.; van Bolhuis, F. *J. Chem. Soc., Chem. Commun.* **1990**, 282.

21. Vriesema, B. K.; Lemaire, M.; Buter, J.; Kellogg, R. M. *J. Org. Chem.* **1986**, *51*, 5169.

22. Stock, T.; Kellogg, R. M.; unpublished results.

23. Cooper, S. R.; Rawle, S. C. *Struct. Bonding* **1990**, *72*, 1.

24. De Simone, R. E.; Glick, M. D. *J. Am. Chem. Soc.* **1976**, *98*, 762.

25. Pelter, A.; Smith, K.; Brown, H. C. *Borane Reagents;* Academic Press; London, 1988.

26. Brown, H. C.; Ravindran, N. *Inorg. Chem.* **1977**, *16*, 2938.

27. For example: Brown, H. C.; Ravidran, N. *Inorg. Chem.* **1977**, *16*, 2938.

28. For leading references, see: Juaristi, E. *J. Chem. Educ.* **1979**, *56*, 438.

29. Desper, J. M.; Gellman, S. H. *J. Am. Chem. Soc.* **1990**, *112*, 6732 and references cited therein.

30. Leonard, N. J.; Milligan, T. W.; Brown, T. L. *J. Am. Chem. Soc.* **1960**, *82*, 4075.
    (b) Leonard, N. J.; Johnson, C. R. *J. Am. Chem. Soc.* **1962**, *84*, 3701.
    (c) Leonard, N. J.; Rippie, W. L. *J. Org. Chem.* **1963**, *28*, 1957.

31. Doi, J. T.; Musker, W. K. *J. Am. Chem. Soc.* **1978**, *100*, 3533.

32. For example: Setzer, W. N.; Coleman, B. R.; Wilson, G. S.; Glass, R. S. *Tetrahedron* **1981**, *37*, 2743.

33. Musker, W. K.; Roush, P. B. *J. Am. Chem. Soc.* **1976**, *98*, 6745.

34. (a) Musker, W. K.; Olmstead, M. M.; Goodrow, M. H. *Acta Crystallogr.* **1983**, *C39*(7), 887.
    (b) Musker, W. K.; Olmstead, M. M.; Kessler, R. M. *Inorg. Chem.* **1984**, *23*, 1764.
    (c) Furukawa, N.; Kawada, A.; Kawai, T.; Fujihava, H. *J. Chem. Soc., Chem. Commun.* **1985**, 1266.
    (d) Jones, II, G.; Malba, V.; Bergmark, W. R. *J. Am. Chem. Soc.* **1986**, *108*, 4214.
    (e) Ryan, M. D.; Swanson, D. D.; Glass, R. S.; Wilson, G. S. *J. Phys. Chem.* **1981**, *85*, 1069.
    (f) Asmus, K.-D. *Acc. Chem. Res.* **1979**, *12*, 436.
    (g) Musker, W. K. *Acc. Chem Res.* **1980**, *13*, 200.
    (h) Doi, J. T.; Goodrow, M. H; Musker, W. K. *J. Org. Chem.* **1986**, *51*, 1026.

35. For example: (a) Hosseini, M. W.; Blacker, A. J.; Lehn, J.-M. *J. Chem. Soc., Chem. Commun.* **1988**, 596.
    (b) Dietrich, B.; Fyles, T. M.; Hosseini, M. W.; Lehn, J.-M.; Kaye, K. C. *J. Chem. Soc., Chem. Commun.* **1988**, 691.

36. General review on cyclic sulfones: Zoller U. In *The Chemistry of Sulphones and Sulphoxides*; Patai, S., Z. Rappoport, and C. Stirling, Eds.; John Wiley; New York, 1988; pp 379–481.

37. Paquette, L. A. *Acc. Chem. Res.* **1968,** *1,* 209.

38. Review of chemistry of cyclic sulphonium salts: Dittmer, D. C.; Patwardhan, B. H. In *The Chemistry of the Sulphonium Group*; Stirling, C. J. M., and S. Patai, Eds.; John Wiley; New York, 1981; pp 387–521.

CHAPTER 15

Crown Thioether Chemistry and Its Biomedical Applications

Stephen R. Cooper

Inorganic Chemistry Laboratory
University of Oxford
Oxford OX1 3QR
England

15.1 Introduction

Explosive development of crown thioether chemistry in the last 10 years has yielded efficient ligand syntheses, insight into conformational factors influencing complexation, and a wealth of information on coordination chemistry. This conceptual infrastructure now permits application of this chemistry to practical ends. Taking recent reviews as its starting point,[1,2,3] this chapter updates selected recent developments and examines potential applications of crown thioethers in nuclear medicine. It then describes our present work on Tc and Re and concludes with a view to likely future developments.

15.2 Availability and Potential Applications

Crown thioethers (Figure 15.1) were first prepared over 50 years ago,[2,3] but low-yield synthetic routes long limited their use as ligands. Indeed, development of reliable high-yield routes provided a major impetus to investigation of their coordination chemistry, most of which has been reported in the last 10 years. As a penetrating example, Ochrymowycz's preparation of 9S3[4] did not attract interest from coordination chemists until 1983;[5] development of high-yield syntheses[6,7] resulted in an explosion of activity.[2,3] Kellogg (see Chapter 14) has lucidly summarized the critical role played by ligand synthesis in the development of crown thioether coordination chemistry.

Crown thioether ligands are prepared by routes that have now appeared in the literature.[7,8,9] All involve reaction of thioether dithiols with dihalides

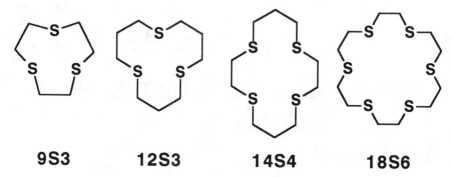

Figure 15.1. The major crown thioether ligands with the abbreviations used here.

or oligo (thioether) dihalides, generally in dimethylformamide containing Cs_2CO_3[9,10,11] as a base of a noncoordinating cation.[12] (Sellmann's beautiful synthesis of 9S3 on a $Mo(CO)_3$-template[6] is an exception in that good yields are obtained without use of Kellogg's cesium-mediated cyclization technique.) Some syntheses require hazardous β-haloethyl sulfides (sulfur mustards) as intermediates, but the low vapor pressure of those used mitigates difficulties in handling. In addition, Menger's recent work provides a safe and efficient means of destroying mustards through microemulsion methods.[13,14] Use of a peristaltic pump instead of a constant addition funnel considerably simplifies addition of the reactants under high-dilution conditions.[15]

With reasonable synthetic routes now available and a basic foundation of coordination chemistry from which to work, the question now turns to the uses of which crown thioethers might be put. Possible industrial applications include use as extractants[16,17] in hydrometallurgical processes [for concentration and recovery of precious metal ions such as Pt(II)] and treatment of industrial effluvia (removal of toxic metal ions). Unfortunately, however, ligand cost now precludes large-scale industrial applications, but further advances in synthetic methodology may obviate this impediment. Practical applications of crown thioethers are at present more likely in the biomedical field. Here, final product cost will be largely determined by expenses incurred in developing the product and securing regulatory approval for it rather than by the cost of the ligands themselves.

15.3 Biomedical applications

15.3.1 Nuclear Medicine: Importance of Tc and Re

Probably the most promising application of crown thioethers lies in delivery of radioisotopes such as ^{99m}Tc in nuclear medicine. Radioimaging agents based upon γ emission by ^{99m}Tc now play a major role in clinical medicine. Closely related to Tc-based diagnostic imaging agents are the corresponding complexes of Re as radiotherapeutic agents. Complexes of [186]Re and [188]Re, especially the former, have been proposed as therapeutic agents of destruction of neoplasms.[18] Both Re isotopes emit β particles with energies (and therefore penetration depths) that make them suitable for therapeutic applications. A further potential advantage, at least in principle, is the possible parallel between analogous imaging ^{99m}Tc complexes and therapeutic [186]Re ones; identification of tumors with the γ-emitting ^{99m}Tc complex could precede their destruction by the β-emitting [186]Re analogue (with the efficacy of radiotherapy followed with the ^{99m}Tc complex).

Development of a new class of Tc chelates may yield substantial dividends to clinical imaging. While ^{99m}Tc imaging agents are now available for organs such as the heart, liver, kidneys, and brain, Clarke and Podbielski[19] have pointed out that no Tc imaging agent is so good that improvement need not be sought. In addition, the nuclear properties of [99m]Tc (specifically $\tau_{1/2}$ and γ energy) are superior to those of other commonly used radioisotopes (e.g., ^{201}Tl, ^{131}I, ^{111}In, and ^{67}Ga). Consequently, significant improvements in diagnostic imaging could result from investigation of new classes of Tc chelates that could potentially replace these isotopes in their current uses—given ligands that confer suitable biodistribution. Thus there is a need for investigation of new ligand classes.

Complexes with predominantly thioether coordination represent one such as-yet-unexplored class. Apart from [Tc(isonitrile)$_6$]$^+$ complexes, clinical applications of ^{99m}Tc largely involve Tc in its higher oxidation states coordinated to oxygen or nitrogen donor groups, but to date Tc complexes of 18N$_2$S$_4$[20] and trans-[TcO$_2$(14N$_2$S$_2$)]$^{+21,22}$ (where 18N$_2$S$_4$ and 14N$_2$S$_2$ are 1,10-diaza-4,7,13,16-tetrathiacyclohexadecane and 1,4-dithia-8,11-diazacyclotetradecane, the diamine analogues of 18S6 and 14S4, respectively) represent the only thioether complexes to be reported. No homoleptic thioether complexes have been reported either for Tc or Re, for [Re(CO)$_3$(9S3)]$^{+23}$ and [Re$_2$Cl$_5$(2,5-dithiahexane)$_2$]^{24} are the only structurally characterized Re complexes involving predominantly thioether coordination. The present research was therefore undertaken to complement existing studies and to develop new facets of Tc and Re coordination chemistry for applications in nuclear medicine.

15.3.2 Potential Utility of Crown Thioethers

Despite the paucity of literature as precedent, thioether chelates of Tc offer considerable potential advantages as radioimaging agents. Previous work established that thioethers avidly bind second- and third-row transition metals, in which they stabilize lower oxidation states.[2,3,25] Thus, strong binding of Tc could be expected. Thioethers do not interact appreciably with biologically important metal ions such as Ca(II), Mg(II), and Zn(II). Furthermore, the ligands themselves are chemically robust with respect to aerial oxidation, hydrolysis, protonation, and dealkylation (but see section 15.4.5).

Moreover, because they do not hydrogen bond, their Tc complexes should differ in biodistribution from, for example, the more widely studied N- and O-based chelates of Tc(IV) and (V). Thus, thioether chelates are likely to concentrate in different locations than currently used Tc radioimaging agents, which will enhance their clinical utility.

Other advantages arise from use of macrocyclic thioethers in particular. Through variations in ring size, crown thioethers can be used to influence the coordination environment—specifically redox potentials, metal–ligand distances, as well as coordination number and geometry—of metal ions.[2,3] Previous work[2,3] encouraged the expectation that crown thioethers would yield a series of chelates well suited for studies on the crucial role of redox potential and coordinative unsaturation in determining biodistribution.[19,26] Deutsch[27] has emphasized the importance of small systematic variations in chelates to elucidate structure–function relationships.

From such studies may emerge principles bearing on the systematic design of future clinical imaging agents, as well as the necessary foundation for examination of functionalized crown thioethers and podands for attachment of biologically relevant targeting groups (e.g., monoclonal antibodies and enzyme substrate analogues).

15.3.3 Redox Potentials and Their Implications for Nuclear Medicine

Factors influencing the biodistribution, and therefore the utility, of a Tc chelate for radioimaging include redox potential and coordinative saturation. In particular, redox potentials in the physiological range can cause the coexistence of several oxidation states, each with different charges and therefore biodistribution properties. Imaging specificity can suffer as a result.

Crown thioethers may solve this problem. A change in crown ring size influences redox potentials through changing the conformational constraints on the metal ions. A striking example of this phenomenon occurs in the $[Rh^{III/II/I}(L)_2]^{3/2/1+}$ system (L = 9S3,[28,29,30] 10S3,[28] 12S3[28]) where it causes the remarkable stabilization of Rh(II) unique to macrocyclic thioethers. Increasing ring size in the series L = 9S3, 10S3, 12S3 shifts both $E(Rh^{III/II})$ and $E(Rh^{II/I})$ potentials to more positive values, by an average of ≈ 60 and 80 mV, respectively, per additional methylene group in the crown (Fig. 15.2).[28] These results emphasize how choice of crown thioether ligand can be used to "tune" redox potential—and therefore control solution chemistry—within the context of a conserved $M(thioether)_6$ coordination sphere.

This capability has obvious implications for applications in nuclear medicine. Deutsch has emphasized the importance of redox effects in influencing biodistribution of chelates.[19,26] In particular, his work shows how analogous Tc and Re chelates sometimes behave differently in the body owing to their different redox potentials and therefore their different oxidation states under physiological conditions. Crown thioether ligands can be used to form a

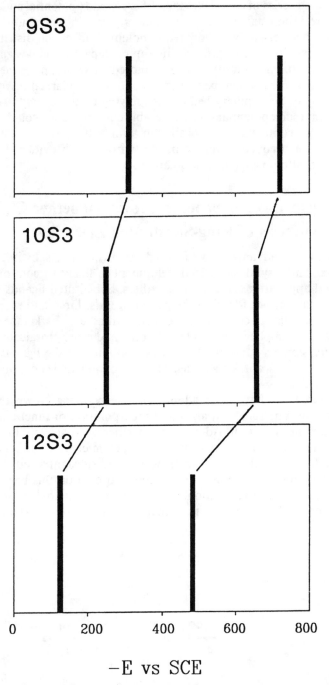

Figure 15.2 Effect of crown thioether ring size on Rh(III/II/I) potentials in [Rh(L)$_2$]$^{3/2/1+}$ complexes (L = 9S3, 10S3, 12S3) (from ref. 28; used with permission).

closely related series of complexes that differ in redox potential and therefore in which oxidation state predominates under physiological conditions. Such a series is ideal for study of how redox potential affects biodistribution.

Another possibility rising from this tuning capability is designing complexes to localize in tumors though a bioreductive trapping mechanism. Tumors are generally hypoxic owing to poor vascularization or simple compression of the capillary bed by the growing tumor mass.[31] A relatively low ambient redox potential results. Suitably chosen ligands could then yield Tc and/or Re complexes that localize in tumors because they undergo reduction and consequent trapping in the tumor,[32] with evident benefit for either diagnostic imaging or radiotherapy.

15.4 Recent Developments in Crown Thioether Chemistry
15.4.1 Synthesis of Ring-Substituted Ligands

Recent synthetic advances have focused upon routes to substituted crown thioethers. This "synthesis-led" development is likely to continue in the biomedical applications of crown thioethers. Substituted ligands offer two attractive features not found in the parent ligands. First, and more importantly, they can be attached to monoclonal antibodies.[33] Parker (see Chapter 4) discusses this application in detail. Second, and more prosaically, suitable substitution confers substantial aqueous solubility on the ligands, thereby removing a significant impediment to use of macrocyclic thioethers in aqueous media.

Two approaches have been adopted for derivatizing crown thioethers: attachment of functionalized arms and incorporation of functionality into the macrocyclic ring itself. Adopting the former approach, Kellogg and co-workers[34] have shown that hydroboration of exomethylene 14S4 yields the corresponding alcohol (Figure 15.3), which has substantial aqueous solubility (≈ 0.05 mol). While no stability-constant data on this ligand are yet available, structural investigation shows that the exomethylene substituent does not alter the conformation from that of the free ligand. This observation

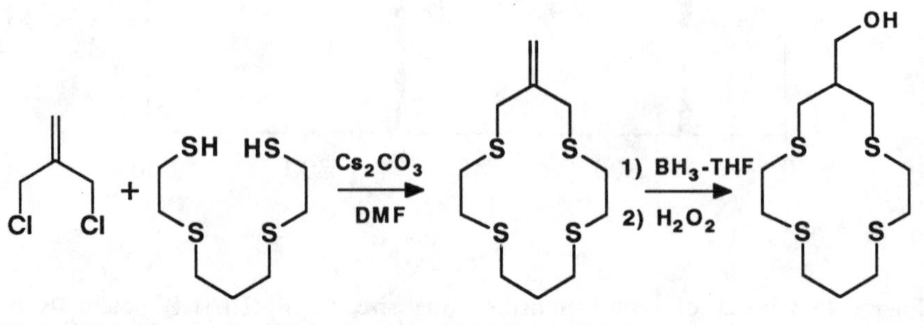

Figure 15.3. Synthesis of 6-hydroxymethyl-14S4 (ref. 34).

encourages speculation that exocyclic substitution may not adversely affect metal binding.

Corresponding results have been found for ring-substituted tridentate ligands. Blower et al. used Sellmann's Mo(CO)$_3$-mediated template synthesis[6] (with use of 3-thiapentane-1,5-dithiol and the appropriate substituted dibromides) to obtain ligands including 2-methyl-,2-methyloxymethyl,2-hydroxymethyl derivatives of 9S3 (Fig. 15.4).[35] Comparison of the Cu(I), Ag(I), and Hg(II) bis complexes with those of the parent ligand indicates that no obvious loss of stability attends ring substitution. Taken together, these two sets of results encourage further exploration of ring substitution, because they establish that it need not interfere with the crucial conformational properties that make 9S3[36] such an especially superlative ligand.

Rorabacher's thermodynamic analysis of ring-substituted 14S4 derivatives, however, illustrates how substituents can either augment or diminish binding affinity. Rorabacher et al. prepared hydroxy-14S4, as well as the *trans* and *cis* isomers of dihydroxy-14S4 by the route shown in Figure 15.5.[37] Their kinetic and thermodynamic studies establish that the introduction of one hydroxy group increases the aqueous binding affinity for Cu(II) (approximately twofold) but that the introduction of two hydroxy groups decreases it (by approximately 3- and 21-fold for the *trans* and *cis* isomers, respectively, all relative to the parent compound).

In similar fashion, Setzer et al. introduced functionality directly into the crown ring of 10S3 and other ligands. They prepared keto-10S3 (9-oxo-1,4,7-trithiacyclodecane) through reaction of 1,3-dichloroacetone with 3-thiapentane-1,5-dithiol (Fig. 15.6)[38,39] and structurally characterized its Ni(II) bis complex as the *meso* isomer.[40] They also found that free hydroxy-12S3[39] crystallizes in a conformation closely resembling that of 12S3 itself,[41] another

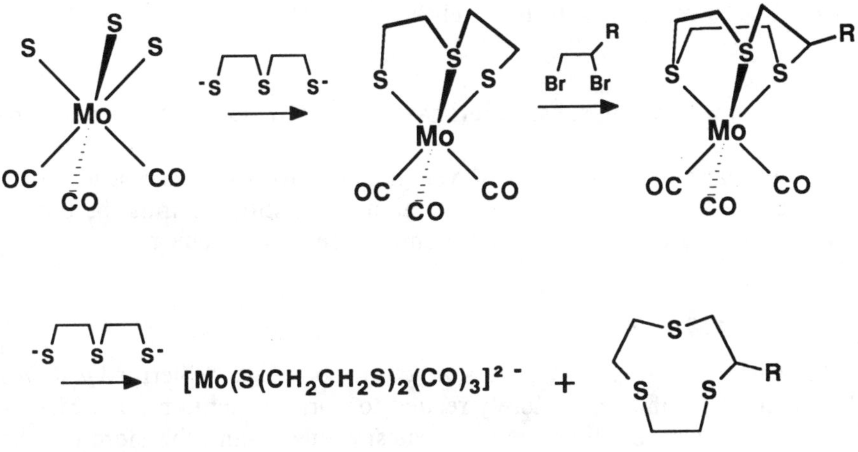

Figure 15.4. Synthesis of ring-substituted 9S3 analogues (ref. 35).

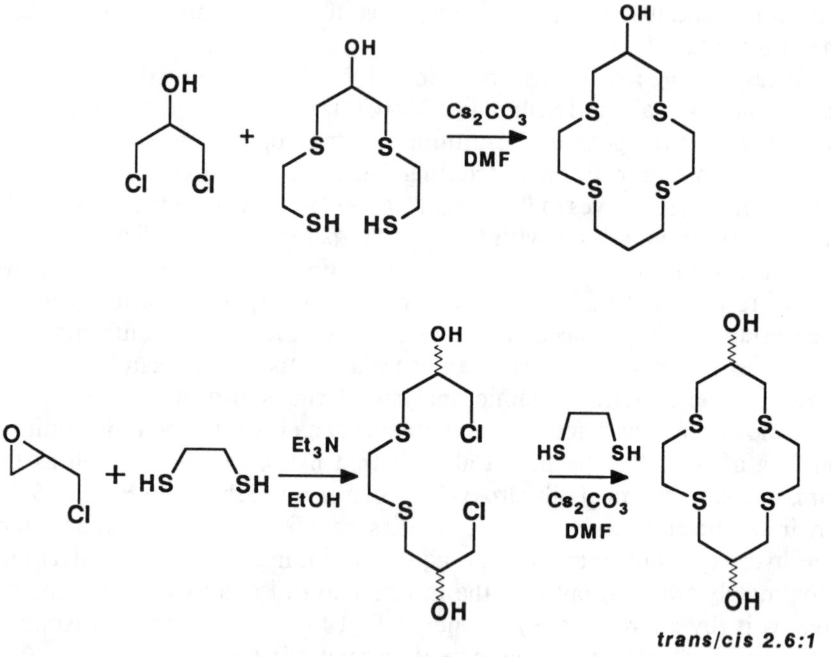

Figure 15.5. Synthesis of 6-hydroxy-14S4 (ref. 37).

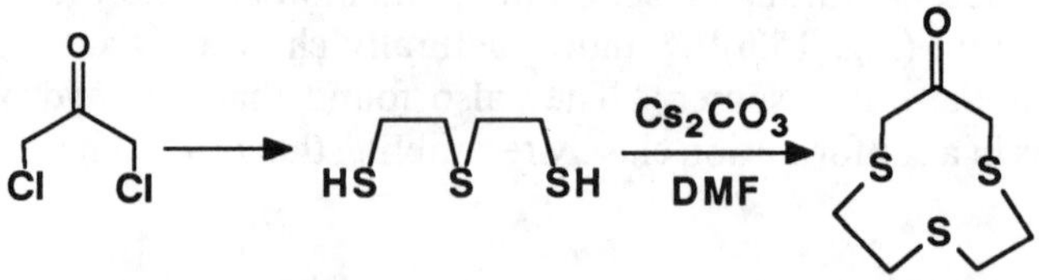

Figure 15.6. Synthesis of keto-10S3 (ref. 39).

example in which ring substitution has had no obvious effect on ring conformation.

In summary, recent synthetic developments now permit access to water-soluble crown thioethers, but introduction of substituents must be undertaken with care to avoid deleterious impact on binding efficacy.

15.4.2 Podands

Although not strictly crown ligands, acyclic tripodal ligands (termed *podands* by Vögtle) are sufficiently closely related to warrant mention here. Macrocyclization invariably presents the greatest obstacle—and therefore has the greatest influence on cost—in synthesis of crown-type ligands. Podands obviate this problem by bringing the arms bearing the donor groups to a common locus, thereby helping to organize them for coordination. We have used the term *supertripodal* to distinguish the hexa-coordinating tripodal ligands discussed here from simple tridentate tripods.[42]

As one example of a crown thioether–related podand, reaction of sodium 2-methylthioethanethiolate with the tritosylate of 1,1,1-tris(hydroxy-methyl)ethane affords ligand L1, 1,1,1-tris((((2-methylthio)ethyl)thio)-methyl)ethane.[43] This ligand strongly binds metal ions such as Co(II)[43] and Ni(II)[44] as well as Ru(II),[45] in each case to produce hexacoordinated [M(L1)]$^{2+}$ species.

In a related system, Sargeson and co-workers[46] recently brought the macrocycle → podand → macrocycle theme full circle to produce a hexathio-cryptand. Reaction of 1,1,1-tris(mercaptomethyl)ethane successively with ethylene oxide and thionyl chloride, followed by cyclization with another equivalent of 1,1,1-tris(mercaptomethyl)ethane under Kellogg's Cs_2CO_3/DMF conditions, yields thiocryptand L2 (Fig. 15.8). Characterization of the cobalt salt established formation of an octahedral low-spin Co(II) complex analogous to those found for 9S3 and 18S6.[5,44,47,48,49]

Recently another class of hexakis(thioether) podands has been developed, derived from conjugation of thiol-bearing arms to a cyclohexane-based framework. Reaction of phloroglucitol tritosylate with 2-methylthioetha-nethiol yields the new podand L3, *cis,cis*-1,3,5-tris(2-methyl-thio)ethyl)thio)cyclohexane (Fig. 15.9).[50] Although structural studies have not yet been completed, spectroscopic results indicate that this expanded version of L1 offers hexakis(thioether) coordination with less "compression" of the MS_3 trigonal pyramid at the closed end of their molecule. Preliminary investigation of its coordinative properties show that this ligand forms stable complexes with a variety of transition metal ions and in many respects behaves similarly to 9S3.

Development of the coordination chemistry of these acyclic thioether ligands is still in a rudimentary phase but offers promise in the same applications proposed for 9S3.

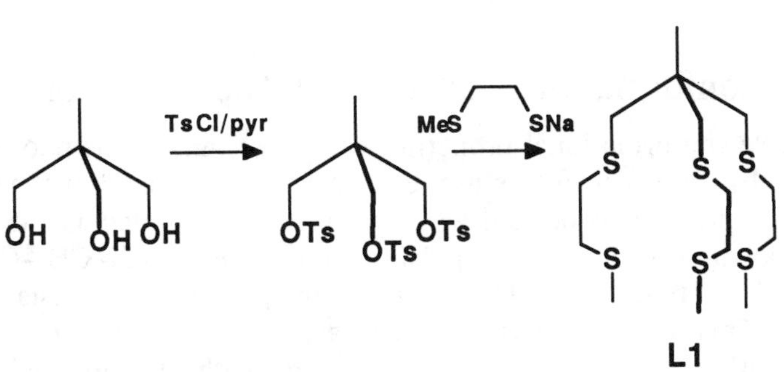

Figure 15.7. Synthesis of L1 (ref. 43).

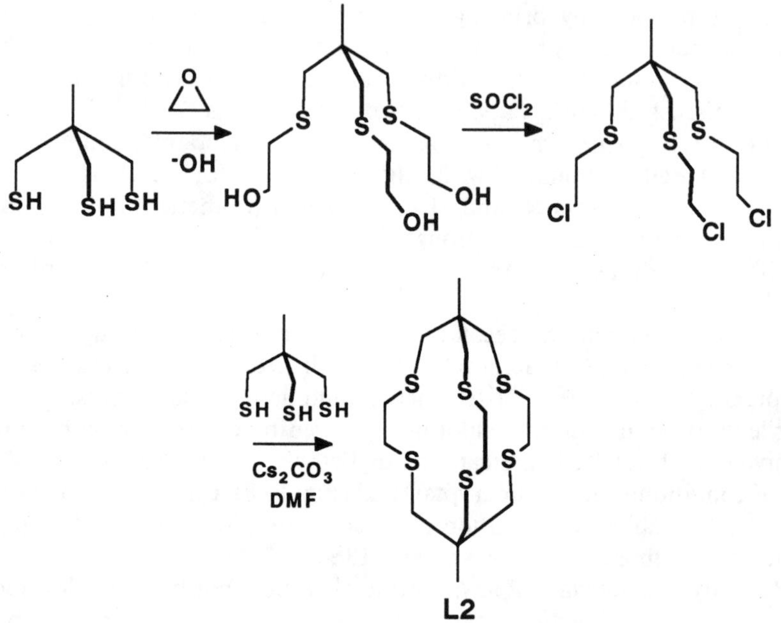

Figure 15.8. Synthesis of L2 (ref. 46).

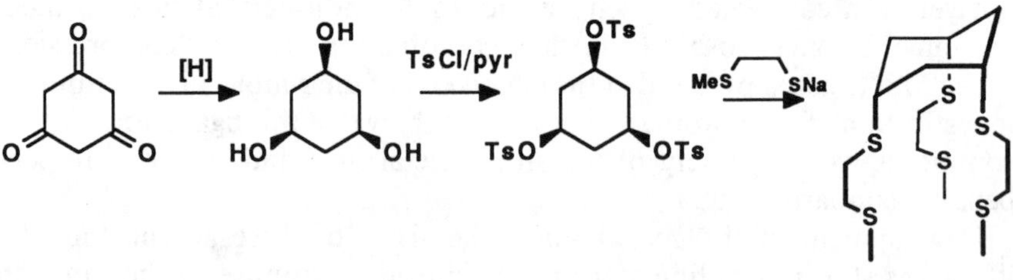

Figure 15.9. Synthesis of L3 (ref. 50).

15.4.3 Conformational Effects and Their Control

Unsuitable conformational properties can make thiacrowns poor ligands[51] and thereby limit their future use. Earlier work indicated that the unfavorable conformations associated with crown thioethers such as 14S4[52] arose from relaxation of H···H 1,4-repulsions in gauche CH_2–X–CH_2–CH_2 units for X = S relative to X = O (because of the greater H···H distance for X = S).[51,53] This relaxation lets thioether-based macrocycles turn themselves "inside out" into a conformation incompatible with chelation. Substitution with *gem*-dimethyl groups should therefore reintroduce those 1,4-repulsions to reverse the conformational preferences and thereby improve binding properties.[54,55] X-ray structural studies confirm that successive *gem*-dimethyl substitution on 14S4 does in fact progressively force the macrocycle toward the "right side in" conformation, and equilibrium measurements reveal concomitant increases in stability constant (Fig. 15.10). Thus, in addition to addressing the origin of the macrocyclic effect,[56] these results clearly provide a strategy for improving the binding ability of thiacrowns with adverse conformational properties.

15.4.4 Microwave Methods in Synthesis

Microwave methods are now becoming more commonplace in both inorganic and organic synthesis. Their use can greatly facilitate preparation of second- and third-row complexes that otherwise resist facile synthesis owing to the kinetic inertness of these metal ions. Microwave heating not only dramatically accelerates reactions (typically about 100-fold) but also often results in products inaccessible by conventional heating methods. For example, reaction of $RuCl_3 \cdot nH_2O$ with excess 9S3 affords $[RuCl_3(9S3)]$, which resists coordination of a second ligand even after several days reflux. Preparation of $[Ru(9S3)_2]^{2+}$ with conventional heating requires the prior preparation of a labile Ru source {e.g., $[Ru(Me_2SO)_6]^{2+}$}. Under microwave heating, the same reaction of $RuCl_3$ proceeds cleanly through to the bis(9S3) complex in minutes.[57] The rapidity of the reaction possibly results in coordination of a second 9S3 before the neutral $[RuCl_3(9S3)]$ complex can precipitate from solution. Clearly, microwave methods may have potential use in preparation of radiopharmaceuticals generally and those derived from crown thioethers in particular.

15.4.5 Coordination Chemistry

An especially significant recent development has been the synthesis of platinum and noble metal complexes in remarkable oxidation states. Isolation of bis(9S3) complexes of Pd(III),[58] Pt(III),[59] Ir(II),[60] and Au(II)[61] extends

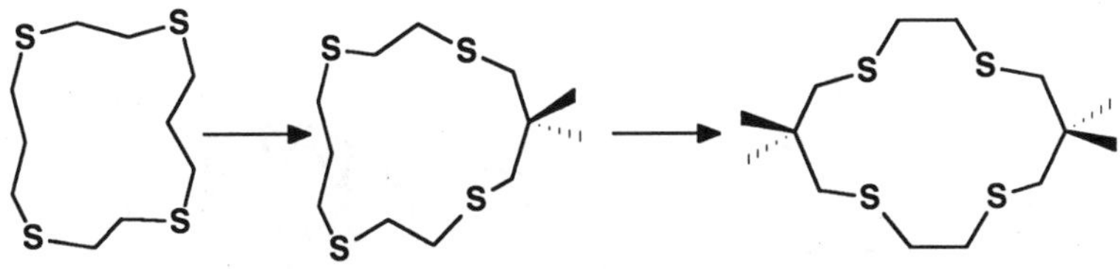

rel. K_{stab} 1   7·3   49

Figure 15.10. Schematic representation of the progressive conformational changes in 14S4 upon successive introduction of one and two *gem*-dimethyl groups (ref. 55).

earlier work demonstrating the existence in solution of mononuclear Rh in the unusual $2+$ oxidation state.[28,29,30] As is certainly true in the Rh(II) case,[28] observation of such unusual oxidation states for the other metal ions probably reflects more the extraordinary complexing ability of 9S3 than merely the influence of thioether coordination. This interplay between oxidation state (and necessarily therefore with redox potential) on one hand and crown conformational properties on the other offers exciting possibilities to nuclear medical applications, as discussed in section 15.3.3.

On a less sanguine note, recent reports indicate that crown thioether complexes of Co, Rh, Ir,[62] and Ru[63] undergo reversible (for M = Co) ring deprotonation and elimination that opens the ring to form an acyclic vinyl thioether thiolate complex (Fig. 15.11). While this reaction itself lacks obvious application, its occurrence may complicate (or perhaps obviate) use of crown thioethers as ligands to high oxidation state species in alkaline media.

Recent work has resulted in the second example of a Re crown thioether complex[23] and the first such coordination (as opposed to organometallic) complex. Reaction of ReO_4^- with 9S3 in the presence of HBF_4 affords $[ReO_3(9S3)]^+$ (Fig. 15.12) as the yellow tetrafluoroborate salt.[64] Structural characterization reveals 9S3 perching upon a trigonal pyramidal $\{ReO_3\}^+$ unit with three mutually *cis* oxo groups and all three S atoms coordinated, the length of the Re–S bonds bearing testimony to their weakness. This complex illustrates the exceptional binding affinity of 9S3, because poor binding at best would be expected between the soft base donor groups and the hard acid (formally $7+$) cation.

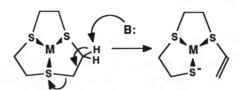

Figure 15.11. Deprotonation of a 9S3 complex to yield the vinyl thiolate complex after cleavage of a C–S bond (ref. 62).

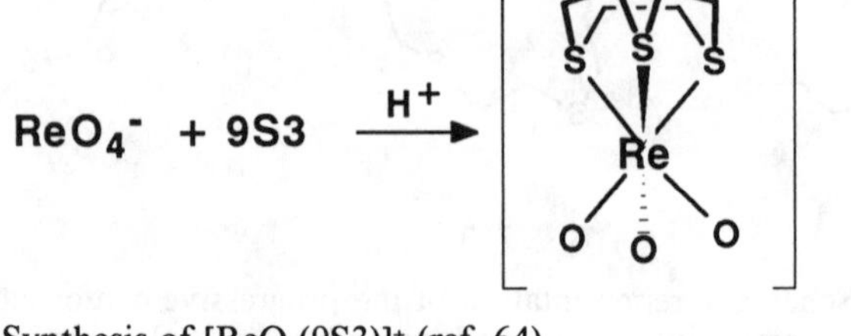

$$ReO_4^- + 9S3 \xrightarrow{\;H^+\;} \left[\begin{array}{c} \end{array} \right]^+$$

Figure 15.12. Synthesis of $[ReO_3(9S3)]^+$ (ref. 64).

15.5 Crown Thioether Chemistry of Tc and Re

While thioether-based chelates have yet to find use in ^{99m}Tc-based radioimaging, recent work has yielded the first examples in which Tc and Re coordinate to crown (macrocyclic) and supertripodal thioether ligands[42] to form homoleptic thioether complexes.

15.5.1 [Re(9S3)$_2$]$^{2+}$

Attention focuses on Re not only because of its own potential nuclear medical applications but also because of its similarity to Tc, from which it differs primarily in its less-oxidizing character (in analogous compounds). Reaction of ReO$_4^-$ with 9S3 in glacial acetic acid in the presence of Sn(II) affords deep red solutions containing [Re(9S3)$_2$]$^{2+}$, from which the dication can be isolated as thin red crystals of its BF$_4^-$ salt.[65] Solid-state magnetic measurements on [Re(9S3)$_2$](BF$_4$)$_2$ give μ_{eff} = 1.6 μ_B at 300 K, decreasing to 1.46 μ_B at 20 K, indicative of an S = $\frac{1}{2}$ system with the substantial orbital contribution expected of a third-row element. FAB MS shows a peak at 634 m/z, corresponding to {[Re(9S3)$_2$](BF$_4$)}$^+$ and thereby substantiating formation of Re(II). Recent X-ray diffraction measurements have confirmed synthesis of the first example of a homoleptic Re(thioether)$_6$ core, with two 9S3 rings coordinated to the metal ion in tridentate fashion at an average Re–S distance of 2.38 Å.

Cyclic voltammetry in MeCn on Pt shows that this complex undergoes a quasireversible one-electron reduction to the 1+ ion around −0.8 V and oxidation to the 3+ ion around 0 V (both versus the saturated calomel electrode, SCE). These results suggest, therefore, that the 3+ and 2+ ions may coexist under physiological conditions.

15.5.2 [Tc(9S3)$_2$]$^{2+}$

The ubiquity of pertechnetate in clinical radioimaging makes it the starting material of choice for crown thioether chelates. Pertechnetate reacts with 9S3 to yield a Tc(II) complex containing a Tc(thioether)$_6$ core, the first such example of a homoleptic thioether complex of this element. Reaction of TcO$_4^-$ with 9S3 in MeCN containing HBF$_4$/Et$_2$O and Sn(II) gives a brown microcrystalline precipitate of [Tc(9S3)$_2$]$^{2+}$ as the BF$_4^-$ salt. (The same complex forms in the absence of Sn(II), in which case the ligand serves as a sacrificial reductant.) This complex is stable to water and can be recrystallized from it. In fact, the synthesis can also be carried on the presence of water. Addition of a saturated solution of NH$_4$PF$_6$ to an aqueous solution of the BF$_4^-$ salt precipitates the PF$_6^-$ salt, for which infrared spectroscopy conforms the absence of the oxotechnetium group. Structural characterization establishes the existence of the first homoleptic TcS$_6$ core, with an average Tc–S distance of 2.38 Å.[66]

Assignment of oxidation state in Tc complexes can be difficult in the absence of structural data. For example, oxidation state assignments can depend on whether or not oxygen atoms bear protons; even X-ray diffraction may not resolve this question unless H atoms can be located. Accordingly, magnetic susceptibility measurements play an important role in the non-structural assignment of oxidation state to complexes for which elemental analytical data may be ambiguous. Deutsch et al.[67] have summarized this application of magnetic measurements. Solution moment measurements by the NMR method confirm the existence of an $S = \frac{1}{2}$ system with μ (298 K) = 1.7 μ_B, reflecting minimal orbital contribution at room temperature. FAB MS further supports formulation of the product as $[Tc(9S3)_2]^{2+}$, with observation of a peak at m/z 546 corresponding to $[Tc(9S3)_2](BF_4)^+$. Thus, all of the data consistently indicate formation of Tc(II) in this homoleptic complex.

Parallel synthetic procedures with the supertripodal compound L1 yield evidence of the analogous Tc(II) complex. Under the conditions described above, L reacts with TcO_4^- to yield a solution identical in appearance with that of the 9S3 homologue, presumably the previously unreported $[Tc(L)]^{2+}$ cation. Efforts are presently underway to isolate and characterize this species.

For clinical applications, two features of these syntheses warrant mention. First, no exogenous reducing agent is required; excess ligand serves as a sacrificial reductant. Second, the synthesis proceeds equally well under aqueous conditions (product characterized by cyclic voltammetric behavior, which is identical with that of the compound crystallized from MeCN). Thus, the coordination chemistry under development could be carried over with minimal modification for use in diagnostic formulations.

Electrochemical characterization of $[Tc(9S3)_2]^{2+}$ by cyclic voltammetry in MeCN on Pt shows a quasi-reversible wave assigned to the Tc(II/I) couple at +0.05 V versus SCE. It also shows an irreversible one-electron oxidation at +1.3 V versus SCE. In addition, a quasi-reversible process appears at −0.66 V versus SCE. On the first scan, this wave has a current one-fourth that of the wave at +0.05 V versus SCE. During subsequent scans, the current associated with this wave at −0.66 V versus SCE grows at the expense of that at +0.05 V versus SCE; by the tenth scan the system reaches either an equilibrium or a steady state. These results suggest that upon reduction of $[Tc(9S3)_2]^{2+}$ to $[Tc(9S3)_2]^+$ (the process at +0.05 V), the electrogenerated $[Tc(9S3)_2]^+$ reacts (perhaps) with CH_3CN to give a product that is redox active in its own right (and which gives rise to the wave at −0.66 V). This species has not yet been characterized.

Perrhenate fails to oxidize 9S3 even in the presence of acid, owing to the generally weaker oxidizing power of Re than Tc in the same oxidation state. Thus, the same reaction carried out with TcO_4^- in place of ReO_4^- yields the brown $[Tc(9S3)_2]^{2+}$ cation, with no evidence yet available for formation of an intermediate oxotechnetium species. Formation of such species may be possible, however, through careful control of pH, because the redox potential of the oxoanions

$$MO_4^- + 8\ H^+ + 5e^- = M(II) + 4\ H_2O$$

decreases with increasing pH. Indeed, solutions of $[Tc(9S3)_2]^{2+}$ slowly become pink on standing and deposit material with strong infrared absorption suggestive of the Tc=O moiety. The nature of this pink material remains to be elucidated.

Synthetic methods worked out for $[^{99}Tc(9S3)_2]^{2+}$ have been successfully extended to no-carrier-added (NCA) conditions using ^{99m}Tc generator eluate (saline) to produce $[^{99m}Tc(9S3)_2]^{2+}$ in high yield. High-pressure liquid chromatographic (HPLC) comparison (by simultaneous in-line optical and γ detection) with authentic $[^{99}Tc(9S3)_2]^{2+}$ confirms that product as $[^{99m}Tc(9S3)_2]^{2+}$, which is stable at the NCA level in saline. Treatment of the product with Sn(II) reduces it to a second complex believed to be $[^{99m}Tc(9S3)_2]^+$, which exhibits identical HPLC behavior with a red complex generated by aqueous $SnCl_2$ reduction of $[^{99}Tc(9S3)_2]^{2+}$. The putative monocation $[Tc(9S3)_2]^+$ is also stable and carrier free in saline. Preliminary biodistribution studies of these two species reveal closely similar behavior, which suggests that the two oxidation states may interconvert *in vivo*. Such interconversion would present both risks and opportunities. On one hand, it may degrade imaging specificity; for example, serum reduction of monocationic Tc(II) diphosphine heart agents to neutral analogues results in ^{99m}Tc wash-out from the myocardial cells.[68]

On the other hand, however, redox interconversion under physiological conditions may offer an intracellular trapping mechanism. In the present case, intracellular oxidation of the mono- to the dicationic form in myocardium, for example, may trap the latter in the heart. If so, this conversion would present an especially effective means of controlling biodistribution.

Judicious choice of crown thioether ligands can shift redox potentials and thereby the Tc redox state prevalent *in vivo*. In summary, through the variety of accessible redox states, the tunable nature of the redox potentials connecting these states, and the simplicity of the system, crown thioether complexes may contribute to a significant breakthrough in ^{99m}Tc-based imaging.

15.6 Prospects for Future Work

Likely directions for future work can be classified into chemical and biological studies. Further chemical investigation can profitably be directed toward characterization of other crown thioether chelates of Tc and Re, with particular emphasis on their redox and ligand substitution behavior. Ring-substituted ligands should receive additional attention to determine quantitatively how functional groups can be introduced without diminishing binding affinity, or, even better, to augment it. Molecular mechanics calculations on free ligands may play an important role in solving this problem, which has clear importance for applications involving attachment of the chelates to other molecules.

Biological studies on how the present Tc complexes distribute themselves in the body may suggest modifications to be made in second-generation crown thioether chelates. Simple examples include use of different ring sizes to shift redox potentials or choosing ring substituents to alter chelate lipophilicity. Biodistribution studies on such a closely related series of complexes with differing redox potentials, degrees of coordinative unsaturation, and molecular morphologies (elongated and compressed cylindrical for bis chelates and 18S6 complexes, respectively, and square pyramidal for 14S4 complexes) may lead to fundamental insight into the factors determining where complexes localize. Last, development of methods to link crown thioether chelates to monoclonal antibodies will further spur interest in their biomedical applications.

In summary, then, crown thioethers seem to have extensive promise for application to nuclear medicine, particularly in conjunction with monoclonal antibodies. Work directed toward this goal is currently in progress.

Acknowledgments

It is a pleasure to thank Mallinckrodt Medical, Inc., for support of this research, Mr. Thomas Mertens (University of Cincinnati) for the FAB MS results, and Messrs. David White and Simba Matondo and Drs. Heinz-Josef Küppers and Phil Blower for their invaluable efforts on this project.

References

1. Cooper, S. R. *Acc. Chem. Res.* **1988**, *21,* 141.

2. Cooper, S. R.; Rawle, S. C. *Struct. Bond.* **1990**, *72,* 1.

3. Blake, A. J.; Schröder, M. *Adv. Inorg. Chem.* **1990**, *35,* 1.

4. Gerber, D.; Chongsawangvirod, P.; Leung, A. K.; Ochrymowycz, L. A. *J. Org. Chem.* **1977**, *42,* 2644.

5. Setzer, W. N.; Ogle, C. A.; Wilson, G. S.; Glass, R. S. *Inorg. Chem.* **1983**, *22,* 266.

6. Sellmann, D.; Zapf, L. S. *Angew. Chem., Int. Ed. Engl.* **1984**, *23,* 807.

7. Blower, P. J.; Cooper, S. R. *Inorg. Chem.* **1987**, *26,* 2009.

8. Wolf, R. E. J.; Hartman, J. R.; Ochrymowycz, L. A.; Cooper, S. R. *Inorg. Synth.* **1989**, *25,* 122.

9. Buter, J.; Kellogg, R. M. *Org. Synth.* **1987**, *65,* 150.

10. Buter, J.; Kellogg, R. M. *J. Chem. Soc., Chem. Commun.* **1980**, 466.

11. Buter, J.; Kellogg, R. M. *J. Org. Chem.* **1981**, *46,* 4481.

12. Dijkstra, G.; Kruizinga, W. H.; Kellogg, R. M. *J. Org. Chem.* **1987**, *52,* 4230.

13. Menger, F. M.; Elrington, A. R. *J. Am. Chem. Soc.* **1990**, *112,* 8201.

14. Menger, F. M.; Elrington, A. R. *J. Am. Chem. Soc.* **1991**, *113,* 9621.

15. Durrant, M. C.; Richards, R. I. *Chem. Ind. (London)* **1991**, 474.

16. Moyer, B. A.; Westerfield, C. L.; McDowell, W. J.; Case, G. N. *Sep. Sci. Technol.* **1988**, *23,* 1325.

17. Moyer, B. A.; McDowell, W. J. U.S. Patent 4,927,610, 1990.

18. Blauenstein, P. *Nouv. J. Chim.* **1990,** *14,* 405.

19. Clarke, M. J.; Podbielski, L. *Coord. Chem. Rev.* **1987,** *78,* 253.

20. Morgan, G. F. E.; Pope, J.; Thornback, J. R.; Theobald, A. E. In *Technetium in Chemistry and Nuclear Medicine*; Nicolini, M., G. Bandoli, and U. Mazzi, Eds.; Cortina International; Verona, 1986; pp 65.

21. Truffer, S.; Ianoz, E.; Lerch, P.; Kosinski, M. *Inorg. Chim. Acta* **1988,** *149,* 217.

22. Ianoz, D.; Mangegazzi, D.; Lerch, P.; Nicolo, F.; Chapuis, G. *Inorg. Chim. Acta* **1989,** *156,* 235.

23. Pomp, C.; Drüeke, S.; Küppers, H.-J.; Wieghardt, K.; Krüger, C.; Nuber, B.; Weiss, J. Z. *Naturforsch.* **1988,** *43b,* 299.

24. Bennett, M. J.; Cotton, F. A.; Walton, R. A. *Proc. R. Soc. London, Ser. A* **1968,** *303,* 175.

25. Murray, S. G.; Hartley, F. R. *Chem. Rev.* **1981,** *81,* 365.

26. Deutsch, E.; Libson, K.; Vanderheyden, J. L.; Ketring, A. R.; Maxon, H. R. *Nucl. Med. Biol.* **1986,** *13,* 465.

27. Konno, T.; Heeg; M. J.; Deutsch, E. *Inorg. Chem.* **1988,** *27,* 4113.

28. Cooper, S. R.; Rawle, S. C.; Yagbasan, R.; Watkin, D. J. *J. Am. Chem. Soc.* **1991,** *113,* 1600.

29. Rawle, S. C.; Yagbasan, R.; Prout, K.; Cooper, S. R. *J. Am. Chem. Soc.* **1987,** *109,* 6181.

30. Blake, A. J.; Gould, R. O.; Holder, A. J.; Hyde, T. I.; Schröder, M. *J. Chem. Soc., Dalton Trans.* **1988,** 1861.

31. Tomlinson, R.; Gray, L. *Brit. J. Cancer* **1955,** *9,* 539.

32. Ware, D. C.; Wilson, W. R.; Denny, W. A.; Rickard, C. E. F. *J. Chem. Soc., Chem. Commun.* **1991,** 1171.

33. Meares, C. F.; Wensel, T. G. *Acc. Chem. Res.* **1984,** *17,* 202.

34. Buter, J.; Kellogg, R. M.; van Bolhuis, F. *J. Chem. Soc., Chem. Commun.* **1990,** 282.

35. Smith, R. J.; Admans, G. D.; Richardson, A. P.; Küppers, H.-J.; Blower, P. J. *J. Chem. Soc., Chem. Commun.* **1991,** 475.

36. Glass, R. S.; Wilson, G. S.; Setzer, W. N. *J. Am. Chem. Soc.* **1980,** *102,* 5068.

37. Pett, V. B.; Leggett, G. H.; Cooper, T. H.; Reed, P. R.; Situmeang, D. A.; Ochrymowycz, L. A.; Rorabacher, D. B. *Inorg. Chem.* **1988,** *27,* 2164.

38. Setzer, W. N.; Cacioppo. E. L.; Grant, G. J.; Glass, R. S. *Phosphorus, Sulfur, Silicon* **1989,** *45,* 223.

39. Setzer, W. N.; Afshar, S.; Burns, N. L.; Ferrante, L. A.; Hester, A. M.; Meehan, E. J., Jr.; Grant, G. J.; Isaac, S. M.; Laudema, C. P.; Lewis, C. M.; VanDerveer, D. G. *Heteroat. Chem.* **1991,** *1,* 375.

40. Setzer, W. N.; Cacioppo, E. L.; Guo, Q.; Grant, G. J.; Kim, D. D.; Hubbard, J. L.; Van Derveer, D. G. *Inorg. Chem.* **1990,** *29,* 2672.

41. Rawle, S. C.; Admans, G.; Cooper, S. R. *J. Chem. Soc., Dalton Trans.* **1988,** 93.

42. Cooper, S. R. *Pure Appl. Chem.* **1990,** *62,* 1123.

43. Thorne, C M.; Rawle, S. C.; Admans, G.; Cooper, S. R. *Inorg. Chem.* **1986,** *25,* 3848.

44. Thorne, C. M.; Rawle, S. C.; Admans, G. A.; Cooper, S. R. *J. Chem. Soc., Chem. Commun.* **1987,** 306.

45. Rawle, S. C.; Cooper, S. R., unpublished work.

46. Osvath, P.; Sargeson, A. M.; Skelton, B. W.; White, A. H. *J. Chem. Soc., Chem. Commun.* **1991,** 1036.

47. Hartman, J. R.; Hintsa, E. J.; Cooper, S. R. *J. Chem. Soc., Chem. Commun.* **1984,** 386.

48. Hartman, J. R.; Hintsa, E. J.; Cooper, S. R. *J. Am. Chem. Soc.* **1986,** *108,* 1208.

49. Wilson, G. S.; Swanson, D. D.; Glass, R. S. *Inorg, Chem.* **1986,** *25,* 3827.

50. White, D. J.; Ashcroft, M. J.; Cooper, S. R., unpublished work.

51. Wolf, R. E. J.; Hartman, J. R.; Storey, J. M. E.; Foxman, B. M.; Cooper, S. R. *J. Am. Chem. Soc.* **1987,** *109,* 4328.

52. DeSimone, R. E.; Glick, M. D. *J. Am. Chem. Soc.* **1976,** *98,* 762.

53. Hartman, J. R.; Wolfe, R. E. J.; Foxman, B. R.; Cooper, S. R. *J. Am. Chem.Soc.* **1983,** *105,* 131.

54. Desper, J. M.; Gellman, S. H. *J. Am. Chem. Soc.* **1990,** *112,* 6732.

55. Desper, J. M.; Gellman, S. H.; Wolf, R. E., Jr.; Cooper, S. R. *J. Am. Chem. Soc.* **1991,** *113,* 8663.

56. Margerum, D. W.; Smith, G. F. *J. Chem. Soc., Chem. Commun.* **1975,** 807.

57. Baghurst, D. R.; Cooper, S. R.; Greene, D. L.; Mingos, D. M. P.; Reynolds, S. M. *Polyhedron* **1990,** *9,* 893.

58. Blake, A. J.; Holder, A. J.; Hyde, T. I.; Schröder, M. *J. Chem. Soc., Chem. Commun.* **1987,** 987.

59. Blake, A. J.; Gould, R. O.; Holder, A. J.; Hyde, T. I.; Lavery, A. J.; Odulate, M. O.; Schröder, M. *J. Chem. Soc., Chem. Commun.* **1987,** 118.

60. Blake, A. J.; Gould, R. O.; Holder, A. J.; Hyde, T. I.; Reid, G.; Schröder, M. *J. Chem. Soc., Dalton Trans.* **1990,** 1759.

61. Blake, A. J.; Greig, J. A.; Holder, A. J.; Hyde, T. I.; Taylor, A.; Schröder, M. *Angew. Chem., Int. Ed. Engl.* **1990,** *29,* 197.

62. Blake, A. J.; Holder, A. J.; Hyde, T. I.; Küppers, H. J.; Schröder, M.; Stoetzel, S.; Wieghardt, K. *J. Chem. Soc., Chem. Commun.* **1989,** 1600.

63. Sellmann, D.; Neuner, H. P.; Moll, M.; Knoch, F. Z. *Naturforsch. [B]* **1991,** *46,* 303.

64. Küppers, H. J.; Nuber, B.; Weiss, J.; Cooper, S. R. *J. Chem. Soc., Chem. Commun.* **1990,** 979.

65. Matondo, S. O. C.; Mountford, P. H.; Watkin, D. J.; Cooper, S. R., unpublished work.

66. White, D. J.; Küppers, H.-J.; Cooper, S. R., unpublished work.

67. Deutsch, E.; Libson, K.; Jurisson, S.; Lindoy, L. *Prog. Inorg. Chem.* **1983,** *30,* 75.

68. Deutsch, E.; Hirth, W. *J. Nucl. Med.* **1987,** *28,* 1491.

CHAPTER 16

From Crowns to Torands and Beyond

Thomas W. Bell
Department of Chemistry
State University of New York
Stony Brook, New York
U.S.A.

Dibenzo-18-crown-6 (*1*, Figure 16.1), the first crown ether discovered by Charles Pedersen,[1] foreshadowed current development of planar crown compounds consisting of multiply fused rings. Despite the reduced basicities of four oxygen atoms in *1* compared with those of 18-crown-6, potassium binding in methanol decreases by only 25% in energetic terms.[2] This effect leads to the notion that planarization of 18-crown-6 could produce exceptionally stable complexes, contrary to the observation that ion encapsulation generally leads to maximum complex stability.[2,3] Planarization also enables cavity rigidification without loss of rapid exchange kinetics, as opposed to encapsulating hosts such as cryptands[3,4] and spherands.[3,5] Bicyclo[3.3.0]octano-18-crown-6 (*2*) represents an early attempt at planarization without reduction in oxygen basicity.[6] This host binds *t*-butylammonium in CDCl$_3$ somewhat more weakly than does 18-crown-6, but the roles of steric interactions and hydrogen-bond directionality are unclear in this comparison. Replacement of oxygen- by nitrogen-binding sites introduces a wide range of possible planar structures. Saturated nitrogen atoms, as in aza-18-crown-6 (*3*), reduce affinities of crown compounds toward alkali metal and alkaline earth ions,[2,3] but unsaturated nitrogen functionalities have larger dipole moments and stronger complexes are expected. Pyrido-18-crown-6 (*4*) binds alkali metal salts somewhat more weakly than does 18-crown-6, but conformational effects may again play a significant role.[2,7] The key pyridocrown is cyclosexipyridine (*5*), for which a synthesis has been reported but complexation results remain absent from the literature; sequestration of sodium by diaryl derivatives of *5* also remains unconfirmed.[7,8] We have investigated the first alkali metal complexes of an all sp^2 hybridized hexaaza-18-crown-6 (*6*) and have introduced a new series of alkylated azacrowns, the torands (e.g., *7* and *8*).[7,9] In torands, the macrocyclic perimeter is com-

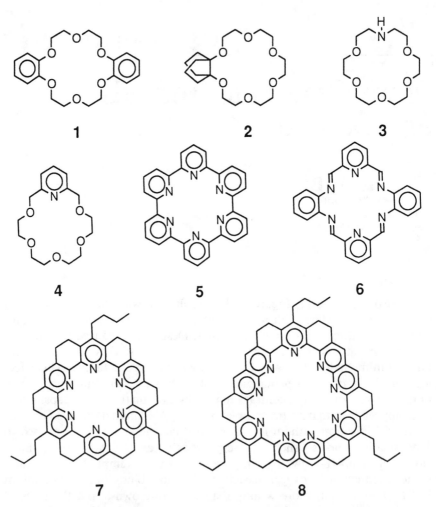

Figure 16.1. Annelated crown ethers (*1* and *2*), an azacrown (*3*), two pyridocrowns (*4* and *5*), a torand model (*6*), and two torands (*7* and *8*).

pletely formed by smaller, fused rings. The ion-binding properties of these planar, toroidal analogues of crown ethers are distinctly different from those of other host classes.

16.1 Torand Models

Torands *7* and *8* were formidable synthetic targets, and our initial studies focused on Schiff base macrocycles, such as *6*, *9*, and *10* (Figure 16.2). At the outset of our work, Drew et al. had reported the syntheses of alkaline earth complexes of *6*, as well as the X-ray structures of 1:1 Cd^{2+} and Pb^{2+}

complexes.[10] These are among the first complexes of "hard" alkaline earth ions with a host containing only "soft" nitrogen donors.[7,11] We found that whereas the condensation of pyridine-2,6-dicarboxaldehyde with benzene-1,2-diamine in the presence of KI gave 6·K[+] in low yield, this complex could be prepared in 56–74% yield by treatment of 6·Sr[2+] with KF (Figure 16.2).[9,12] Free ligand 6 is prepared by reaction of 6·K[+] with [2.2.2]cryptand or excess 18-crown-6; then 6·Na[+] may be synthesized by complexation of 6. Titration of 6·K[+] and 6·Na[+] with 18-crown-6 in DMSO showed that both complexes are stronger than those of 18-crown-6. The solvent might significantly influence this comparison, but solubility limitations dictated the choice of DMSO for NMR and potentiometric competition experiments. Macrocycle 6 is the first all-nitrogen analogue of a crown ether shown to bind alkali metals more strongly, but a puzzling decrease in size selectivity is also observed.

Comparison of X-ray crystal structures and ^{1}H NMR spectra of 6 and 6·Sr[2+] indicates that torand model 6 undergoes a conformational change upon complexation. As in complexes of Pb[2+] and Cd[2+], in 6·Sr(CF$_3$SO$_3$)$_2$ the ligand undergoes a saddle deformation of the planar, roughly circular conformation shown in Figure 16.1. The metal atom is approximately centered in the cavity, with six Sr–N contacts ranging from 2.65–2.82 Å and three "axial" Sr–O contacts of 2.52–2.95 Å.[12] The crystal structures and molecular modeling studies indicate that the planar conformation of this 18-membered macrocycle provides a cavity that is too large for Sr[2+], Pb[2+], and Cd[2+], and the unusual [113]Cd NMR chemical shift may be a consequence of this size mismatch.[12] The cavity radius of the planar ligand best matches 9-coordinate K[+] (1.55 Å)[13]; saddle distortion tilting the benzene and pyridine rings to

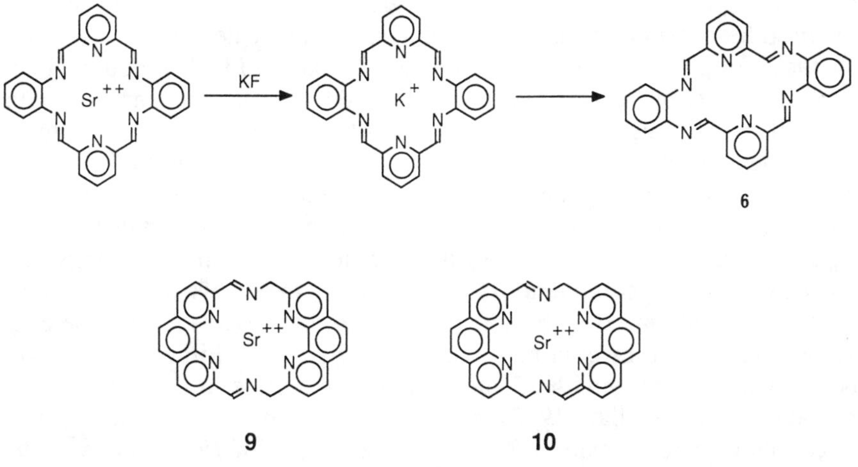

6

9   10

Figure 16.2. Preparation of torand model 6 as a free ligand and strontium complexes of bis(phenanthroline) torand models (9 and 10).[12,14]

opposite sides of the best plane reduces the hole size in response to smaller metals. Molecular mechanics calculations indicate, however, that the planar conformation of 6 is destabilized by dipole–dipole and steric interactions.[12] The approximately elliptical conformation of 6 shown in Figure 16.2 best describes the solid-state structure, and ¹H NMR data are consistent with this as the minimum energy conformation in solution.[12] Hence, 6 is not completely preorganized, and some of the complexation energy must be used to orient all six nitrogen dipoles toward the cavity.

The flexibility observed in 6 led us to search for more rigid, yet easily synthesized, torand models. We condensed 1,10-phenanthroline-2,9-dicar-boxaldehyde with 2,9-bis(aminomethyl)-1,10-phenanthroline in the presence of strontium triflate and obtained a 1:2 mixture of Schiff base macrocycle complexes 9 and 10 in 39% yield.[9,14] We were unable to remove strontium from these complexes without apparently destroying the macrocycles. The structures of 9 and 10 were supported by NMR and analytical data and by reduction to a bis(phenanthroline) macrocycle containing saturated bridges.[14] Generally poor solubility limited the usefulness of 9 and 10 as torand models, but their isolation gave further evidence that planar, rigid azacrowns would form strong complexes with alkali metals and alkaline earth ions.

16.2 Tri-*n*-butyltorand 7

When we began our synthetic work in 1982, torand 7 was one of several targets having different degrees of saturation and different substituents.[15] The 1,2,3,4,5,6,7,8-octahydroacridine ring system proved sufficiently easy to functionalize, and the 9-*n*-butyl derivatives display a good balance between crystallinity and solubility in organic solvents. Our current synthesis of tri-*n*-butyl torand 7 is shown in Figure 16.3. By modifying methods reported for other octahydroacridine derivatives,[16] we developed large scale syntheses of intermediates *11,12,* and *13.*[17] N-oxide *13* is then doubly functionalized, first by heating in acetic anhydride, then by condensation with benzaldehyde in the same pot. Hydrolysis of the resulting acetate produces benzylidene alcohol *14.* We now use the Albright-Goldman method[18,19] to oxidize alcohol *14* to ketone *15* because this method is more economical and simpler to perform on a large scale than previously reported methods.[9,16] Another recent key advance is our development of a method for preparing the new Mannich compound *16* from ketone *15.*[20] This Mannich intermediate is quaternized with methyl iodide to form *17,* which can be converted to enone *18* by elimination using triethylamine. Either *17* or *18* can be condensed directly with *15* and ammonium acetate in DMSO, yielding heptacyclic intermediate *19.* This new method improves the overall yield of coupling *15* to *19* from 30% to 70%. Ozonolysis of *19* gives one of the required macrocyclization intermediates (*20*) containing two of the three octahydroacridine moieties of target 7. The smaller fragment is prepared

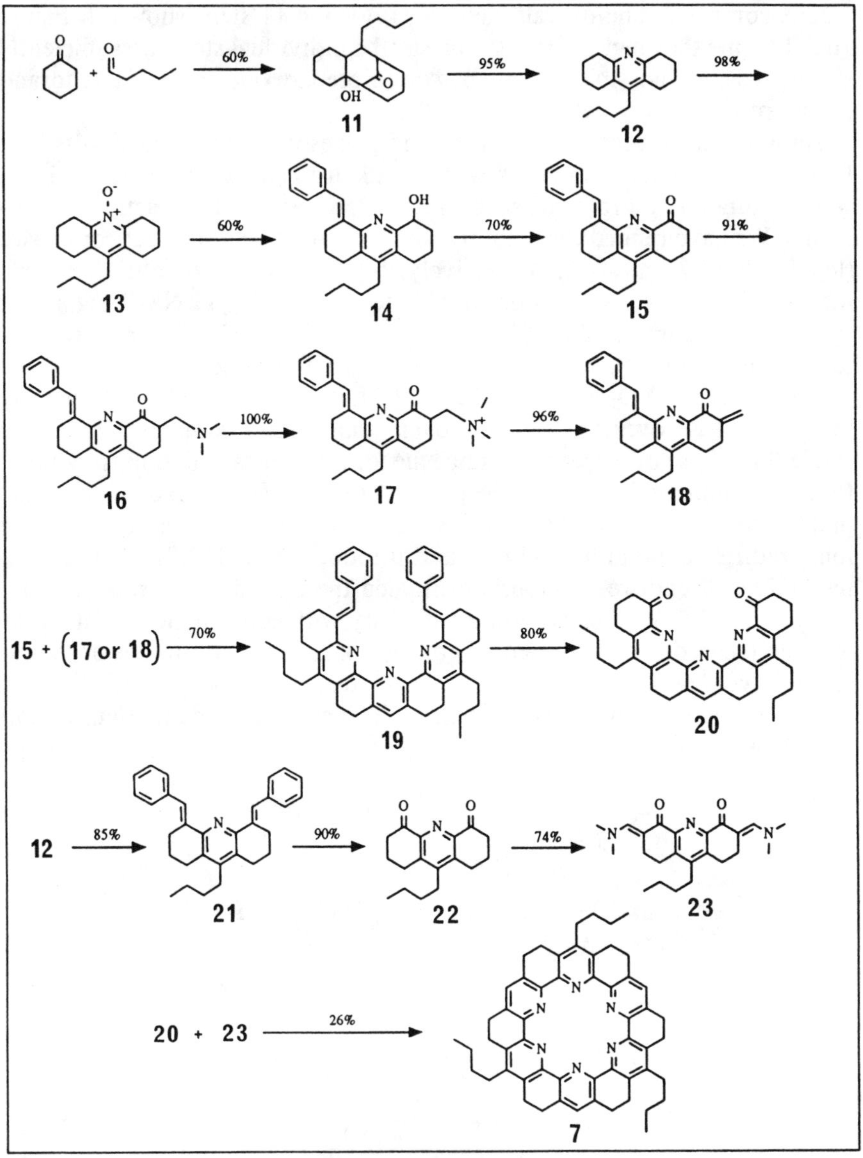

Figure 16.3. Current synthesis of tri-*n*-butyltorand *7*.

from *12* by benzaldehyde condensation, ozonolysis, and reaction with Bred-ereck's reagent. Reaction of *20, 23,* and trifluoromethanesulfonic (triflic) acid in acetic acid apparently gives a bis(pyrylium) macrocycle, which is converted to *7* in situ by reaction with ammonium acetate. Tri-*n*-butyltorand *7* is currently isolated as its salt with triflic acid rather than as its calcium triflate complex.[9] The overall yield is 3% for the 11 steps shown in Figure 16.3. Despite the length of this synthesis, the individual steps are sufficiently simple that it is possible for one person to prepare one gram of this torand in one month.

Torand 7 forms remarkably strong complexes with alkali metal salts, and log K_s values of 1:1 complexes with all alkali metal picrates exceed 11 in water-saturated chloroform, according to the results of extraction experiments. We have refined the stability constants of the Na[+] and K[+] complexes (log K_s = 14.7 and 14.3, respectively) by ^{1}H NMR competition experiments.[9,17,20] Previously, we were unable to resolve the ^{1}H NMR signals of free [2.2.1]cryptand and [2.2.1]cryptand·Li[+] at 300 MHz. Recently we performed this competition at 600 MHz and found log K_s = 13.4 for 7·Li(picrate) in D$_2$O-saturated CDCl$_3$.[20] Thus, 7 exhibits slight selectivity for Na[+] and K[+] over Li[+]. Our discovery that torand complexes are more stable than those of cryptands refutes the idea that encapsulation is required for strong binding. Moreover, the poor selectivity of 7 contradicts the usual preference for the metal that best fits the cavity because the hexacoordinate ionic radii of potassium, sodium, and lithium are 1.38, 1.02, and 0.76 Å, respectively,[13] and we previously estimated the torand cavity radius as approximately 1.3 Å. Effective ionic radii vary with coordination number, so we then focused on crystallography of the alkali metal complexes of torand 7. We have determined the X-ray crystal structures of the lithium, potassium, and rubidium picrate complexes, which are illustrated in Figure 16.4.[9]

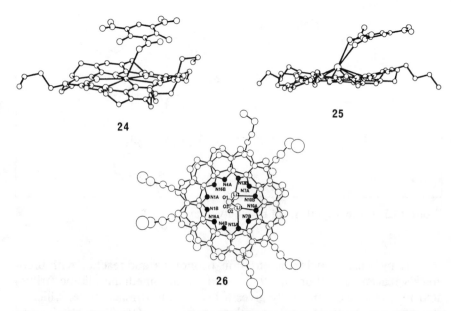

Figure 16.4. Crystal structures of 7·K(picrate) (24), 7·Rb(picrate) (25) and 7·Li(picrate) (26).[9]

In the potassium and rubidium complexes (*24* and *25*), the metal atoms are centered with respect to the cavity of *7*, which adopts the staggered pseudo-D_{3d} conformation predicted by molecular modeling.[9] Potassium is clearly the best fit, since it is only 0.23 Å from the best N plane with K–N distances ranging from 2.73–2.85 Å. Potassium has a 7th contact to picrate, so the cavity radius of *7* should be revised to almost 1.46 Å, the ionic radius of 7-coordinate potassium. The torand adopts the same conformation in *7·*Rb⁺; rubidium lies 1.06 Å for the best N plane, and picrate is bidentate. In stark contrast to these 1:1 complexes, *7·*Li(picrate) (*26*) crystallizes as a 2:2 complex containing three water molecules.[9] This structures is apparently unique for crystals containing lithium picrate, because no anion coordination is observed. The two torands are coaxially stacked and threaded by a hydrated dilithium chain (H_2O-Li⁺-H_2O-Li⁺-H_2O). Each lithium ion binds asymmetrically to two of the six nitrogens of each torand, and two water molecules complete the tetrahedral coordination sphere.

Comparison of the solid-state structures of *7·*Li⁺ and *7·*K⁺ does not offer a clear explanation for the low size-selectivity of torand *7*, and it remains to be determined whether this effect will be generally observed in planar crown compounds. It is possible that in rigidly planar systems, entropy-enthalpy compensation exerts a leveling effect.[21] Lithium in *7·*Li⁺ has a relatively low coordination number equal to 4,[22] suggesting that binding of small ions to *7* may release more solvent molecules than during binding to more flexible macrocycles. Despite the relatively large cavity size of *7*, we have also found that complexes are formed with metal ions throughout the periodic table. Our spectroscopic studies have shown that *7* binds Ca^{2+}, Ba^{2+}, Sc^{2+}, Mn^{2+}, Fe^{3+}, Co^{2+}, Ni^{2+}, Cu^{2+}, Zn^{2+}, Ag^+, Cd^{2+}, Eu^{3+}, Tb^{3+}, Tl^+, Pb^{2+}, and UO_2^{2+}, and 1:1 complexes of many of these metals have been characterized. Torand *7* and its metal complexes partition selectively into CH_2Cl_2 or $CHCl_3$ relative to water. Metals and other ions bind rapidly to the open-faced cavity of *7* and many can be extracted into water by protonation of the torand using aqueous acid.

16.3 An "Expanded" Torand (*8*)

Combining the 9-*n*-butyl-1,2,3,4,5,6,7,8-octahydroacridine building block with the Friedlander condensation of *o*-aminoaldehydes,[23] we have synthesized a series of new naphthyridine-containing macrocycles and clefts.[9,24] "Expanded" torand *8* and U-shaped receptor *31* are both prepared from ketone *15* (Figure 16.5). The key reagent is 4-aminopyrimidine-5-carboxaldehyde. Hydrolysis of the pyrimidine ring of the condensation product (*27*) gives a new *o*-aminoaldehyde (*28*), which can be either condensed with another equivalent of ketone *15* or directly ozonized to ketoaminoaldehyde *30*. Ozonolysis of the former product gives diketone *29*, which can then be converted to U-shaped receptor *31* by reaction with 2-aminonicotinaldehyde. Intermediate *30* contains both moieties needed for Friedlander con-

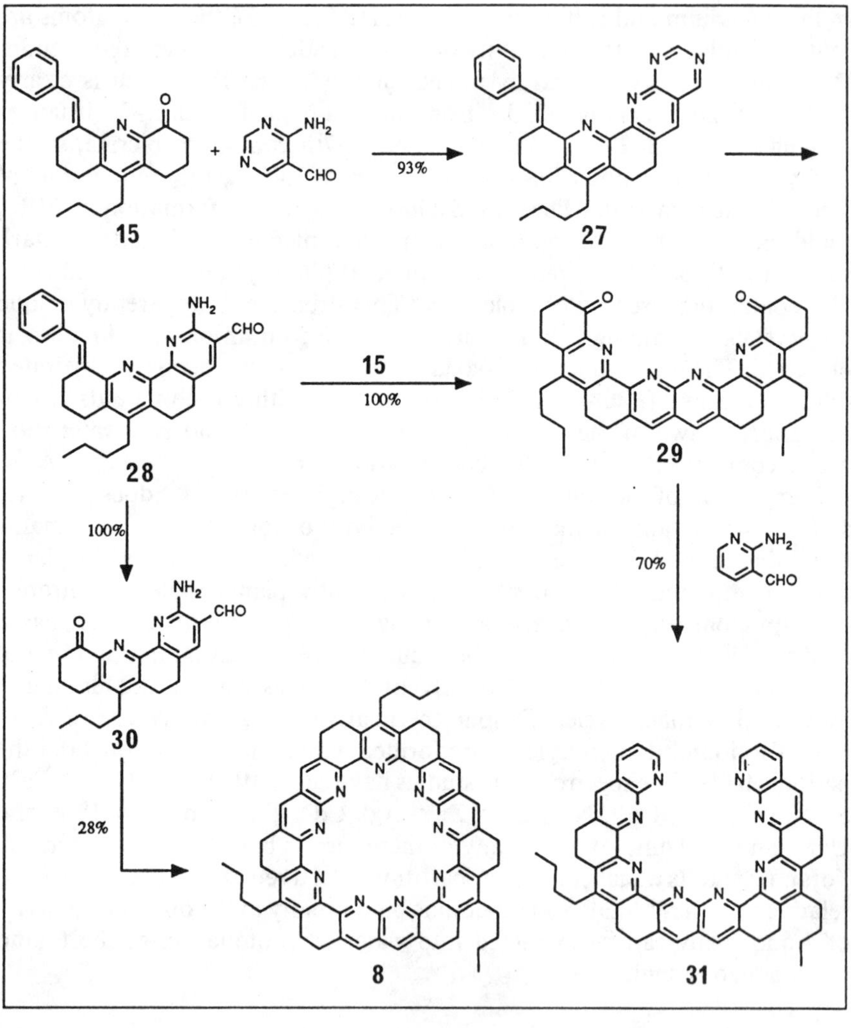

Figure 16.5. Synthesis of "expanded" torand 8 and U-shaped host 31.[9]

densation, and CsOH-templated trimerization gives expanded torand 8 in up to 50% yield.[9] Although 8 is a larger torand than 7, it can be prepared by this method in higher overall yield and in fewer steps. The cesium complex of 8 is much weaker than alkali metal complexes of 7, and the free ligand is obtained simply by boiling in water.

We recognized that receptors 8,29, and 31 could form nearly ideal, planar hydrogen-bonded complexes with appropriate organic guest molecules. Diketone 29 solubilizes urea in chloroform, forming the strongest known complex of urea with a neutral, metal-free receptor.[24] Torand 8 and U-shaped receptor *31* both form stable complexes with guanidinium chloride. The torand is more preorganized and forms a stronger guanidinium complex; however, we have performed molecular mechanics calculations indicating that *31* has a helical shape as a free ligand. Thus, *31* binds guanidinium by an "induced-fit" mechanism, and the complexation reaction may be followed by UV-visible spectroscopy.[9] Naphthyridine-containing systems have not yet been extended to produce full-turn helices, but this was recently accomplished in the polypyridine series related to torand *7*.[25]

16.4 Future Research and Applications

Torands are a relatively new class of cation receptors with unusual properties, and they present unique opportunities for practical applications. In order to realize their full potential, chemists will need to (1) better understand the energetics and kinetics of torand complexation, (2) further simplify torand synthesis, (3) devise methods for flexible control of torand substituents, and (4) expand the repertoire of heterocycles and hydrogen-bonding groups that may be incorporated into the structures of torands.

Stability constants of a wide range of torand complexes with alkaline earth salts, transition metals, heavy metals, lanthanides, and actinides need to be determined. Influences of counterions and solvents must also be evaluated. The ^1H NMR competition method we have used to measure the stabilities of $7 \cdot Li^+$, $7 \cdot Na^+$, and $7 \cdot K^+$ is too cumbersome for this purpose, and alternative methods are required. Fluorescence and potentiometric titration are good possibilities for more rapidly surveying the binding ability of *7* throughout the periodic table. Polypyridines are electrochemically active, and the effects of bound metals on torand redox potentials must be investigated. These studies could lead to applications of torands as reversible extractants or as optical or electrochemical sensors for various metals. The complexes of Eu^{3+} and Tb^{3+} are of particular interest for applications as luminescent labels and luminophores in electroluminescent display devices. If sufficiently inexpensive and nontoxic, water-soluble torand complexes of Gd^{3+} would be of interest as NMR imaging agents. Even without modification, torand *7* may be surface active because of its planarity and lipophilicity. Thin films composed of torands and prepared at liquid/gas or solid/liquid interfaces should display unique and potentially useful properties. The solid-state structure of $7 \cdot Li^+$ also suggests that it will be possible to assemble linear-chain polymetallic complexes of stacked torands to prepare materials of interest for ion and electron conductivity.

We are currently developing triaryltrialkyltorands (*32*) in an approach to discotic liquid crystals (Figure 16.6).[20] So far, we have been unable to prepare hexaalkyltorands, which would be expected to align coaxially to form columnar discotic mesophases, as shown in diagram A of Figure 16.6. An alternative approach is to attach three long chains to the para positions of each aryl group in triaryltrialkyltorands to form discotic liquid crystals as

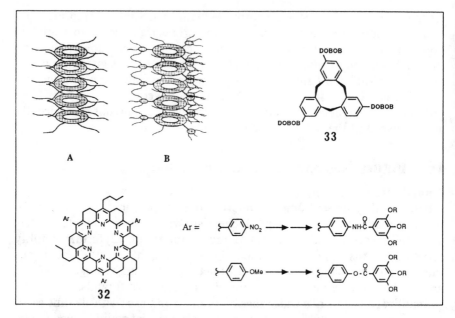

Figure 16.6. Approaches to columnar discotic liquid crystals.[20]

shown in diagram B of Figure 16.6. This approach using the 3,4,5-tris(*p*-*n*-dodecyloxybenzyloxy)benzoyloxy (DOBOB) group induces cyclotricatechy-lene (*33*) to form columnar mesophases that are stable between room temperature and 150 °C.[26] Syntheses of tris(*p*-nitrophenyl) and tris(*p*-methoxyphenyl) torands are being developed for conversion to DOBOB derivatives, as shown in Figure 16.6. The resulting tubular torand mesophases constitute an attractive approach to synthetic ion channels. Among other possibilities, electrodes coated with tubular mesophases may respond selectively toward ions small enough to bind to the channels.

Synthetically feasible target torands having cavities of different sizes and shapes and incorporating different heterocycles include expanded systems *34,35,* and *36* (Figure 16.7). It should be possible to obtain triangular torand *34* by direct analogy with the facile Friedlander synthesis of *8* (Figure 16.5). In target *34,* the naphthyridine rings of *8* are replaced by anthyridines, offering the possibility of reduction to dihydroanthyridines or oxidation to anthyridinone moieties. This expands the set of potential guests that may be complexed by ionic and hydrogen-bonding interactions. Pyrrole rings are incorporated into *35* and *36,* producing "mixed heterocycle" torands having a combination of hydrogen-bond acceptor and donor sites. Deprotonation of the pyrrole rings would produce highly delocalized polyanions that should form strong complexes with many metals. These torands are similar to the smaller, more flexible "expanded" porphyrins[27] of current interest for their unusual coordination behavior and potential applications, including pho-

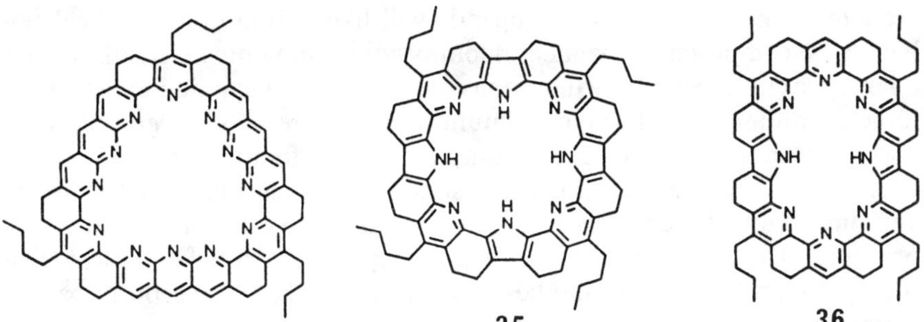

Figure 16.7. New target torands.

todynamic therapy. Anionic complexes of *35* with lanthanides such as $(35 \cdot Eu)^-$ are of interest for developing NMR shift reagents for Na^+ and K^+, as well as NMR imaging agents based on this principle.[28]

The exceptionally large cavities of *34–36* offer the possibility of binding multiple cationic guests. Under acidic conditions, these torands will be polyprotonated, having several inward-directed hydrogen-bond donors. These cationic hosts are of interest for recognition of anionic guests, including halides, carbonate, sulfate, and phosphate. Torands *8* and *34* might also afford the first complexes of metal–metal bonded clusters encircled by macrocyclic ligands. Intensive work on transition metal clusters has been driven in large part by the promise of unusual catalytic activity by species intermediate in structure between homogeneous and heterogeneous catalysts. It has been recognized[29] that this promise remains largely unrealized, often because the "cluster compound breaks down under reaction conditions to give highly reactive mononuclear entities that are responsible for the catalysis." Thus, stabilization of the cluster "throughout reaction is a problem central to the establishment of cluster catalysis." Large torands such as *8* and *34–36* offer an attractive solution to this problem, because they may stabilize clusters of two to six metal atoms, forming soluble complexes in which one or more substrates can bind axially without steric interference by the ligand. This approach is superior to the use of solid supports,[29] because nuclearity might be specifically controlled and catalytic complexes could be characterized in solution by spectroscopic methods.

We should not neglect to mention that our research program on torands has also spawned a new family of "hexagonal lattice" receptors containing hydrogen-bonding clefts. Molecules composed entirely of fused rings are highly preorganized prior to macrocyclization and exhibit useful complexation properties. Thus, diketone *20* forms alkali metal complexes of comparable strength to those of crown ethers. Related tridentate, quaterdentate, and quinquedentate ligands may be solubilized in water or attached to polymers to sense, sequester, or immobilize various metals. Helical analogues are already known,[25] and these ligands will likely form double-helical polymetallic complexes. Expanded, naphthyridine-containing clefts may bind a large variety of small organic molecules and ions, including ureas, guanidines, amidines, and alkylammonium ions. We are now integrating these "artificial receptors" with dye chromophores and fluorophores to produce analytical reagents and chemical sensors. This novel approach to clinical and environmental analysis is a practical alternative to biomolecular approaches. The selectivities of artificial receptors are still far poorer than those of enzymes and antibodies, but hexagonal lattice receptors permit closer coupling between the binding event and signal transduction.

Our discovery of a new 14-membered azamacrocycle that undergoes ion-selective tautomerism illustrates the promise of n-butyloctahydroacridine derivatives in analytical chemistry (Figure 16.8).[30] We condensed diketone *22* with hydrazine to form *E,Z*-dihydrazone *37*. Macrocycle *38* precipitated as a yellow-orange solid from reaction of *37* with *22*. In chloroform and methylene chloride, *38* has a unique bis(vinylhydrazone) structure rather than the expected bis(azine) structure, as shown in Figure 16.8. Upon complexation of alkali and alkaline earth salts, *38* undergoes tautomerism to the bis(azine) form with dramatic changes in the UV-visible absorption spectrum. Displaying general selectivity for smaller cations such as Li^+ and Mg^{2+}, the 380 nm band of *38* disappears when CH_2Cl_2 solutions are exposed to the solid chloride and nitrate salts of these metals.

Torands and hexagonal lattice hosts are logical outgrowths of attempts to rigidify crown compounds. The exquisite control of molecular architecture

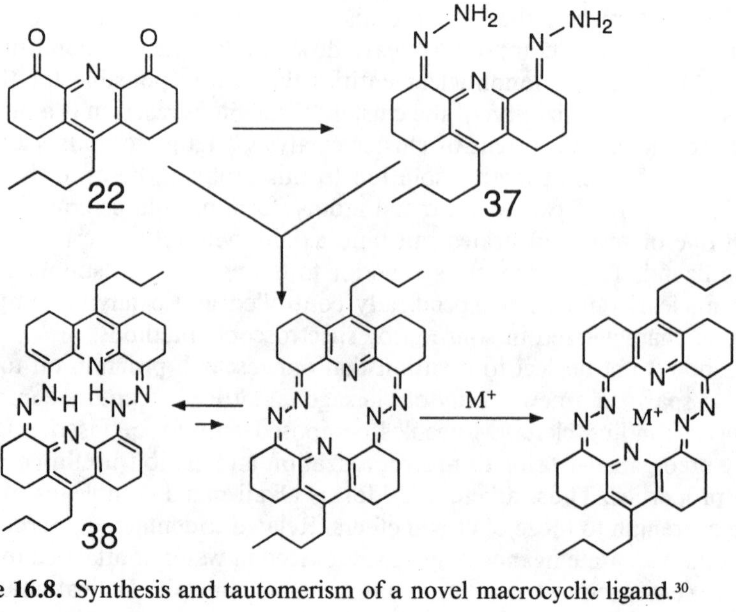

Figure 16.8. Synthesis and tautomerism of a novel macrocyclic ligand.[30]

offered by fused-ring systems assures their continued interest both in fundamental and applied chemistry.

Acknowledgment

My collaborators, whose names are listed in the following references, are warmly thanked for their essential contributions to the research program described here. Grants from the National Institutes of Health (GM32937 and AI30945), the Petroleum Research Fund, and the New York Science and Technology Foundation through the Center for Biotechnology are gratefully acknowledged.

References

1. (a) Pedersen, C. J. *J. Am. Chem. Soc.* **1967,** *89,* 2495.
   (b) Pedersen, C. J. *Aldrichim. Acta.* **1971,** *4,* 1.
   (c) Pedersen, C. J. *Angew. Chem., Int. Ed. Engl.* **1988,** *27,* 1021.

2. Izatt, R. M.; Bradshaw, J. S.; Nielsen, S. A.; Lamb, J. D.; Christensen, J. J. *Chem. Rev.* **1985,** *85,* 271.

3. Vögtle, F. *Supramolecular Chemistry* (translation by M. Grognuz); Wiley; New York; 1991; p 9.

4. (a) Lehn, J.-M. *Angew. Chem., Int. Ed. Eng.* **1988,** *27,* 89.
   (b) Lehn, J.-M. *Angew. Chem., Int. Ed. Eng.* **1990,** *29,* 923.

5. (a) Cram, D. J. *Angew. Chem., Int. Ed. Eng.* **1986,** *25,* 1039.
   (b) Cram, D. J. *Science (Washington, D.C.)* **1988,** *240,* 760.

6. Bell, T. W.; Cheng, P. G.; Newcomb, M.; Cram, D. J *J. Am. Chem. Soc.* **1982,** *104,* 5185.

7. Bell, T. W.; Sahni, S. K. In *Inclusion Compounds Vol. 4: Key Organic Host Systems*; Atwood, J. E.; Davies, E.; MacNicol, D., Eds.; Oxford Univ. Press; Oxford, 1991; p 329.

8. (a) Newkome, G. R.; Lee, H.-W. *J. Am. Chem. Soc.* **1983,** *105,* 5956.
   (b) Toner, J. L. *Tetrahedron Lett.* **1983,** *24,* 2707.

9. (a) Bell, T. W.; Firestone, A. *J. Am. Chem. Soc.* **1986,** *108,* 8109.
   (b) Bell, T. W.; Firestone, A.; Hu, L.-Y.; Guzzo, F. *J. Inclusion Phenom.* **1987,** *5,* 149.
   (c) Bell, T. W.; Firestone, A.; Ludwig, R. *J. Chem. Soc., Chem. Commun.* **1989,** 1902.
   (d) Bell, T. W.; Firestone, A.; Liu, J.; Ludwig, R.; Rothenberger, S. D. In *Inclusion Phenomena and Molecular Recognition*; Atwood, J. L., Ed.; Plenum; New York, 1990, p 49.
   (e) Bell, T. W.; Liu, J. *Angew. Chem., Int. Ed. Eng.* **1990,** *29,* 923.
   (f) Bell, T. W.; Cragg, P. J.; Drew, M. G. B.; Firestone, A.; Kwok, D.-I. A. *Angew. Chem., Int. Ed. Engl.* **1992,** *31,* 345.
   (g) Bell, T. W.; Cragg, P. J.; Drew, M. G. B.; Firestone, A.; Kwok, D.-I. A. *Angew. Chem., Int. Ed. Engl.* **1992,** *31,* 348.

10. Drew, M. G. B.; Cabral, J. de O.; Cabral, M. F.; Esho, F. S.; Nelson, S. M. *J. Chem. Soc., Chem. Commun.* **1979,** 1033.

11. Majestic, V.; Newkome, G. R. *Top. Curr. Chem.* **1982,** *106,* 79.

12. (a) Bell, T. W.; Guzzo, F. *J. Am. Chem. Soc.* **1984,** *106,* 6111.
    (b) Guzzo, F., M. S. Thesis, SUNY, Stony Brook, 1985.
    (c) Bell, T. W.; Guzzo, F. *J. Chem. Soc., Chem. Commun.* **1986,** 769.
    (d) Bell, T. W.; Guzzo, F. *Ann. N. Y. Acad. Sci.* **1986,** *471,* 291.
    (e) Marchetti, P. S.; Bank, S.; Bell, T. W.; Kennedy M. A.; Ellis, P. D. *J. Am. Chem. Soc.* **1989,** *111,* 2063.
    (f) Bell, T. W.; Guzzo, F.; Drew, M. G. B. *J. Am. Chem. Soc.* **1991,** *113,* 3115.

13. Shannon, R. D. *Acta. Crystallogr. A.* **1976,** *32,* 751.

14. Hu, L.-Y., Ph.D. Thesis, SUNY, Stony Brook, 1988.

15. For approaches to hexaazakekulenes via the dibenzo[*b*,*j*][1,10]phenanthroline ring system, see: (a) Ransohoff, J. E. B.; Staab, H. A. *Tetrahedron Lett.* **1985,** *26,* 6179.
(b) Bell, T. W.; Hu, L.-Y.; Patel, S. V. *J Org. Chem.* **1987,** *52,* 3847. See also ref. 14.

16. (a) Gill, N. S.; James, K. B.; Lions, F.; Potts, K. T. *J. Am. Chem. Soc.* **1952,** *74,* 4923.
(b) Pitha, J.; Plesek, J.; Horak, M. *Coll. Czech. Chem. Commun.* **1961,** *26,* 1209.
(c) Tilichenko, M. N. *Uch. Zap. Sarat. Gos. Univ.* **1962,** *75,* 60.
(d) Pitha, J.; Tilichenko, M. N.; Kharchenko, V. G. *Zh. Obshch. Khim.* **1964,** *34,* 1936.
(e) Stankevich, E. I.; Vanags, G. *Khim. Geterotsikl. Soedin., Akad. Nauk. Latv. SSR* **1965,** *2,* 305.

17. (a) Bell, T. W.; Firestone, A. *J. Org. Chem.* **1986,** *51,* 764.
(b) Bell, T. W.; Rothenberger, S. D. *Tetrahedron Lett.* **1987,** *28,* 4817.
(c) Firestone, A., Ph.D. Thesis, SUNY, Stony Brook, 1988.
(d) Bell, T. W.; Cho, J.-M.; Firestone, A.; Healy, K.; Liu, J.; Ludwig, R.; Rothenberger, S. D. *Organic Syntheses* **1990,** *69,* 226.

18. Albright, J. D.; Goldman, L. *J. Am. Chem. Soc.* **1967,** *89,* 2416.

19. Liu, J., Ph.D. Thesis, SUNY, Stony Brook, 1990.

20. Ludwig, R., Ph.D. Thesis, SUNY, Stony Brook, 1992.

21. (a) Inoue, Y.; Liu, Y.; Hakushi, T. In *Cation Binding by Macrocycles*; Inoue, Y., and Gokel, G. W., Eds.; M. Dekker; New York, 1990.
(b) Inoue, Y.; Amano, F.; Okada, N.; Inada, H.; Ouchi, M.; Tai, A.; Hakushi, T.; Liu, Y.; Tong, L.-H. *J. Chem. Soc., Perkin Trans. 2* **1990,** 1239.

22. Olsher, U.; Izatt, R. M.; Bradshaw, J. S.; Dalley, N. K. *Chem. Rev.* **1991,** *91,* 137.

23. Caluwe, P. *Tetrahedron* **1980,** *36,* 2359.

24. Bell, T. W.; Liu, J. *J. Am. Chem. Soc.* **1988,** *110,* 3673.

25. Bell, T. W.; Jousselin, H. *J. Am. Chem. Soc.* **1991,** *113,* 6283.

26. Malthete, J.; Collet, A. *J. Am. Chem. Soc.* **1987,** *109,* 7544.

27. (a) Sessler, J. L. *Comments Inorg. Chem.* **1988,** *7,* 333.
(b) Vogel, E. *Pure & Appl. Chem.* **1990,** *62,* 557.

28. Szklaruk, J.; Marecek, J. F.; Springer, A. L.; Springer, C. S., Jr. *Inorg. Chem.* **1990,** *29,* 660.

29. Whyman, R. In *Surface Organometallic Chemistry: Molecular Approaches to Surface Catalysis*; Basset, J.-M., et al., Eds.; Kluwer; Dordrecht, Netherlands, 1988.

30. Bell, T. W.; Papoulis, A. T. *Angew. Chem., Int. Ed. Engl.* **1992,** *31,* in press.

Index